Modern Transportation Systems and Networks

Modern Transportation Systems and Networks

Edited by **Samuel Morgan**

Willford Press

New York

Published by Willford Press,
118-35 Queens Blvd., Suite 400,
Forest Hills, NY 11375, USA
www.willfordpress.com

Modern Transportation Systems and Networks
Edited by Samuel Morgan

International Standard Book Number: 978-1-68285-176-0 (Hardback)

Printed in China.

Contents

Preface

I am honored to present to you this unique book which encompasses the most up-to-date data in the field. I was extremely pleased to get this opportunity of editing the work of experts from across the globe. I have also written papers in this field and researched the various aspects revolving around the progress of the discipline. I have tried to unify my knowledge along with that of stalwarts from every corner of the world, to produce a text which not only benefits the readers but also facilitates the growth of the field.

Transportation is an important sector in every nation. The advancements in science and technology have constantly aided the progress of this field. The latest findings in this field are compiled in this all inclusive book. Modern intelligent transport systems, transportation infrastructure, urban transit systems, traffic planning and management, interdisciplinary transportation research, environmental impacts of transportation, etc. are some of the areas discussed in this book. It will serve as a resource guide for students, researchers, engineers and professionals as well.

Finally, I would like to thank all the contributing authors for their valuable time and contributions. This book would not have been possible without their efforts. I would also like to thank my friends and family for their constant support.

Editor

Influence of filler–bitumen ratio on performance of modified asphalt mortar by additive

Hongsheng Qiu · Ximing Tan · Shu Shi · Heng Zhang

Abstract When the filler–bitumen ratio of asphalt mortar changes, its adhesion and viscoelasticity will also change, as well as its mechanical performances, such as fatigue durability at normal temperature and low-temperature ductility. Thus, the filler–bitumen ratio directly affects the asphalt mortar's performance. This paper tested the physical indexes of the No. 70 matrix asphalt mortar modified by additive Sasobit (SB) and Sasowam (SW) through dynamic shear rheometer and bending beam rheometer under different temperature conditions, and comprehensively analyzed the high-temperature anti-rutting and fatigue performance, low-temperature crack resistance performance, and ductility of asphalt mortar. The results show that ore powder not only can increase the anti-rutting factor but also can increase the aging resistance of asphalt. SB has better performances than SW at high temperatures. As for the filler–bitumen ratio of asphalt mortar with additive SB, the recommended value is between 0.8 and 1.2, and the value may be a little larger for that with SW.

Keywords Filler–bitumen ratio · Asphalt mortar · Anti-rutting factor · Small beam specimens · Creep rate

1 Introduction

Asphalt is often the choice of highway pavement material in the world. With the traffic growth, the overloaded vehicles will cause the early damage of asphalt pavement, including high-temperature rutting, low-temperature cracks, subsidence, potholes, etc. Mortar theory assumes that asphalt mixture is a dispersed system of multilevel interspace reticulate structure with coarse aggregate as the dispersed phase, and asphalt mortar is the first decentralized system of asphalt mixture, with fine aggregate as the dispersed phase scattered in the asphalt medium [1]. The structure, composition, and performance of asphalt mortar have direct influence on the road performance of the asphalt pavement. A lot of research about asphalt mortar [2–6] has been done in terms of the standard Marshall test, immersion Marshall test, rutting test, and low-temperature bending test of beam specimens. Zhang et al. [7] studied the asphalt mortar property with SHRP equipment dynamic shear rheometer (DSR) and bend beam rheometer (BBR), tested and analyzed the effect of filler–bitumen ratio on performance of mortar at high and low temperatures in 2004; and they recommended that the filler–bitumen ratio should not exceed 1.4 for high-temperature performance and not exceed 1.0 for low-temperature performance. But they did not give the minimum value of filler–bitumen ratio, and did not consider additive in asphalt. It has been concluded that modified asphalt by additive can significantly improve the asphalt pavement performance [8–10]. It is still important to explore a proper filler–bitumen ratio to control the asphalt performance.

In this paper, we conducted tests on asphalt mortar with additive SB (Sasobit) and SW (Sasowam) through DSR and BBR. By analyzing test data, the high-temperature anti-rutting and low-temperature crack resistance performance, and the fatigue durability of asphalt mortar are explored. Finally, a filler–bitumen ratio of asphalt mortar is recommended.

2 Test scheme introduction

According to the strategic highway research program (SHRP) [11], No. 70 matrix asphalt and limestone ore

H. Qiu (✉) · X. Tan · S. Shi · H. Zhang
School of Transportation, Wuhan University of Technology, Wuhan 430063, China
e-mail: 13377889059@163.com

Table 1 Test results of indexes of ore powder and its comparison with technical requirements

	Density (g/cm^3)	Hydrophilic coefficient	Penetration rate of ore for different siel sizes of screen (mm)			
			0.6	0.3	0.15	0.075
Test result	2.730	0.76	100	99.1	95.8	78.4
Technical requirements	$\nless 2.5$	$\ngtr 1$	100	–	90–100	75–100

powder are used as raw materials to produce the aggregate gradation. The density and hydrophilic coefficient of the ore powder are listed in Table 1.

First, we placed the limestone powder into different sizes of screen mesh to measure penetration rate of the limestone. The obtained results are compared with the technical requirements [1], as shown in Table 1. Then, the ore powder was added into No. 70 matrix asphalt to have asphalt mortar specimens with four filler–bitumen ratios of 0.8, 1.0, 1.2, and 1.4.

3 High-temperature and fatigue performance of asphalt mortar

The DSR test is generally used to evaluate the high-temperature characteristics of asphalt mortar [1, 3, 5]. When flat spacing is twice larger than the maximum diameter of rubber cement and the size of the added grain is not more than 0.250 mm, the DSR can be used to test rubber paste properties [12–14].

The DSR in Fig. 1a is used to characterize the viscous and elastic properties of materials by measuring complex material shear modulus (G^*) and phase angle (δ) [15–17]. DSR can form a piece of thick film between retainer plate and the vibration plate, which are retained under a certain load, and the vibration plate rotates at an angular velocity and angle, as shown in Fig. 1b.

The shear stress τ, shear strain r, complex modulus G^*, and phase angle of specimens are calculated by

$$\tau = \frac{2T}{\pi r^3}, \quad \gamma = \frac{\theta r}{h}, \quad G^* = \frac{\tau_{\max} - \tau_{\min}}{\gamma_{\max} - \gamma_{\min}}, \quad \delta = 2\pi f \cdot \Delta t,$$

where T is the maximum torque; r the vibration plate radius; h the sample height; θ the vibration plate rotation angle; $\tau_{\max}$, $\tau_{\min}$, $\gamma_{\max}$, and $\gamma_{\min}$ the maximum or minimum shear stress and strain; Δt the lag time.

Under the normal asphalt pavement temperature and traffic load, the viscous and elastic properties of asphalt can be obtained by testing complex shear modulus G^*, and the relative index δ between recoverable and non-recoverable deformations. $G^*/\sin\delta$ is an anti-rutting factor, indicating distortion resistances of the asphalt material. At the design maximum pavement temperature, the higher the $G^*/\sin\delta$, the smaller the deformation of the asphalt. Thus, the higher $G^*/\sin\delta$ is more conducive to resist rutting. In addition, $G^*\sin\delta$ stands for the capability of the cementing material to resist fatigue [17].

In this experiment, the prepared asphalt mortar was added with SB and SW of 3 % in mass fraction, respectively, as the original specimens; and the original specimens were subjected to aging treatment in rolling thin film oven. We used the original and aging specimens in the DSR test under different temperature conditions. DSR strain is 12 %, and angular velocity $\omega = 10$ rad/s. By applying DSR test, asphalt mortar's fatigue performance was evaluated, that is, rubber paste's rheological properties under the normal temperatures. In Superpave Binder Specifications, fatigue performance is evaluated by applying a pressure to the aging asphalt [18]. SHRP

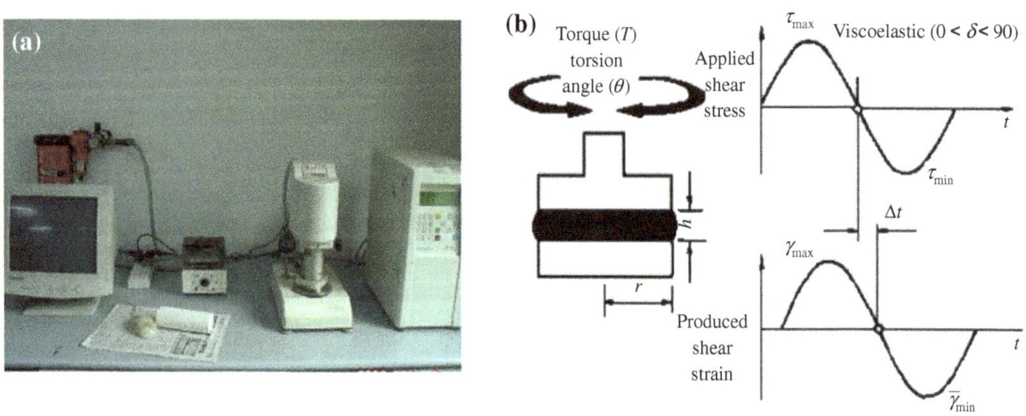

Fig. 1 SHRP dynamic shear flow meter and principle diagram. **a** Dynamic shear rheometer. **b** Dynamic shear rheometer working principle

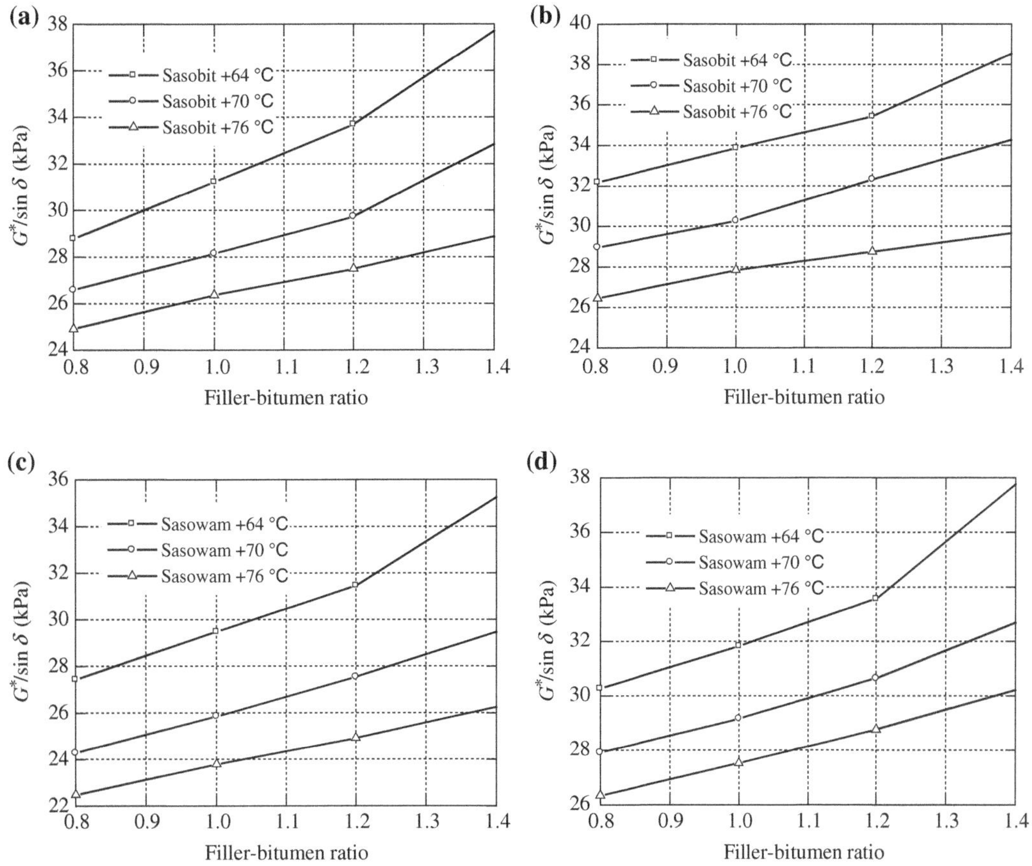

Fig. 2 Relationship between filler–bitumen ratio and rutting factor. **a** Original specimens with additive SB. **b** Aging specimens with SB. **c** Original specimens with SW. **d** Aging specimens with SW

specifies that $G^*\sin\delta$ should not exceed 5,000 kPa in the anti-fatigue cracking performance test [11]. Under different temperatures, the relationship between filler–bitumen ratio and rutting factor can be obtained for the original and aging specimens, as shown in Fig. 2; and the relationship between filler–bitumen ratio and fatigue resistance for the specimens at normal temperatures are shown in Fig. 3.

From Fig. 2, we can reach the following conclusions:

(1) Anti-rutting factor of asphalt mortar increases with filler–bitumen ratio. For the specimens with SB and SW, the increase of $G^*/\sin\delta$ is the largest at 64 °C. When filler–bitumen ratio increases from 0.8 to 1.4, $G^*/\sin\delta$ increases from 28.78 to 37.71, 27.43 to 35.24 for SB and SW, respectively, corresponding to an increment of 31 % and 28.5 %. Thus, the high-temperature resistance performance of crack of asphalt mortar with additive SB is better than that with additive SW.

(2) As the temperature and the filler–bitumen ratio increases, the increment of $G^*/\sin\delta$ reduces gradually. The filler–bitumen ratio plays an important role in determining the high-temperature performance of asphalt and prevents the asphalt short-term aging. It is necessary to increase filler–bitumen ratio.

From Fig. 3, the following conclusions are drawn:

(1) The increment of $G^*\sin\delta$ is 13 % when the filler–bitumen ratio increases from 0.8 to 1.2. It increases by 20 % when the filler–bitumen ratio increases from 1.2 to 1.4. Obviously, filler–bitumen ratio has a great effect on the fatigue cracking of cementing material.

(2) $G^*\sin\delta$ increases when filler–bitumen ratio increases from 0.8 to 1.4, and the increase rate is relatively stable.

4 Test analysis of asphalt mortar's performance at low temperature

Bending beam rheometer in Fig. 4 is an instrument which evaluates the performance of asphalt material at low temperatures [19–21]. It puts asphalt beam specimen on the supporting frame and then loads on it to test the creep stiffness modulus (S) and creep slope (m) of asphalt material. The asphalt binder with smaller S and larger

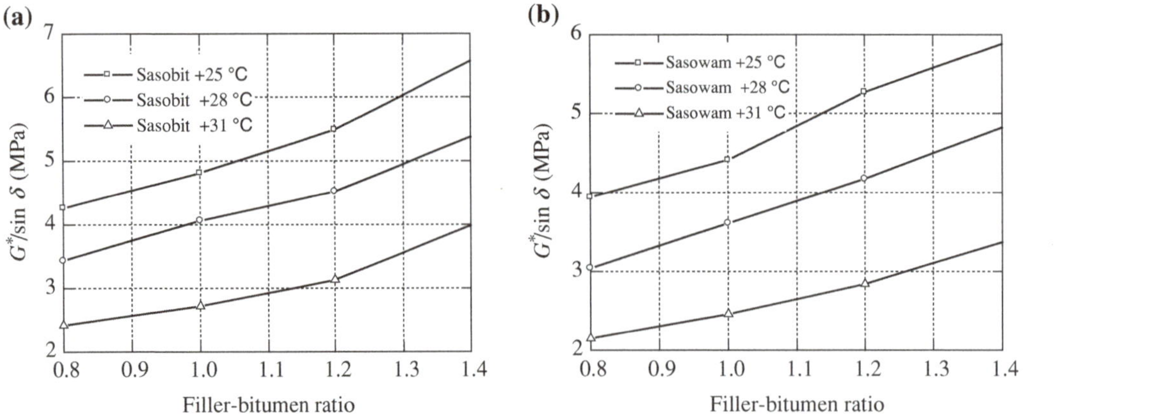

Fig. 3 Relationship between the filler–bitumen ratio and $G^*/\sin(\delta)$. **a** Relationship between the filler–bitumen ratio and $G^*\sin(\delta)$ of SB. **b** Relationship between the filler–bitumen ratio and $G^*\sin(\delta)$ of SW

m shows better crack resistance than that with larger *S* and smaller *m* at low temperatures.

Bend beam rheometer is recommended to study the low-temperature performance of the asphalt mortar by many researches for its advantages such as accurate data and good reproducibility [20]. In the process of BBR test, there are three important prerequisites, including time–temperature equivalent relationship, the influence of physical hardening, and linear viscoelasticity, which should be met by BBR test [22].

In this paper, we use the BBR to measure the bending creep stiffness modulus *S* and creep rate *m* at −18, −12, and −6 °C, respectively, and the results are shown in Figs. 5 and 6.

From Fig. 5, we obtain the following conclusions:

(1) The low-temperature resistance performance of asphalt mortar becomes worse with the increase of filler–bitumen ratio, especially when the ratio increases from 1.2 to 1.4. Its performance falls fast at low temperature, and the decrease of temperature also has a great influence on the stiffness modulus of rubber paste.

(2) The creep rate decreases gradually with the increase of filler–bitumen ratio, and the crack resistance decreases rapidly at low temperature when the filler–bitumen ratio exceeds 1.2.

From Fig. 6, it is seen that with the increase of filler–bitumen ratio, creep stiffness modulus increases; that is, the low-temperature anti-cracking performance of the mortar becomes worse with the increase of filler–bitumen ratio. At the same time, the temperature also has a great impact on the rubber paste's stiffness modulus when it decreases. Amplitude of creep rate does not reduce significantly. Hence the filler–bitumen ratio of asphalt mortar with additive SW can be a little larger than that of SB.

5 Ductility test analysis for asphalt mortar

The interaction of asphalt and additive not only has great influence on the formation and the quantity of the asphalt

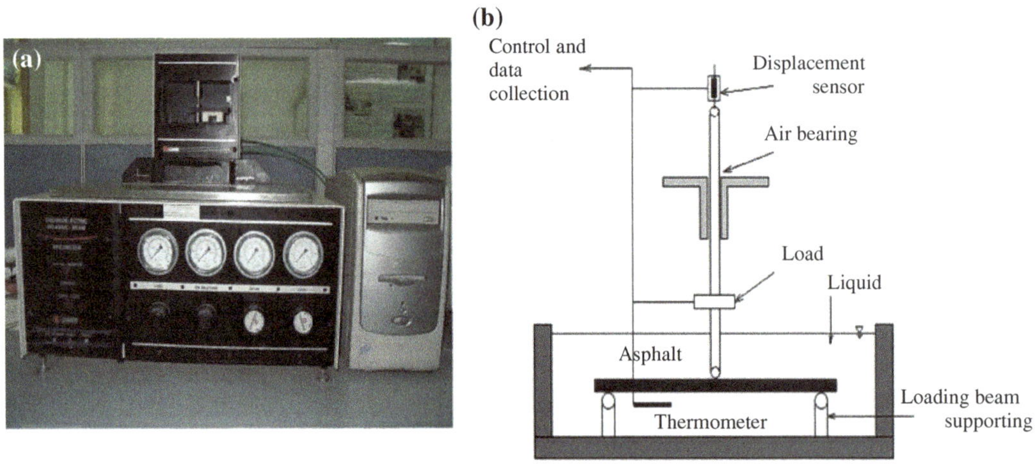

Fig. 4 Bending beam rheometer and its principle diagram. **a** Photo. **b** Work principle

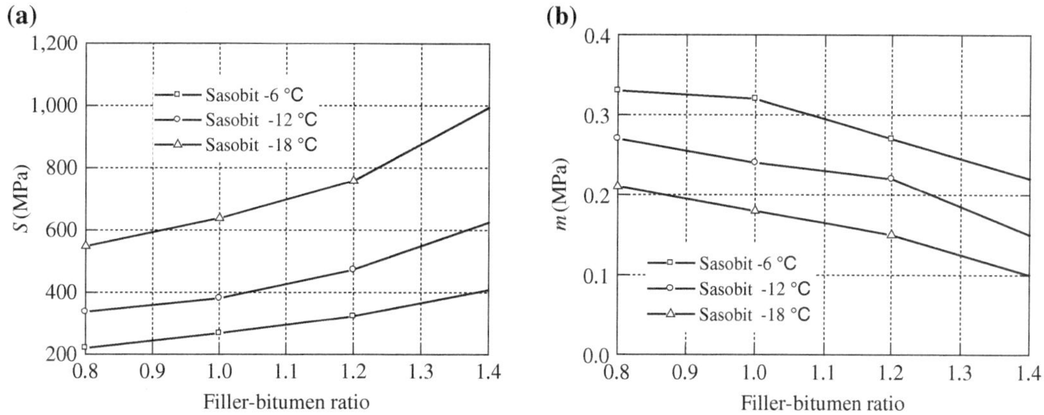

Fig. 5 BBR test results of SB modified asphalt mortar. **a** Bending creep stiffness modulus versus filler–bitumen ratio. **b** Creep rate versus filler–bitumen ratio

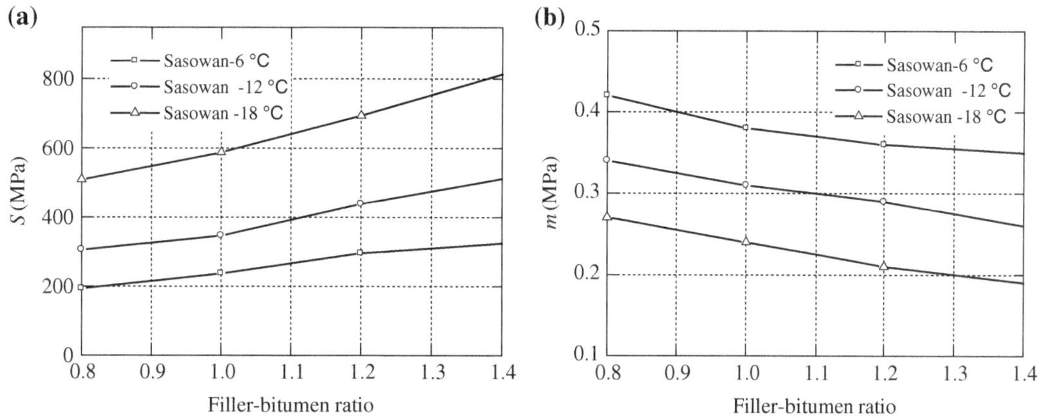

Fig. 6 BBR test results of SW modified asphalt mortar. **a** Bending creep stiffness modulus versus filler–bitumen ratio. **b** Creep rate versus filler–bitumen ratio

Table 2 Test results of modified asphalt's ductility (cm)

Additive	Temperature (°C)	Filler–bitumen ratio			
		0.8	1.0	1.2	1.4
SB	25	4.26	4.81	5.49	6.57
	28	3.43	4.06	4.52	5.38
	33	2.42	2.72	3.13	3.98
SW	25	3.94	4.41	5.27	5.88
	28	3.04	3.61	4.17	4.82
	33	2.15	2.46	2.84	3.37

mortar's structure, but also has great influence on its viscosity and low-temperature brittleness [22–24]. The ductility of asphalt mortar, compared to asphalt mixture, is more accurate to reflect the low-temperature ductility. This paper discusses ductility value of the asphalt mortar at a certain tensile speed (5 cm/min) under the different filler–bitumen ratios and different temperature conditions. Test results of ductility values of the asphalt mortar are shown in Table 2 and Fig. 7.

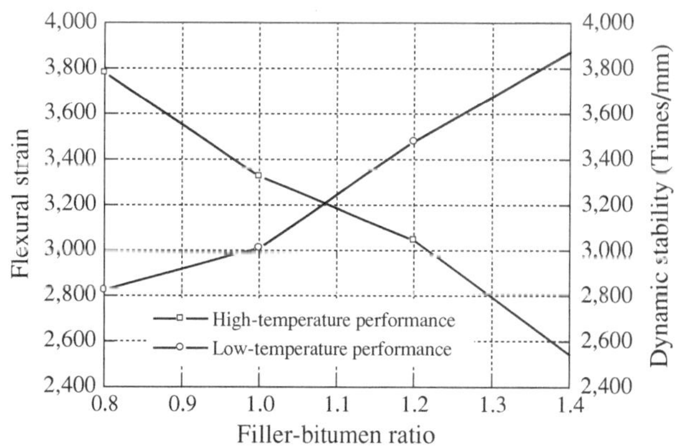

Fig. 7 Test results of SB modified asphalt mixture at high and low temperatures

Table 2 shows that the ductility of the asphalt mortar decreases with the increase of filler–bitumen ratio. Hence, the filler–bitumen ratio should not be too great. Here we suggest that it should be less than 1.2.

From Fig. 7 one can see that the bending strain at low temperature decreases gradually with the increase of filler–bitumen ratio, and the low-temperature performance reduces sharply when filler–bitumen ratio continues to increase from 1.2 to 1.4. However, the high-temperature performance increases gradually with the increase of filler–bitumen ratio. Compared to the low-temperature performance, the high-temperature performance has smaller increase amplitude when filler–bitumen ratio continues to increase from 1.2 to 1.4. According to the high-temperature and low-temperature performance results, the filler–bitumen ratio of asphalt mortar with additive SB should be less than 1.2.

6 Conclusions

(1) According to the above research results, we know that rutting factor increases gradually with the filler–bitumen ratio and has a largest amplitude at 64 °C. The increment of rutting factor decreases with temperature. The filler–bitumen ratio of asphalt mortar with additive SB has a larger effect on anti-rutting factor at low temperature than that of SW. The increment of anti-rutting factor after aging treatment becomes smaller than that of before aging, which shows that ore powder can prevent asphalt aging. Thus, ore powder not only can increase the anti-rutting factor but also can increase the aging resistance of asphalt.

(2) According to two additives' performances and different testing temperatures, the influence of SB on anti-rutting factor is greater than that of SW. Hence, SB has better performances than SW at high temperatures.

(3) With the increase of ore powder's dosage, anti-rutting factor increases gradually, and the increasing amplitude is less affected by temperature. When the filler–bitumen ratio reaches 1.2, the fatigue factor of the asphalt mortar with SB increases significantly; while for SW it is not so obvious. Compared to SB, the fatigue factor of the specimens with SW has smaller amplitude increase. Hence, the filler–bitumen ratio of SB should be less than 1.2.

(4) According to low-temperature performance test, when the filler–bitumen ratio of the asphalt mortar with SB continues to increase above 1.2, creep stiffness modulus increases rapidly, while creep rate and ductility decrease rapidly in the ductility test. Hence, filler–bitumen ratio for that of SB should be less than 1.2, while that of SW can be larger than 1.2 since SW has better performances than SB at low temperatures.

(5) The influence of additive on asphalt mortar performance can be obtained through the high-temperature and low-temperature and ductility tests on warm mix asphalt mortar. The recommended range of filler–bitumen ratio of the asphalt mortar with SB is between 0.8 and 1.2.

References

1. Zhang ZQ, Wang YC (2006) Influence of asphalt mortar on hot mix asphalt performance at high and low temperature. J Chang'an Univ Nat Sci Ed 26(2):1–5 (in Chinese)
2. Feng H (2008) Research on experimentation characteristic of asphalt mortar with viseo-elasticity theory. Dissertation, University of Science and Technology, Changsha (in Chinese)
3. Brice D, Herve DB, Pierre C et al (2007) Linear viscoelastic properties of bituminous materials: from binders to mastics (with discussion), J Assoc Asph Paving Technol 76:455–494
4. Brown ER, Haddock JE, Crawford C (1996) Investigation of stone matrix asphalt mortars. J Transp Res Board 1530: 95–102
5. Gubler R, Liu Y, Anderson DA et al (1999) Investigation of the system filler and asphalt binders by rheological means. Association of Asphalt Paving Technologists, 68(2):284–304
6. Witczak MW, Kaloush K, Pellinen T (2002) Simple performance test for super-pave mix design. Transportation Research Board, National Research Council. National Academy Press, Washington, pp 172–180
7. Zhang ZQ, Zhang WP, Li P (2004) Ratio of filler bitumen of asphalt mixture. J Chang'an Univ Nat Sci Ed 24(5):7–10 (in Chinese)
8. Kim YR, Little DN (2004) Linear viscoelastic analysis of asphalt mastics. J Mater Civil Eng 16(2):122–132
9. Yousefi AA, Kadi AA, Roy C (2000) Composite asphalt binders: effect of modified RPE on asphalt. J Mater Civil Eng 12(2): 113–123
10. Neguleseu II, Daly WH, Mohammad LN et al (2000) Recycling of polymer modified asphalt pavement—a binder study. Transportation Research Board 79th Annual Meeting, Washington, DC, 9–13 Jan 2000
11. American Association of State Highway and Transportation Officials (1993) AASHTO guide for design of pavement structures. AASHTO, Washington
12. National Cooperative Highway Research Program, Guide for mechanistic empirical design of new and rehabilitated pavement structures final report.
13. American Association of State Highway Transportation Officials, method of test for determining rutting susceptibility of asphalt paving mixtures using the asphalt pavement analyzer.
14. Ministry of transportation (2008) Technical specifications for construction of highway asphalt pavements. China Communications Press, Beijing (in Chinese)
15. Huang YB, Qin L (2005) Methods to improve asphalts and discussion of improved asphalts capacity. J Guangxi Univ Nat Sci Ed 30(1):122–124 (in Chinese)

16. Chen HX, Yuan YJ, Zhang ZQ et al (2005) Influencing factors in dynamic shear rheometer tests. Central South Highw Eng 30(2):142–164 (in Chinese)
17. The Federal Aviation Administration (2008) Hot mix asphalt mixture calculates the manual. People's Communication Press, Beijing (in Chinese)
18. Duan SR, Ren RB (2001) Application of dynamic shear rheometer (DSR) evaluation of pavement performance. Central South Highw Eng 4:81–83
19. Lin XX (2002) The discussion on the Superpave composition ratio characteristic. East China Highw 2:45–47 (in Chinese)
20. Feng IL (2006) Rheological analysis of asphalt tested by bending beam rheometer. J Wuhan Univ Technol 30(2):205–208 (in Chinese)
21. Li YW, Wang JS, Huang WY et al (2011) Warm mix asphalt pavement construction technology. China Architecture &Building Press, Beijing (in Chinese)
22. Zhang YL (2011) Asphalt pavement construction technology and quality control. People's Communication Press, Beijing (in Chinese)
23. Li YS (1995) Creep and relaxation of 4 kinds of rock under uniaxial compression tests. Chin J Rock Mech Eng 14(1):39–47 (in Chinese)
24. Zheng NX, Zhang X, Li A (2004) Discuss of problems in SHRP performance grading system. J Highw Transp Res Dev 9:55–59 (in Chinese)

2

Evaluation of alternative criteria for determining the optimal location of RWIS stations

Tae J. Kwon · Liping Fu

Abstract This article presents a comprehensive framework for determining the location of road weather information system (RWIS) stations over a regional road network. In the proposed methodology, the region is divided into a grid of equal-sized zones which are considered as the minimum spatial unit for allocating a candidate set of RWIS stations. These zones are ranked according to a set of pre-specified criteria that reflect the needs for, and potential benefits from, real-time road weather information, including road surface temperature variability, precipitation, network traffic, and collision patterns. A case study based on the existing RWIS network in the province of Ontario was conducted to illustrate the major features of the proposed method and evaluate the implications of alternative location selection criteria. The findings of the study suggest that it is feasible to develop a systematic process for locating RWIS stations using an integrated location criterion to capture multiple factors being considered in practice. The study has also revealed the need to establish quantitative models for estimating the benefit of real-time information from RWIS stations, which is the foundation of a cost–benefit-based RWIS location optimization model.

T. J. Kwon (✉) · L. Fu
Department of Civil and Environmental Engineering, University of Waterloo, 200 University Ave West, Waterloo, ON N2L 3G1, Canada
e-mail: tkwon@uwaterloo.ca

L. Fu
e-mail: lfu@uwaterloo.ca

L. Fu
School of Transportation and Logistics, Southwest Jiaotong University, Chengdu, China

Keywords RWIS · Road surface temperature · Accident rate · Traffic volume · Location selection criteria

1 Introduction

Accurate and timely information on road weather and surface conditions (RSC) in winter seasons is a necessity for road authorities to optimize their winter maintenance operations and improve the safety and mobility of the traveling public. One of the primary tools for acquiring this information is a road weather information system (RWIS). The term RWIS is generally used to describe a system consisting of many environmental sensors placed at the roadside or in the roadway, connected together for collecting real-time local atmospheric and pavement conditions information [1]. Data collected by these series of sensors are subsequently examined and/or manipulated, and utilized in generating site-specific road weather and condition forecasts. With such information, hazardous highway segments are flagged to alert maintenance personnel for prompt maintenance operations, which in turn could effectively reduce salt usage and, thus, detrimental effect on the environment [2]. Furthermore, RWIS is essential for some innovative snow- and ice-control strategies such as direct liquid application (DLA, i.e., de-icing and anti-icing), and fixed automated spray technology (FAST) [3]. More importantly, some prior studies have shown that localized road weather and condition information from RWIS can produce direct benefits of reduced costs because of, for instance, efficient use of labor, materials, and equipment for WRM operations [4]. Moreover, there are also some indirect benefits of using RWIS information. For instance, motorists can save their travel time and reduce fuel consumption from the enhanced traffic signal timing, and can also be warned of real-time hazards from dynamic warning systems [5].

Acknowledging these potential and perceived benefits of RWIS, many transportation agencies have invested millions of dollars in establishing their current RWIS network and continue expanding their network for better winter maintenance decision support and traveler information provision. While effective in providing real-time information on road weather and surface conditions, RWIS stations are expensive to install and operate, and therefore can only be deployed at a limited number of locations. Considering the vast road network varied conditions of which need to be constantly monitored locally and regionally during winter weather events, RWIS stations must be placed strategically so that they can collectively provide informative inputs required for accurate estimation of the road weather and surface conditions of the whole highway network.

There are a few studies conducted previously, which have attempted to determine the location of RWIS stations at the regional level. Eriksson and Norrman [6] undertook a study on optimally locating RWIS stations in Sweden where they identified conditions hazardous to road transport as a criterion for locating RWIS stations. In their study, they identified 10 different slipperiness types using one winter season of RWIS data, and linearly regressed each type with location attributes including latitude, longitude, elevation, distance to coast , etc. With the resulting regression model, they mapped out the occurrences of each slipperiness type over the entire study area. Candidate RWIS sites were recommended based on the estimated slipperiness counts and four different landuse groups. Although the proposed method in that study seems to provide a good reference for analysis of station locations with respect to various locational attributes and landuse types, it is a heuristic approach considering only one location criterion—road weather condition. In addition, those authors did not provide much explanations/justifications as to how their four landuse groups such as forest/water, open/water, forest, and open areas were determined. Such categorization schemes can be deemed subjective and thus scientifically less persuasive.

World Weather Watch [7] conducted a study involving a climatological analysis for the purpose of optimizing the RWIS station locations. The study extensively reviewed micrometeorological variations by investigating local physiography, topography, temperature, and snowfall precipitation amount within the study area. The study also took into account the hotspots identified by the maintenance personnel which require regular monitoring. By combining all those factors, a list of high-risk sites are identified as the recommended locations for new RWIS stations in the region. A similar yet more inclusive study was carried out by Alberta Department of Transportation in collaboration with AECOM and Telvent, in which they proposed a sound approach to determine the location and density of RWIS stations, by identifying and analyzing the RWIS-deficient regions (RDR), and by following general budget guidelines, respectively [8]. Their approach consists of two components namely macro and micro assessments. In the macro-assessment phase, the authors took several factors into account when determining the RDR such as traffic loads, accident rates, climatic zones, availability of meteorological information, and discussions with regional road maintenance personnel, and key stakeholders. In the micro-assessment phase, among the selected subset of potential new RWIS locations, a final site was selected by conducting detailed field visits to ensure their suitability and feasibility (e.g., appropriate sensor selection/configuration, budget, access to power, etc.) for installing a new RWIS station.

These past studies, while logical in methodology, lack scientific and systematic formulation of justification on how all those factors/criteria are put together to determine the potential high-risk sites. More importantly, both studies had failed to provide a clear linkage between the considered criteria and the purposes of RWIS stations. For instance, the latter study included accident rates as one of their potential location selection criteria, but did not establish a solid rationale as to why such a criterion should be given a priority when choosing a new location. In other words, what is the relationship between the accident rate and the availability of a RWIS station? Without a valid justification/explanation on why each location selection criterion is considered and utilized, incorporation of such selection criteria in the studies cannot be legitimized.

Currently, majority of provincial and municipal transportation agencies do not have a systematic approach for determining the suitable RWIS locations, but rather rely heavily on the experience of regional/local maintenance personnel. According to the survey conducted for synthesizing the current best practices on how RWIS stations are deployed, most respondents claimed to consider different weather zones, which are the areas with historically heavy snow, high wind, low-lying road segments, etc. As for the non-weather-related factors, they would mainly consider collision rate, traffic loads, frequency of WRM operations, and road class. All the information is put together through a series of face-to-face meetings with key stakeholders and field experts to narrow down various candidate locations to a manageable size and decide based on the budget availability. Finding a resolution through this process is laborious and time-consuming one thereby requiring greater manpower.

As described above, past efforts and current practices for determining the location of RWIS stations have several limitations. In particular, what constitutes a good RWIS location? What location selection criteria should be included in the analysis for determining the potential RWIS stations? What are the pros and cons of these criteria used for locating RWIS stations? How easily can the RWIS station location scheme be applied, adopted, and utilized by the RWIS network expansion planners? These questions represent the main concern of this study.

The main objective of this study is to construct a systematic approach/framework for determining the candidate RWIS station location. The article is organized as follows. The first section presents an introduction to the topic under review. The second section discusses several criteria for identifying a "good" location of RWIS stations that are also used as a basis of location-ranking criteria in this study. The third section describes the proposed methodology to be undertaken in this study. Analysis and evaluation of considered criteria using a case study based on Ontario RWIS network are presented in the fourth section. The last section is composed of conclusions and recommendations for future study.

2 Regional location selection criteria

It is critical to clearly define the criteria that can be used to measure the "goodness" of a location for installing a RWIS station. To identify the criteria, we must start with determining the values or benefits of RWIS stations. As discussed previously, RWIS stations are installed to collect real-time RSC data, and their values are reflected in the use of these RWIS data, including improved mobility and safety (i.e., beneficial for motorists), and reduced winter road maintenance (WRM) costs and salt usage (i.e., beneficial for agency and environment). These benefits are the basis for the criteria defined in the following section.

2.1 Weather-related factors

Intuitively, RWIS stations should be installed in areas that experience severe yet, less-predictable weather patterns, which thus need real-time monitoring the most. Therefore, it is important to analyze the spatial distribution patterns of some critical weather variables such as temperature and precipitation. For example, the variability of surface temperature (VST) and mean surface temperature (MST) are important factors to be considered as they can provide a definite measure of how much the surface temperature would vary over time and space. Late November to early December is the time of year with the highest probabilities of having black-ice or frost. The higher elevation and greater distance away from large water bodies can both contribute to generating colder surface temperatures. This gradually leads to a period of longer winter months with higher likelihood of having frost on road surface, and thus exposes motorists to great danger.

Snowfall amount is another important factor as it makes logical sense that a RWIS station needs to be situated in areas where snowfalls occur frequently. This is particularly true when, having a better monitoring capability can intuitively increase mobility and safety by performing a more prompt WRM operation [4]. Hence, these three weather factors are included in the analysis for selecting a good location of a RWIS station.

2.2 Traffic-related factors

The benefits received by RWIS can intuitively increase by placing them in locations where a greater number of traveling publics can be served. For this reason, many transportation agencies are inclined to incorporate macro-level traffic criteria such as collisions, traffic volume, and highway type (HT) [1, 2]. A recent study conducted by Greening et al. [9] showed that a well-maintained RWIS network would in fact reduce the accident rates by a significant amount, which in turn would bring huge savings.

Notwithstanding the fact that they ignored other confounding factors such as a general vehicle technology advancement, which could also have led to the reduction of accident rates, their study clearly demonstrated that the use of RWIS information could potentially prevent accidents by prompt WRM operations. Furthermore, the survey dedicated to providing the current practices of deploying a RWIS system showed that more than 60 % of DOTs who participated would also consider highway class along with collision rate and traffic volume. Their intension for taking highway class into account is similar to considering traffic volume in the context of providing benefits to more number of road users. As such, the three traffic-related factors mentioned above are included as location selection criteria in this study.

3 Methodology

In this section, the methodology undertaken to carry out the analysis is briefly described.

Figure 1 shows a flowchart of the proposed method for analyzing and evaluating different location selection criteria for choosing an ideal RWIS station location. As shown in the figure, the first step is to divide the region under investigation into a grid of equal-sized zones (i.e., cells) which are assumed to be the minimum spatial unit for allocating a candidate set of RWIS locations. Once the grid covering the entire study area is created, then major road segments, including highways and arterials, are superimposed onto the grid in such a way that only the cells containing the road segments can be selected for further analysis. This step helps one single out only the candidate locations of RWIS stations from all cells, reducing the degree of complexity by having only the "important" cells in the analysis and evaluation phases. Six different criteria are prepared, processed and/or modeled (a point that will be discussed later in this article), and are then used to assign an averaged unique value to each cell in the grid such that each cell can be readily ranked to determine "more" or "less" important sites. Finally, these ranked location criteria are evaluated using Ontario's current RWIS network.

4 Case study

4.1 Ontario RWIS network

Ontario Ministry of Transportation (MTO) installed its first RWIS station in 1991 as a research and development project. Over the years, the number of stations has increased, mostly in the southern part of Ontario where the road network is the most dense. Since then, the benefits of RWIS being an effective tool for proactively deploying WRM operation have been recognized [2]. As such, MTO installed additional 85 stations between 2001 and 2004 in a series of three phases. Phase 1 (25 stations) primarily focused on installing along the key economic corridors (i.e., Hwy 401, 400, and 417), whereas Phases 2 and 3 (60 stations) focused on installing stations in the areas where larger gaps previously existed in both the northern and the southern parts. MTO has been expanding the network each year and currently has a total of 140 RWIS stations as depicted in Fig. 2. Figure 2 shows black pushpins and red lines representing the location of individual RWIS stations and the Ontario highway network, respectively. The figure also shows a grid of equal-sized cells each having an area of 50 $\times$ 50 km^2. This spatial resolution was decided based on the survey, which revealed that the most provinces and states would typically keep a distance of 40–50 km between two stations. However, it is important to note that such a criterion is not a requirement, and can vary according to the needs.

As described earlier, the road network is used as a constraint to extract only the matching cells as the other cells that do not match cannot have a RWIS station (e.g., a cell covering, for instance, the lake is not considered as a potential candidate site).

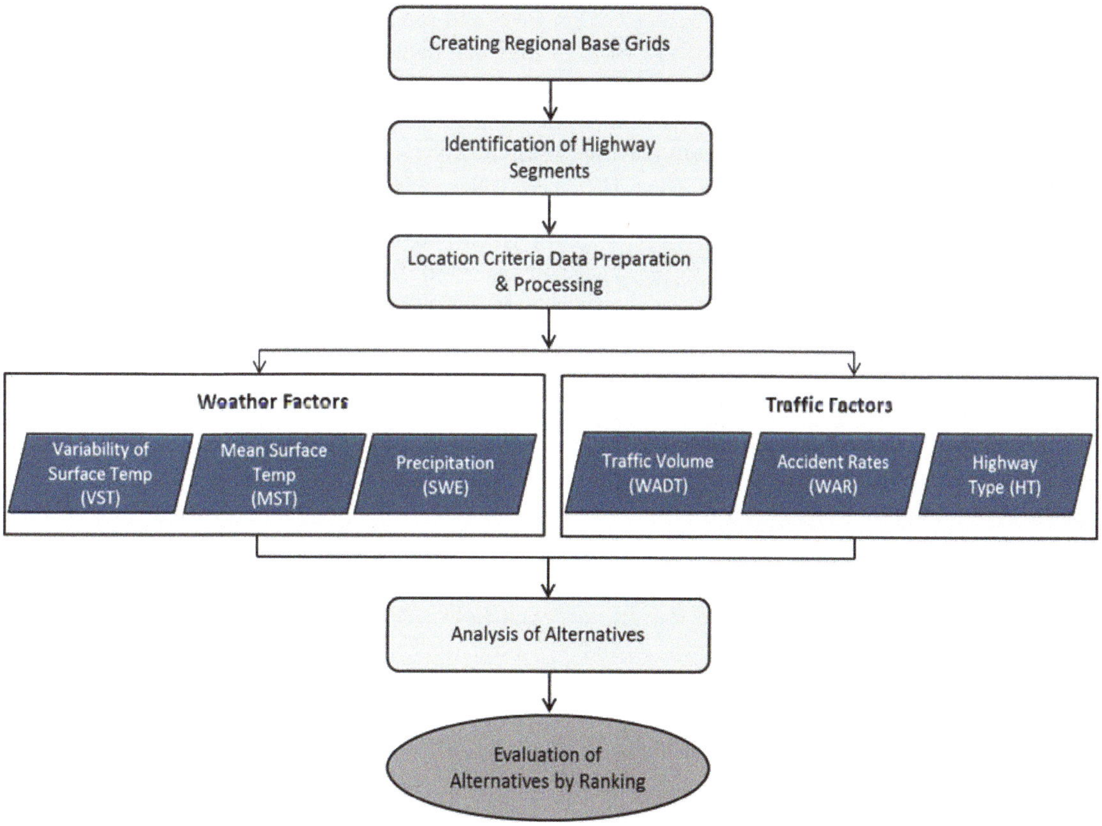

Fig. 1 Flowchart of evaluating alternatives for choosing a RWIS station location

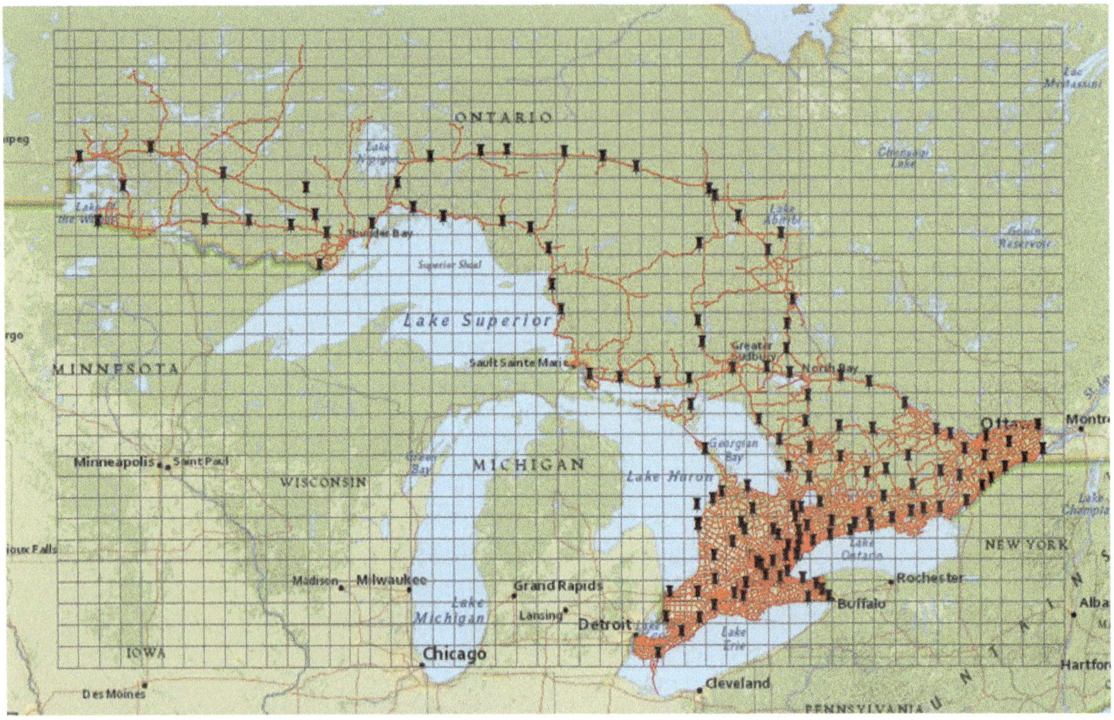

Fig. 2 Current Ontario RWIS Network with 140 stations

4.2 Data preparation and analysis of the location selection criteria

4.2.1 Modeling of VST and MST

The two dependent variables of interest are VST and MST, which can be defined by calculating the standard deviation of surface temperature, and the average of minimum surface temperature, respectively, using multiple-year data. The environmental variables (i.e., auxiliary information) used in this study include latitude (*lat*), longitude (*long*), elevation (*elev*), distance from water (d_w), and relative topography (RT_1, RT_3, and RT_5). The justification for choosing such variables are that latitude is expected to affect the spatial variation of surface temperature, whereas longitude may capture the influence of winds. Elevation in meters above mean sea level could be linked to the VST (e.g., the higher the elevation, lower the temperature), and distance from large water bodies, the Euclidean distance in kilometers, represents the degree of continentality [6]. Finally, relative topography is included to describe the exposure, and is calculated by taking the difference in elevation between each station location and an average of pixels within the respective radius range (e.g., 1, 3, and 5 km). Since the monthly variation of surface temperature can vary significantly from one month to another, the two dependent variables, VST and MST, are modeled on a monthly basis. Owing to data unavailability and deficiency, 3-year surface temperature data collected during 2006–2008 from a total of 45 Ontario RWIS stations for the month of January are used for modeling. ArcGIS 10.1 is used as a base platform for this study, where digital elevation model (DEM) with 1-km spatial resolution as well as water layers including lakes and sea are utilized to obtain the aforementioned auxiliary information. Once all the required information has been obtained, SPSS is used to perform the multiple linear regression (MLR) analysis with all variables being tested at the 5 % significance level. The resulting equations for two dependent variables are as follows:

$$VST = 0.403(lat) + 0.076(long) + 0.161(dis_w) - 0.011(RT_5) - 5.974, \quad (1)$$
$$R^2 = 72.2\%$$

$$MST = -2.398(lat) - 0.518(long) - 0.016(elev) + 0.296(dis_w) - 0.049(RT_3) + 61.937, \quad (2)$$
$$R^2 = 88.3\ \%$$

To construct the continuous surface to be used later for evaluating different alternatives, a similar approach is undertaken to extract all the required auxiliary information from all available cells, where the above calibrated equations are used to calculate both VST and MST values for each cell. Figure 3 shows the VST and MST maps. Note that only the cells that contain roads are used for analysis as previously described.

From Eqs. (1) and (2), it can be clearly seen that all regression coefficients make intuitive sense. For instance, latitude, longitude, and distance to water have a positive correlation with VST implying that as values of each

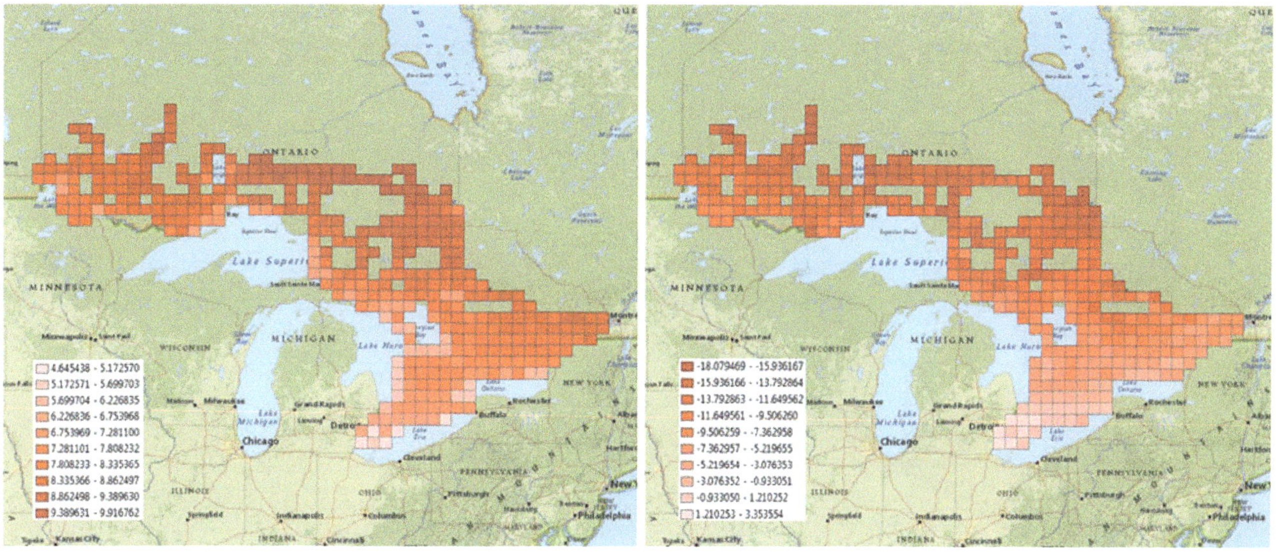

Fig. 3 Processed VST (*left*) and MST (*right*) Maps

parameter increase, so does the VST. It is true particularly during colder months that surface temperature would likely vary at a greater range at high latitude regions. Furthermore, VST is likely higher as it moves into deeper continents which are typically more mountainous causing a larger temperature variation. These phenomena are explained graphically in Fig. 3 (left) that VST cells in southern regions and/or nearby lakes tend to exhibit less variation compared with other cells. As for the MST, all the regression coefficients except for distance to water are found to be negative. This is observed since the minimum surface temperature would drop as it moves to the northern regions with higher elevations. Notice that MST can vary by a significant amount (~ 20 °C) between the northernmost and southernmost cells in Fig. 3 (right). A positive correlation of distance to water represents that as it moves away from water, temperature would likely drop because of the lack of significant water bodies nearby [10].

4.2.2 Processing and integration of other location selection criteria

4.2.2.1 Precipitation (snowfall) Distribution of precipitation amounts, particularly of snowfall amounts must thoroughly be investigated to determine the regions where heavy snowfalls are likely to occur so that recommendations for RWIS stations can be made accordingly. This can be done by analyzing the long-term historical snowfall observations. Daymet is an online weather data archive where daily surface weather and climatological summaries are available for public use [11]. Snow-water-equivalent (SWE), which describes the amount of water contained within the snowpack expressed in kg/m^2, is estimated such that biases in shortwave radiation estimates related to multiple reflections between the surface and atmosphere are reduced [12]. The average annual summary maps for SWE are obtained for periods from 2001 and 2005 covering the entire North America. Since these files came in a raster format, 5-year map is generated by taking a sum of all available SWE layers using ArcGIS 10.1. Once all the maps have been combined and averaged, each cell is assigned with the corresponding SWE value (i.e., sum of all SWE data within each cell) for the entire grid. In Fig. 4, it can be seen that central regions seem to have the most snowfalls, and the amounts are being gradually diminished as it moves to outer regions. The farthest southern regions appeared to have the least amount of snow.

4.2.2.2 Traffic volume (WADT), accident rate (WAR), and highway type (HT) It was emphasized earlier that a RWIS station should be located in places where traffic volumes and accident rates are high such that the benefits to road users can be maximized. This reasoning can also be applied to HT where higher classes of highways should be given a priority when installing a RWIS station. For this reason, winter average daily traffic (WADT), winter accident rate (WAR), and HT have been collected and processed to be implemented as RWIS location selection criteria. Traffic information management is centered at the MTO-provided data for the 2000–2010 provincial highways winter traffic volumes and accident numbers along with highway class, all of which are geocoded with linear highway-referencing system (LHRS). LHRS can be defined as a location-referencing method to easily identify a specific location with respect to a known point [13]. MTO currently has a total of 2,588 geocoded locations across the province as shown in Fig. 5a. With these geocoded locations, WADT data have been mapped onto the grid where the data are averaged and

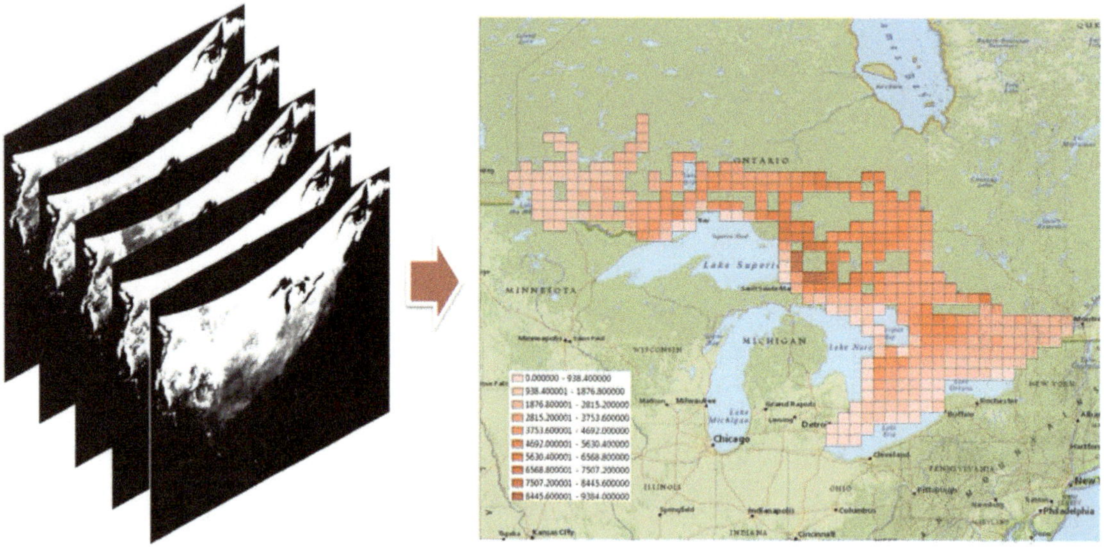

Fig. 4 Five annual average SWE maps (*left*) and the 5-year Average SWE map (*right*)

assigned to the corresponding cells. WAR used in the analysis is defined as the number of reportable accidents occurring during winter months on a particular highway section for every million vehicle kilometers (MVK) traveled on that section during the same period, and can be expressed as

$$WAR = \frac{Accident\ Number\ (WAN)}{\left(\frac{WADT \times 212 \times Section\ Length\ in\ km)}{1{,}000{,}000}\right)} \quad (3)$$

Representative section lengths for all geocoded points are used when calculating WAR. As for HT, four different types that are currently being used and have been classified by MTO are used. Following the similar approach of the other traffic dataset, HT data have been first geocoded using the LHRS, and then the averaged values are assigned to each cell. Figure 5b–d depicts the processed WAR, WADT, and HT, respectively. Note that all these three figures are classified by 10 equal intervals. As can be seen in the figure, WAR and WADT data are appeared to share some common traits that there are many number of "high-risk" cells in the southern region for having relatively heavier traffic loads and higher accident rates. This makes a logical sense that increase in exposure would likely increase the number of accidents. On the other hand, northern regions consist of many low-valued cells indicating that they are less important when considering traffic as a location criterion. Similar conclusions can be drawn by analyzing the HT figure that a great number of high-class highways are situated in the southern region suggesting the needs of RWIS stations.

4.3 Evaluation of alternatives

In this section, different alternatives are evaluated by applying them to "relocate" Ontario's existing 140 RWIS stations and comparing the results to their current locations. For each alternative, the candidate cells are sorted based on their values of the given selection criterion in a descending order and the top 140 cells are identified accordingly as the new best locations based on that alternative. The percent of cells (locations) matched is calculated as a performance measure.

Before the evaluation, a correlation analysis has been carried out to resolve any possible occurrence of multicollinearity, which could essentially affect and thus produce biased results when two or more factors are added. Spearman's coefficient of rank correlation, denoted as ρ is a nonparametric measure of statistical dependence between any two ranked variables, which are assumed to have a monotonic relationship. Its product ranges between -1 and 1, indicative of strong negative and positive correlations, respectively. As such, this method was used to assess the correlation with other location selection criteria under analysis. The equation for Spearman's rank correlation coefficient is expressed as [14]

$$\rho = 1 - \frac{6 \sum d_i^2}{n(n^2 - 1)} \quad (4)$$

where d_i is a difference in paired rank; and n is a total number of cases, which in our case is 377 (i.e., a total number of cells in the grid). Table 1 summarizes the Spearman's rank correlation coefficients for all the variables considered in this study.

As anticipated (from a visual inspection of the earlier figures), VST and MST, and WAR and WADT are found to be moderately correlated. Such findings are reasonable since, for example, the rate of accident would vastly be influenced by the number of travelers, or vice versa. The

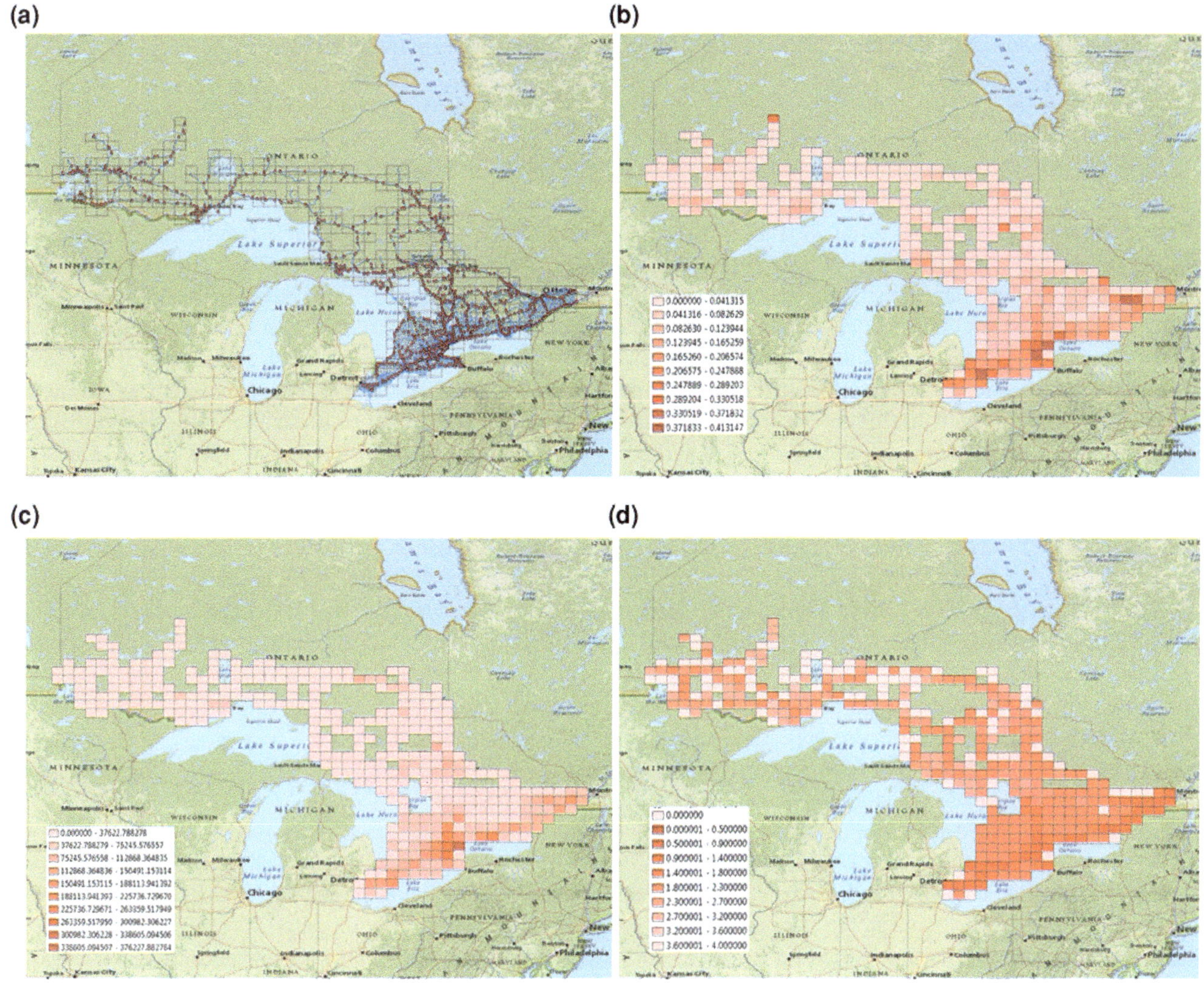

Fig. 5 **a** LHRS, **b** WAR, **c** WADT, and **d** HT

fact that VST and MST are positively correlated indicates that when MST is lower, its variability would be higher. It is interesting to see some degree of correlation between WADT and MST. This is probably because the low MST values are present mostly in the northern regions whereas a majority of high WADT values appear in the southern regions. Hence, two weather factors of VST and SWE, and two traffic factors of WAR and HT are considered for the evaluation.

4.3.1 Alternative 1: weather factors combined

Two pre-defined weather factors namely VST and SWE are used to evaluate the current RWIS network in the province of Ontario. VST values in each cell have been added to the corresponding cell of SWE. Note that both factors are normalized with a range between 0 and 1 to eliminate any potential bias. Figure 6 describes the result of the combined location selection criteria. It also shows the current

Table 1 Spearman's rank correlation coefficients

	VST	MST	SWE	HT	WAR	WADT
VST	1	0.643638	0.194449	−0.42406	−0.4318	−0.43333
MST	–	1	0.298177	−0.46556	−0.50343	−0.54187
SWE	–	–	1	−0.12457	−0.16166	−0.1729
HT	–	–	–	1	0.39289	0.351165
WAR	–	–	–	–	1	0.555482
WADT	–	–	–	–	–	1

Ontario RWIS stations that have been superimposed on the map. Highlighted cells represent the top 140 cells that are recommended as potential RWIS station locations.

As can be seen in the figure, a map generated by combining two weather factors suggests a potential RWIS site to be in the middle to upper part of the region, where VST and SWE are also found to be significant in those regions. A percent of matching is found to be 24 % by having 33 cells matched with the actual locations of the existing RWIS stations. Notice that there are many cells in the central regions (i.e., highlighted cells) where no RWIS stations are there to monitor the highly varying weather conditions with historically heavy snowfall events. From this analysis, it can be stated that the current RWIS location setting is less susceptible to capturing the variability of weather conditions.

4.3.2 Alternative 2: traffic factors combined

A second alternative contains the two predetermined traffic-related factors namely WAR and HT. Figure 7 illustrates the proposed 140 locations of RWIS stations when considering the traffic factors only. Notice that the map now focuses more on the areas where high accident rates/highway class exist. This alternative suggests that almost all southern parts of the province should have RWIS stations installed while most parts in the northern region are left uncovered. Ninety-three of the 140 existing RWIS stations (66 %) would be located at the same sites based on this alternative. Such a high matching rate should not be viewed as an indication that this location criterion is the best; instead, it should rather be considered as an indication that these factors are most heavily weighted in Ontario's current RWIS location practice.

4.3.3 Alternative 3: weather and traffic factors combined

A third alternative is proposed by combining both weather and traffic factors to balance out the deficiencies and limitations of alternatives 1 and 2. Figure 8 shows the proposed 140 locations where RWIS stations are recommended to be sited when considering all four factors previously explained. A visual inspection of the result shows that the identified cells are well distributed over the entire province. The percent of matching for alternative 3 is found to be smaller (i.e., 57 %) than that of alternative 2. However it is worthwhile to emphasize that such a number is based on the evaluation with respect to the current locations of Ontario RWIS stations, and thus does not provide an absolute measure of the performance. While the current RWIS setting is able to provide a reasonably good coverage on both traffic and weather factors, the large difference between the two suggests that the RWIS stations may not have been located optimally.

5 Conclusions

In this study, a general framework for choosing the location of a RWIS station at a regional level has been developed. Several location selection criteria are proposed on the basis of scientific and/or rational justifications that they would, to a certain extent, help us delineate good RWIS locations.

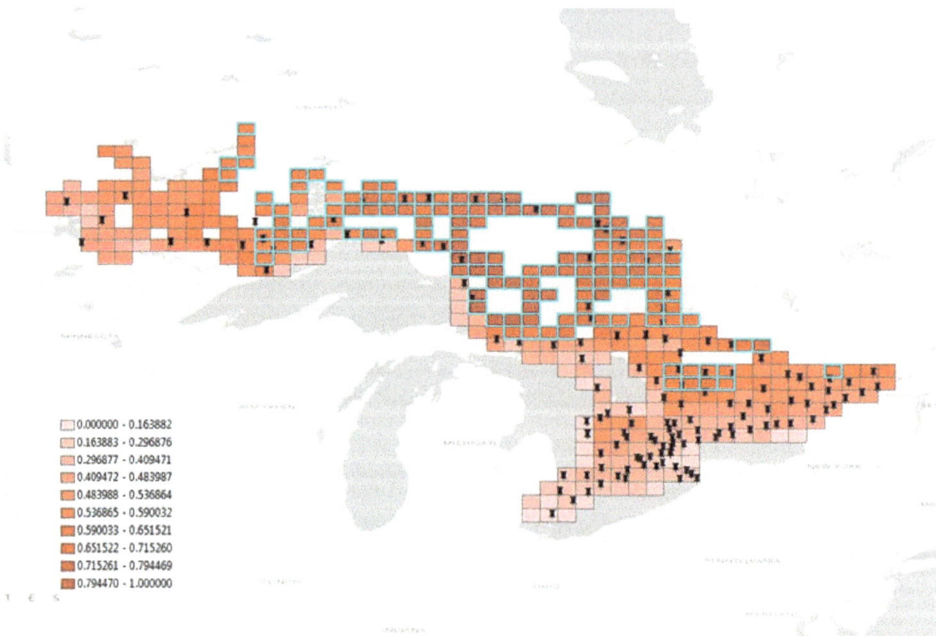

Fig. 6 Alternative 1 with the two weather factors combined

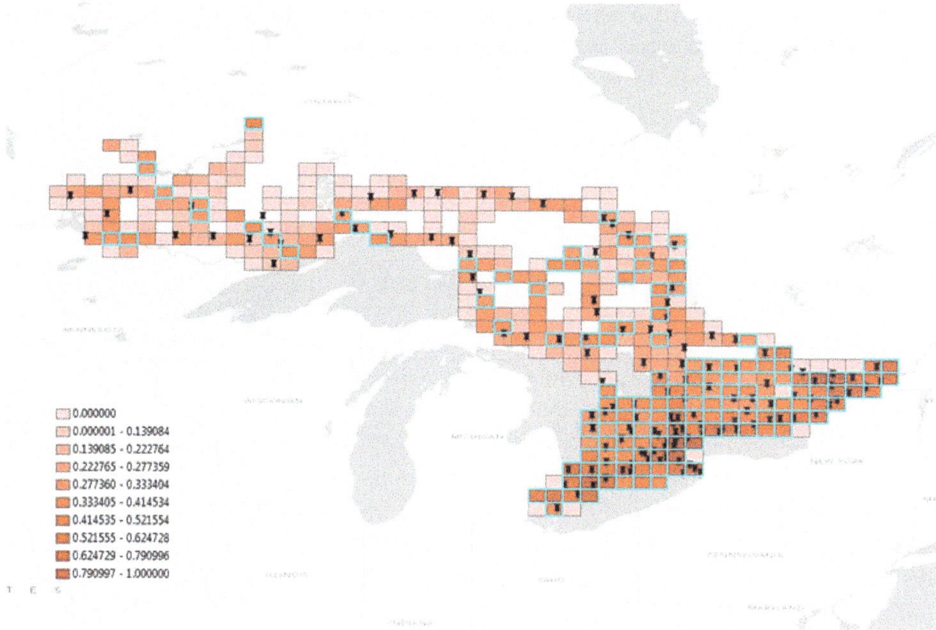

Fig. 7 Alternative 2 with the two traffic factors combined

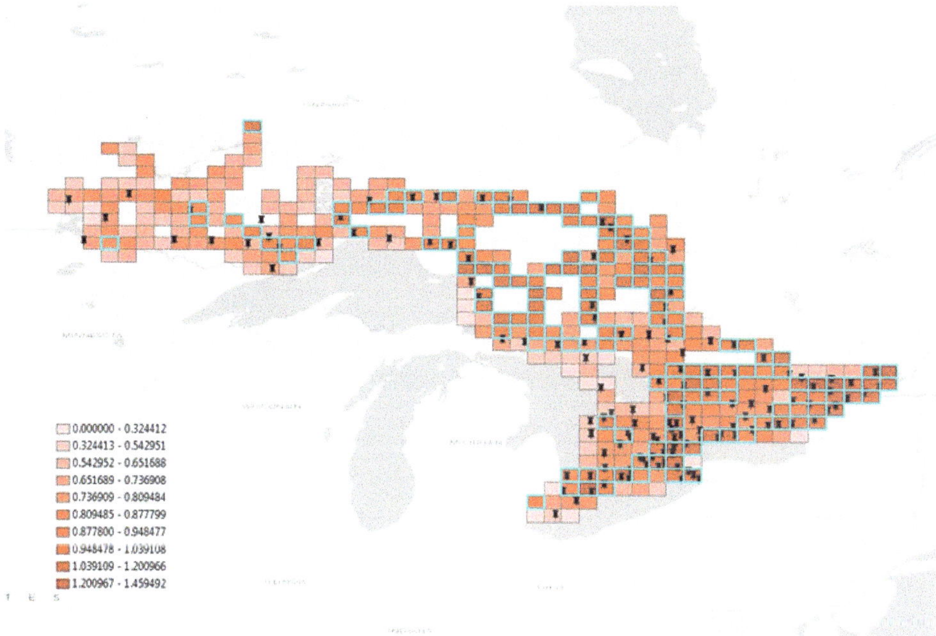

Fig. 8 Alternative 3 with all four factors combined

Three weather factors considered include VST, MST, and snow water equivalent (SWE), of which VST and MST are modeled by taking into account the several locational attributes. WADT, WAR, and HT are also considered under the rationale that serving a higher number of traveling publics would provide more benefits. After confirming the multicollinearity between the considered variables, a total of three location selection alternatives have been developed: alternatives 1 and 2 take into account two weather factors, VST and SWE; and two traffic factors, WAR and HT, respectively. Alternative 3 is a combination of alternatives 1 and 2. These alternatives are used to evaluate the current Ontario RWIS network. A percent of matching is utilized as an evaluation metric for benchmarking the current location setting. The findings have revealed that the alternative 1 is more focused on the northern region comprising of highly varying weather conditions with 24 % matching rate, while the alternative 2 is more focused on the southern region with heavy traffic loads with 66 % matching rate. The high matching rate of the alternative 2 indicates that the current RWIS network has been set up in such way that it predominantly considers the need of covering the road network. The alternative 3 with 57 % matching rate seems to balance the limitations of the first two alternatives by showing the potential candidate RWIS locations across the whole province. It is unknown how much of weight needs to be put on each of the criterion discussed here to have an optimized RWIS network, but it is clear that the proposed framework is easy to apply when planning a RWIS network expansion by introducing different weights to individual criteria based on their importance.

This research can be extended in several directions. First, VST and MST models can be improved by utilizing a geostatistical interpolation technique such as kriging. Several studies have found that kriging would provide a better estimation than regression, especially when variables are spatially dependent on each other [15, 16]. Second, the use of LHRS needs to be improved since the cells that do not have a reference point will have no data. Third, methodical guidelines need to be established for determining a number of RWIS stations to be allocated within a cell. This is particularly important for DOTs wanting to install more than one RWIS station within, for instance, as used herein minimum spatial unit of 50 km^2 to enhance and extend their monitoring capability and spatial coverage. Finally, it is of necessity to establish quantitative models for estimating the benefit of using information from RWIS stations, with which a cost-benefit based RWIS location optimization model can be developed.

Acknowledgments The authors wish to thank Max Perchanok, Heather McClintock, Mary Anne Grepsma, and Bill Harrett at the Ontario Ministry of Transportation. This research was funded by the Aurora Program, and was partially funded by National Sciences and Engineering Research Council of Canada (NSERC) and Ontario Ministry of Transportation (MTO).

References

1. Garrett JK, Boyce B, Krechmer D (2008) Implementation and evaluation of RWIS ESS siting guide. Federal Highway Administration Road Weather Management Program
2. Buchanan F, Gwartz SE (2005) Road weather information systems at the ministry of transportation. Transportation Association of Canada, Ontario
3. Canadian Strategic Highway Research Program (2000) Anti-Icing and RWIS Technology in Canada
4. Ye Z, Strong S, Fay L et al (2009) Cost benefits of weather information for winter road maintenance. Final report, Iowa Department of Transportation
5. Boselly SE (2001) Benefit/cost study of RWIS and anti-icing technologies. National Cooperative Highway Research Program, TRB
6. Eriksson M, Norrman J (2001) Analysis of station locations in a road weather information system. Meteorol Appl 448(8):437–448
7. World Weather Watch (2009) Climatological assessment for the purposes of siting road sensors. Final Report
8. Mackinnon D, Lo A (2009) Alberta Transportation Road Weather Information System (RWIS) expansion study. Alberta Transportation
9. Greening K, Johns D, Bridge P et al (2012) A study to determine the effects of employing a well maintained RWIS network on accident rates on major highways in the US state of Idaho. Standing International Road Weather Commission, Helsinki
10. Ahrens CD (2009) Meteorology today: an introduction to weather, climate, and the environment. Brooks/Cole, Cengage Learning, Belmont
11. Thornton PE, Thornton MM, Mayer BW et al (2012) Daymet: annual surface weather on a 1 km grid for North America, Acquired on 16 December 2012 from Oak Ridge National Laboratory Distributed Active Archive Center, Oak Ridge.
12. Thornton PE, Hasenauer H, White MA (2000) Simultaneous estimation of daily solar radiation and humidity from observed temperature and precipitation: an application over complex terrain in Austria. Agric For Meteorol 104(4):255–271
13. Baker W, Blessing W (1974) Highway linear reference methods, synthesis of highway practice 21, National Cooperative Highway Research Program. National Academy Press, Washington
14. Steel GDR, Torrie JH (1980) Principles and procedures of statistics: a biometrical approach, 2nd edn. McGraw-Hill, New York
15. Hengl T, Heuvelink GBM, Stein A (2003) Comparison of kriging with external drift and regression-kriging. International Institute for Geo-Information Science and Earth Observation (ITC), Enschede
16. Mesquita S, Sousa AJ (2009) Bioclimatic mapping using geostatistical approaches: application to mainland Portugal. Int J Climatol 29(14):2156–2170

3

Examining the effect of speed, roadside features, and roadway geometry on crash experience along a rural corridor

Donald C. Watson Jr. · Ahmed Al-Kaisy · Nathan D. Anderson

Abstract This paper presents a current investigation into crash experience along a 15.7-mile rural corridor in southwest Montana with the aim of better understanding crash causal factors along the corridor. The study utilized ten years of crash data, geometric data, and observed free-flow speed data along the corridor. A systematic approach was used where every tenth of a mile was described in term of the crash experience, speed, alignment, and roadside features. Using bivariate and multivariate statistical analyses, the study investigated the crash experience along the corridor as well as some of the underlying relationships which could explain some of the crash causal factors. Results show a strong association between crash rates and horizontal curvatures even for flat curves that can be negotiated at speeds above the posted speed limit, per the highway design equations. Higher crash rates were also found to be associated with the difference between the observed free-flow speeds and the speed dictated by the curve radius or sight distance as per the design equations. Further, results strongly support the safety benefits of guardrails as evidenced by the lower crash rates and severities. The presence of fixed objects and the steepness of side slopes were also found to have an effect on crash rates and severities.

D. C. Watson Jr.
289 North Lake Lulu Drive, Winter Haven, FL 33880, USA

A. Al-Kaisy (✉)
Western Transportation Institute, Montana State University, 213 Cobleigh Hall, Bozeman, MT 59717, USA
e-mail: aalkaisy@ce.montana.edu

N. D. Anderson
16304 South Yancey Lane, Spangle, WA 99031, USA

Keywords Crashes · Rural · Analysis · Alignment · Speed · Roadside

1 Introduction

The highway system in the U.S. is critical for the economic and social development in this country as it allows for people and goods to travel efficiently in rural and urban areas as well as between towns and cities. However, this high level of mobility for people and goods comes at a remarkable price as suggested by the recent crash statistics. Vehicle collisions cost the U.S. billions of dollars each year, and resulted in an estimated 34,080 fatalities in 2012 [1]. Therefore, improving safety on the highway system has been one of the priorities at the national level. For the traffic safety professional community, it is of critical value to understand the association between crash occurrence and other related variables such as highway geometry, roadside features, driver characteristics, and environmental factors. To this end, two different approaches are typically employed. Site-specific crash analyses at high-crash locations are used with the objective of identifying the underlying crash contributory factors at those locations. The second type of analysis is more aggregate in nature, and crashes are characterized on a highway network with the objective of discerning trends and patterns between crash occurrence and any of the variables mentioned earlier. While the first type of analysis is more microscopic in that the analysis focuses on a single site, the latter is more macroscopic as the analysis addresses the whole network of the highway system at the city, state, or national level. The data required for each analysis type are different in terms of the level of detail and the availability of accurate location information.

The research presented in this paper is concerned with a new approach in analyzing crash data along an extended stretch of a rural corridor that is known to have higher-than-average crash experience. The approach used in this study is macroscopic in that it examined the whole corridor systematically without focusing on any specific crash or any particular site. However, the approach can also be viewed as microscopic in that detailed crash-related information at a relatively high resolution was used where every tenth of a mile was characterized as an individual segment within the corridor.

2 Background

In this section, the most important studies that attempted to establish the relationship between crash experience and any of the underlying variables as related to roadway and traffic attributes are presented. Further, the analysis of non-intersection highway segments in regards to segment length is also presented. Anderson and Krammes [2] discovered that curves with a degree of curvature greater than 4 had higher accident rates. These curves required speed reductions, while curves with lower values did not. Similar results were found by Lamm et al [3]. Conversely, a study on roads in Virginia observed a decrease in sideswipe and run-off-the-road crashes (ROR) with an increase in curvature [4]. Radius was also shown to affect accident rates as Caliendo and Lamberti [5] noticed a decrease in accident rates with the increase in radius for radii between 200 and 500 m. Cenek et al. [6] observed this relationship for a wider range of radii, while Hauer [7] observed this relationship for all radii. Hauer also found that curves with large deflection angles are more hazardous than those with smaller values. A few other studies evaluated other geometric variables such as lane width, shoulder width, pavement type, skid resistance, annual average daily traffic (AADT), spiral transitions, and passing behavior [6, 8–14]. Another two articles described the relationship between speed and curvature [15, 16]. Another study by Tate and Turner found that the difference between the negotiation speed and design speed on curves was strongly related to the injury crash rate [17]. Negotiation speed was defined as the 85th percentile speed of free-flow vehicles (those with headways greater than 6 seconds). Results showed that as the difference in speed exceeded 15–20 km/h, crash rates increased significantly and that this relationship deteriorated when non-injury crashes were included. Other studies compared accident rates to standard deviation of speed, mean speed reduction, and difference between operating speed and speed limit [2, 4, 18, 19]. These studies derived similar conclusions.

The relationship between operating speed and accident frequency was also evaluated. One study found that higher operating speeds generally led to fewer accidents [19]. This was attributed to the influence of geometric features on drivers' speed. While curves pose more risk, they are negotiated at lower speeds. Another study found that accident severity was affected by operating speed with a 1 % increase in the average operating speed resulting in a 0.074 % decrease in the number of minor injuries and a 0.095 % increase in the number of fatalities [20]. Other research found that higher speed limits increased the probability of a more severe accident and that accident severity increased outside level and straight roadways [21, 22].

In regards to the segment length used for crash analysis outside intersection-related sites, one study found that there is no definitive length which performs better than any other and that the length of segment used depends solely on the type of research being done [23]. Rather than focusing solely on fixed length segments, a study by Koorey [24] concluded that variable length segments perform better in crash analysis. Because fixed segments are based on length and variable segments are based on geometric and environmental features, the latter more accurately represents roadway characteristics. Similar conclusions to those of Koorey were reported by Mayora and Rubio [25], who used both fixed and variable length segments to investigate crash rate prediction in Spain's two lane rural roads. They found that fixed length segments did not perform as well in correlation analysis as compared to variable length segments. For both segment types, access control, sight distance, and design consistency were found to have the highest correlation to crash rates. Sight distance was also found to be significant in a study done by Caliendo and Lamberti [5] on four lane median separated roads.

There are good reasons to believe that, despite the many studies in the literature that attempted to understand the relationships between crash occurrence and traffic and roadway features, those relationships are not fully understood with a reasonable level of certainty. The current study is an effort to contribute to the existing limited knowledge using the US-191 corridor case study presented in this paper.

3 Study site

The site investigated in this study consists of a 15.7-mile stretch of the US-191 highway running through the Gallatin Canyon (between mileposts 48.3 and 64) in southwest Montana. This highway segment was designed in 1954 to accommodate the natural contours of the Gallatin River, creating many sharp curves with limited sight distance for drivers at a few locations. This particular stretch was chosen because it was identified as one of the high crash rural corridors in the state of Montana [26]. Currently, it has a posted speed limit of 60 mph with advisory speed limits on some of its restricting curves. In 2008, the AADT at this corridor was in the order of 6300 vehicles per day [27].

4 Study design

4.1 Approach

The crash corridor in this study involved several restrictive curves, non-restrictive curves, and tangent segments. Restrictive curves contain limited sight distance and/or small radii that restrict travel speed relative to the speed limit, while non-restrictive curves have sufficient sight distance and greater radii that do not restrict travel speed. Tangent segments exist between horizontal curves and constitute the majority of corridor by distance. This study attempted to investigate crash experience at the corridor level using a mesoscopic approach. Specifically, while no focus was given to any particular location within the corridor and the whole corridor was treated uniformly, the analysis involved portioning the corridor into very short segments, each of which was associated with several descriptors using spatial, geometric, and speed data. The unit length used for the segment was one tenth of a mile, which is considered very short given the rural context of the corridor. An important consideration in selecting this length was the resolution of the spatial information for crash data obtained from the Montana Department of Transportation (MDT). Using crash data and segment descriptors, the analysis in this study attempted to identify the underlying relationships between crash experience on one hand and road geometry and speed characteristics on the other hand.

4.2 Data collection and processing

Significant amounts of data were needed in this investigation including data on crash experience, horizontal alignment, roadside features, and prevailing free-flow speed.

- *Crash data* Ten years of crash data for the study corridor were obtained from the MDT and included the location of the crash up to 0.1 mile accuracy, crash attributes, drivers' attributes, and limited environmental information. A total of 356 crashes occurred in the study corridor for the period between January 1, 2000 and December 31, 2009.
- *Horizontal alignment* Detailed engineering design drawings for the study corridor were obtained from the MDT and used to extract information on horizontal alignment related to this study.
- *Roadside features* Information on roadside features was gathered by the research team using detailed field observations. Four aspects of roadside features were of interest to this study: side slope, presence of guardrail, presence of fixed objects, and the relative location of the valley/river (or otherwise mountain wall) to the roadside. Only those objects within 30 feet of the travel lane were included in the study.
- *Speed data* Speed data was collected at 44 locations along the study corridor for the two directions of travel. Speed radar guns were used to collect speed observations of vehicles traveling at headways of six seconds or greater to ensure that vehicles were not following (interacting) with the lead vehicles. The appropriate procedures were followed when using the equipment to attain high accuracy of speed measurements. Specifically, the distance from the road's centerline and between the vehicle and observation point was measured in the field, especially on tangent segments. This information was used later to adjust for the effect of the firing angle between the direction of motion and the radar line, as deemed applicable. On most curved sections, the device was aimed at vehicles at the appropriate angle, thus requiring no further adjustment. Further, the research team tried to take speed observations without being seen by drivers to avoid any effect on the observed speeds. The sample size for each data set was estimated using a level of significance of 0.1, speed standard deviation of 5 mph, and a tolerance of $\pm$ 1.5 mph. The type of vehicle was also included in the speed study field sheet using three main classes: cars, recreational vehicles, and heavy vehicles including trucks and buses.

The objective of conducting these extensive speed measurements was to establish a free-flow speed profile along the study corridor. In establishing the speed profile, it was assumed that vehicle speed on tangents and very flat curves remained relatively constant outside the acceleration and deceleration areas. Using this free-flow speed profile, deviation from the speed dictated by radius, sight distance, or speed limit, whichever was lowest, was derived and used in the analysis. The term used to describe this deviation is ΔV.

4.3 Study crash indicators

This study investigated crash experience on the US-191 corridor using four different crash indicators; crash frequency, severity index, combined safety index, and crash rate per mile distance. The crash frequency is the number of crashes occurring during the study period for all crash types or for a specific crash type such as the frequency of PDO, injury, or fatal crashes. The severity index takes into account the severity of crashes regardless of its frequency. To calculate the severity index, weights are to be applied to the different severity-type crashes. In this study, weights applied to the PDO, injury, and fatal crashes are 1, 5, and 15, respectively. The severity index is found by multiplying the frequency of each type of crash by its assigned weight, summing all values, and dividing the sum by the total frequency of crashes. The combined safety index takes both frequency and severity into consideration in describing crash experience. Again, this index utilizes different weights for each type of crash, i.e., PDO, injury, and fatal crashes. The weights used in this study are the same as those used for the severity index discussed earlier. To calculate the combined safety index, frequency of the PDO, injury, and fatal crashes are multiplied by their assigned weights and summed together into a single index value. Finally, crash rate by distance (in miles) was also utilized in the analyses involved in this study.

5 Study results

5.1 Preliminary examination of crash data

Out of the 356 crashes investigated in this study, 86 injury crashes and 6 fatal crashes occurred in the study corridor over the ten-year study period. Of all the drivers involved in these accidents, around 74 % were male drivers while the remaining 26 % were female drivers. The highest proportion of drivers involved in crashes belongs to the age group 20–29 years, followed by the age groups 30–39, and 40–49 years, respectively. Trucks claimed 13 % of the total crashes in the corridor. In terms of light condition, 61 % of crashes occurred during daylight, 34 % during dark, and the remaining 5 % of crashes occurred during dawn and dusk. Ice, snow, and slush on pavement were associated with 43.5 % of the crashes investigated by this study. Figure 1 shows the breakdown of corridor crashes by crash type over the ten-year study period. More than half of all crashes were ROR, with animal-vehicle collisions, and rear end crashes being the next most common types of crashes, respectively.

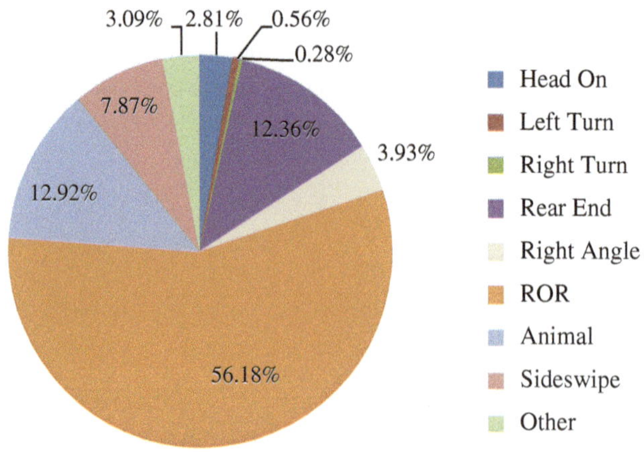

Fig. 1 Crashes on US-191 corridor broken down by crash type

Table 1 shows the number of vehicles involved in crashes broken down by crash severity for all crashes investigated by this study. The most important trend shown in this table is the fact that multiple-vehicle crashes represent higher percentages of fatal and injury crashes compared with the PDO crashes. The relationship between the number of vehicles involved in a crash and crash severity was found significant using the Pearson χ^2 test shown in Table 1. Using crash location relative to the corridor, the frequencies of PDO, injury, and fatal crashes along the corridor are shown in Fig. 2. As clearly shown in this figure, some segments have higher frequencies, severities, or both when compared to other segments in the corridor. Specifically, the segment between milepost 61 and 62 had the highest frequency among all other segments in the corridor. However, three out of the six fatal crashes that took place in the corridor over ten years occurred in the segment between mileposts 52 and 53.

5.2 Crash-alignment relationship

Each 0.1 mile segment of the study corridor was assigned an alignment descriptor which describes the most restrictive alignment within the segment. Tangent, non-restrictive, and restrictive horizontal curvatures were assigned to each segment. Non-restrictive curvatures are defined as those values that would safely allow vehicles to negotiate curves at speeds equal or greater than the posted speed limit. The restrictive curves, on the other hand, would force drivers to travel at speeds lower than the posted speed limit if they wanted to safely negotiate the curve. Those determinations are made using the well-known highway geometric design equations. Figure 3 illustrates the relationships between degree of curvature and crash frequency, crash severity, and the combined safety index. While the scatter plots do not clearly show a strong relationship between the variables, a relatively fuzzy trend can be discerned between the degree of curvature on one hand and crash frequency and the combined safety index on the other hand in both directions of travel. No specific trend can be discerned for the relationship between degree of curvature and crash severity. This is somewhat logical in that sharper curves may increase the likelihood of vehicles leaving the travel way (and consequently involved in a crash) while the severity of the crash is more dependent on other factors such as roadside design features.

Table 1 Number of crashes by severity type and number of vehicles involved in crash

	PDO	Injury	Fatality	Pearson χ^2 test
One-vehicle crash	195 (73.9 %)	49 (57 %)	2 (33.4 %)	$\chi^2 = 23.239$
Two-vehicle crash	65 (24.6 %)	28 (32.6 % %)	4 (66.6 %)	$P < 0.001$
Three-vehicle crash	4 (1.5 %)	9 (10.4 %)	0 (0.0 %)	
Total	264 (100 %)	86 (100 %)	6 (100 %)	

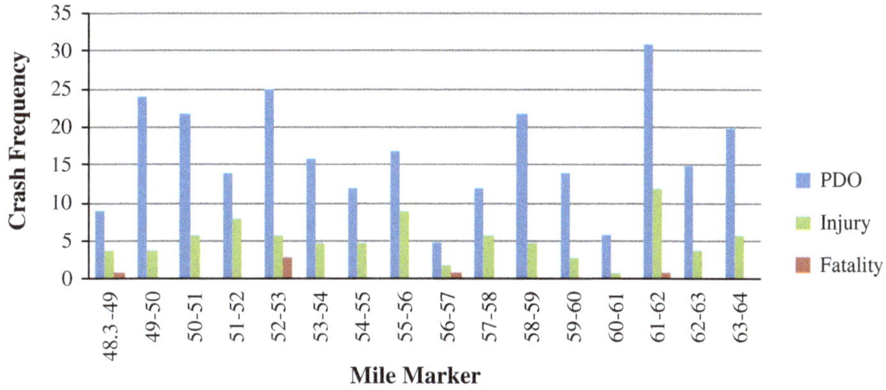

Fig. 2 Frequency of PDO, injury, and fatal crashes per 1-mile segment along study corridor

To further assess the fuzzy trends shown in Fig. 3, crash rate per mile per year was established for the three types of segments; those involving restrictive curves, those involving non-restrictive curves, and tangent segments in the two directions of travel. The results are shown in Table 2. Two important trends are clearly exhibited in this table. The first trend is that the crash rate is higher for restrictive curves, followed by non-restrictive curves, and tangent segments, respectively. This trend is consistent across crash type by severity in the two directions of travel. The other trend that can easily be discerned in this table is the fact that crash rates in the northbound direction are higher than those in the southbound direction. One possible explanation is that higher average speeds were observed in the northbound direction as compared to the southbound direction.

5.3 Examination of the crash-speed relationship

The speed variable that is used in this study (ΔV) is the difference between prevailing free-flow speed and speed dictated by radius or sight distance on restrictive curves and the difference between free-flow speed and speed limit on non-restrictive curves and tangents. It is believed that this variable is mainly a function of highway alignment. Figure 4 shows scatter plot of free-flow speed and ΔV with respect to the degree of curvature within the study corridor in the two directions of travel. The general trend shown in this figure is that the higher the degree the curvature, the lower the free-flow speed. This is consistent with the expectation that horizontal curvatures restrict the selected free-flow speed on high speed highway facilities. In regards to ΔV, it is clear in this figure that higher ΔV values are generally associated with sharper curves, i.e., higher degrees of curvature.

To investigate the crash experience as a function of ΔV, crash rate per mile per year was established for a three-tier ΔV classification for each severity-type crash in the two directions of travel. The results of this analysis are presented in Table 3. A closer examination of the crash rates in this table reveals a clear trend; the greater the value of ΔV, the greater the crash rate. This trend is consistent throughout the table with one minor exception. Despite this observed consistency, the relationship between ΔV and crash severity was not found significant as suggested by the Pearson χ^2 results shown in Table 3.

5.4 Examination of crash relationship with roadside safety features

Four highway roadside features that are thought to have relationships with safety were investigated in this study. These features involve the presence of guardrail, side slope, the presence of fixed objects, and the presence of valley/river or the mountain wall adjacent to the travel lane. Figure 5 shows the average crash frequency, severity index, and combined safety index value for the aforementioned variables in the two directions of travel.

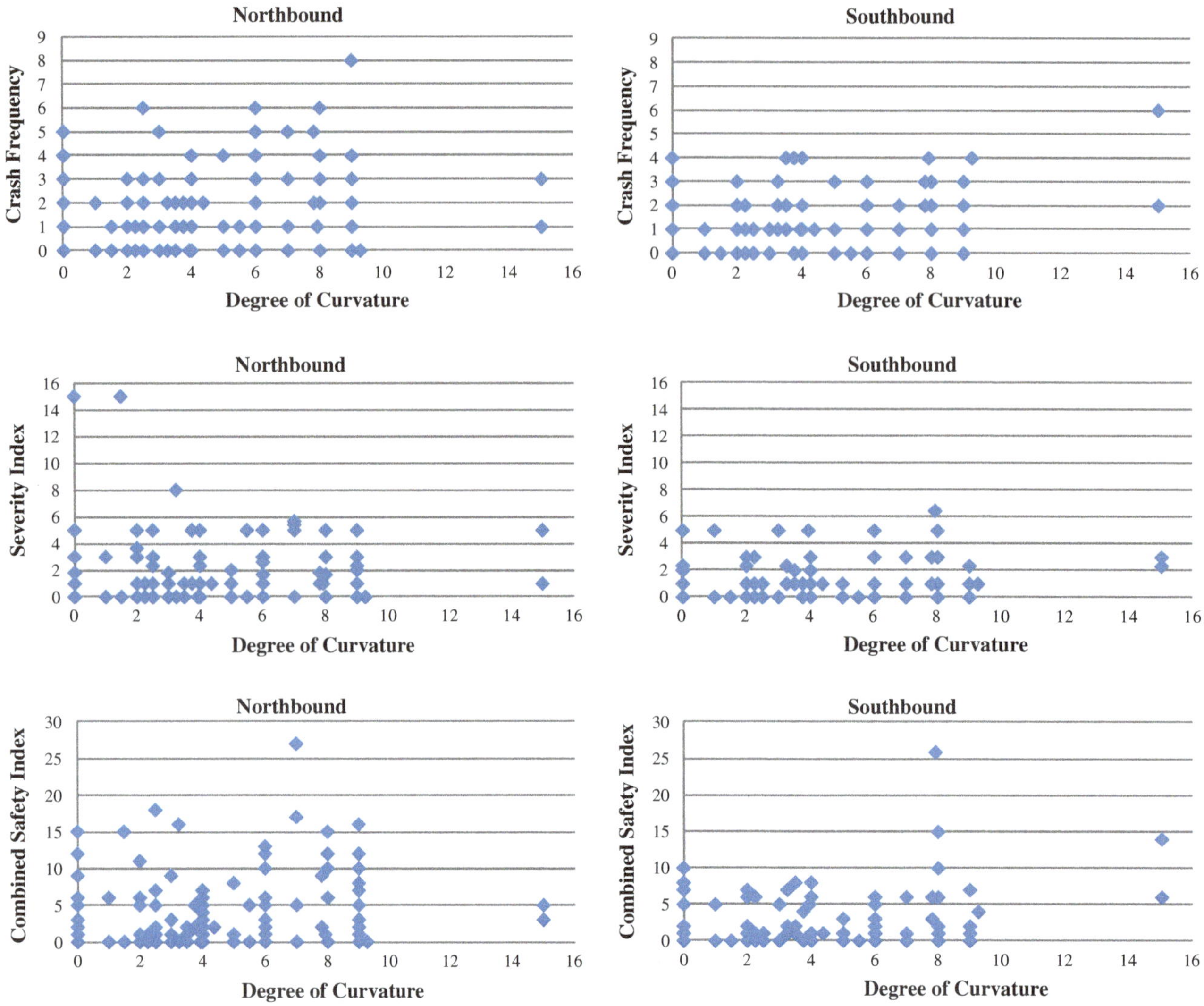

Fig. 3 Crash frequency, severity index, and combined safety index for various degrees of curvature in the two directions of travel

Table 2 Crash rate per mile per year on US-191 based on direction, severity, and alignment

	Restrictive Curves	Non-restrictive Curves	Tangents	Total
North PDO	1.584	0.827	0.471	0.945
South PDO	1.158	0.735	0.358	0.740
North injury	0.670	0.294	0.151	0.364
South injury	0.345	0.202	0.113	0.217
North fatal	0.041	0.037	0.019	0.032
South fatal	0.020	0.000	0.000	0.006
Total	3.819	2.096	1.112	2.304

In regards to the presence of guardrail, the overall trend that could be discerned from Fig. 5 is that segments with no guardrail installations had higher average values for crash indicators as compared to segments with guardrail installations. However, the effect of the amount of guardrail installations on the crash indicators may not be completely consistent. To more accurately assess this trend, the average safety indicators were calculated for segments with and without guardrail installations and the results are shown in Table 4. As clearly shown in this table, guardrail installations are consistently associated with a reduction in crash frequencies and severities, and consequently in the combined safety index. *t* tests were conducted at the 90 % confidence level to determine the significance of these reductions. Only differences in the crash frequency and combined safety index for the northbound direction were found significant as shown in Table 4.

Side slope is also thought to have an effect on the likelihood of vehicles recovering from an errant maneuver and on the severity of ROR. Figure 5 shows that moderate side slopes are associated with the highest crash indicators in both directions of travel. This is somewhat unexpected

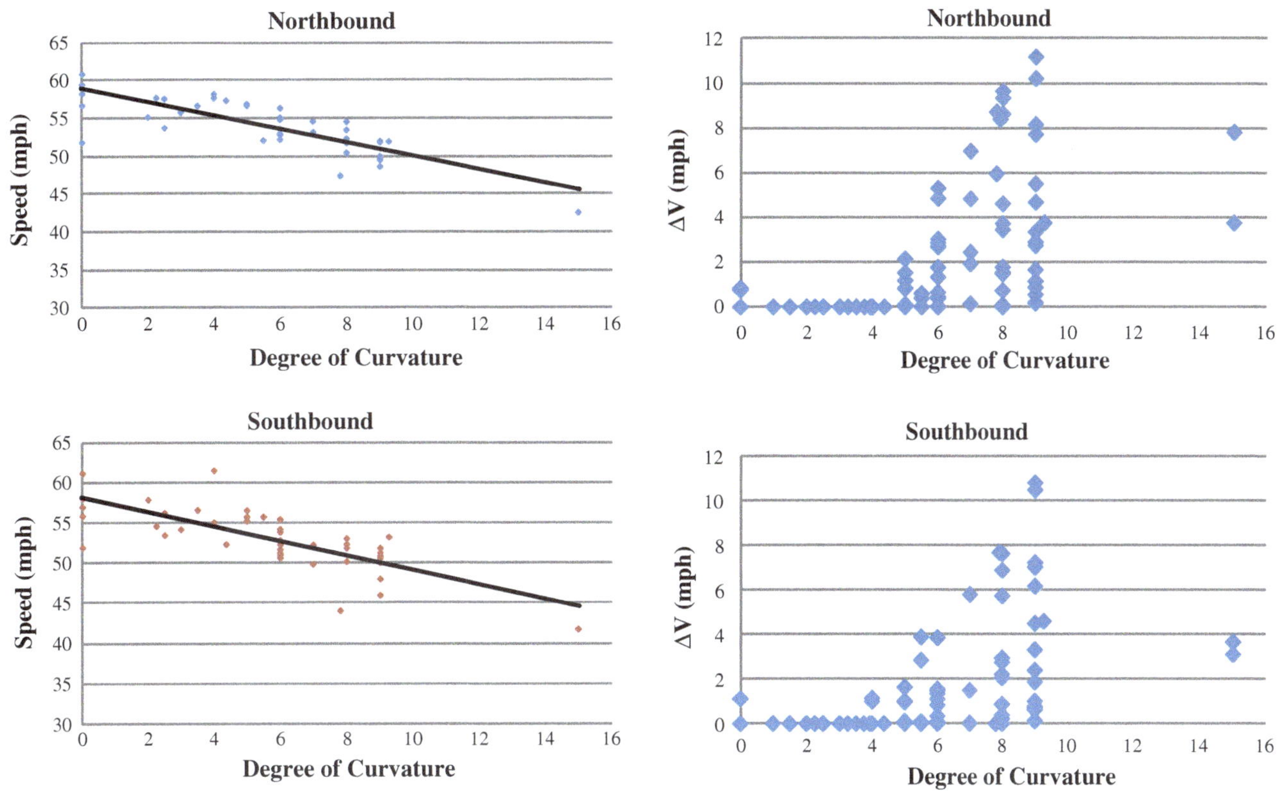

Fig. 4 Scatter plots of free-flow speed and ΔV versus degree of curvature

Table 3 Crash rate per mile per year as a function of ΔV for various crash types by severity

	$\Delta V = 0$	$0 < \Delta V < 4$	$4 \leq \Delta V \leq 12$	Pearson χ^2 test
NB—PDO	77 (0.828)	48 (1.043)	23 (1.211)	$\chi^2 = 0.492$
NB—Injury and Fatal	29 (0.376)	22 (0.478)	11 (0.579)	$P = 0.781$
SB—PDO	68 (0.701)	37 (0.787)	11 (0.786)	$\chi^2 = 4.629$
SB—Injury and Fatal	16 (0.165)	11 (0.234)	8 (0.571)	$P = 0.099$

as steeper side slopes should pose more hazards to the traveling public and contribute to the number and severity of crashes. To investigate this unexpected trend, a breakdown of study segments by guardrail installation and side slope was established as shown in Table 5.

As shown in Table 5, the guardrail installation is over-represented on segments with steep side slopes, a matter that is expected. This is believed to explain the lower average crash indicators for segments with steep slopes as guardrails prevent most vehicles from leaving the road and thus contribute to lower crash frequency and severity. The relationship between side slope and percent guardrail installation was found significant as suggested by the Pearson χ^2 results.

In regards to the fixed objects at the roadside, the research hypothesis is that the presence of fixed objects may contribute to the frequency and severity of crashes. A closer look at Fig. 5 shows a general consistency with this logical hypothesis, with the northbound direction being more consistent as compared to the southbound direction. To more quantitatively assess the effect of fixed objects on crash experience, the average crash indicator values for segments with and without fixed roadside objects were calculated and presented in Table 6.

This table clearly shows that segments with no fixed roadside objects consistently exhibited lower crash indicators compared with those that have fixed objects. This supports the logical hypothesis about the hazards posed by the presence of fixed objects at the roadside. While the mean crash indicators supported the aforementioned hypothesis, *t* tests conducted at the 90 % confidence level found the difference in crash indicators for sections with and without roadside objects significant only for the combined safety index in the northbound direction.

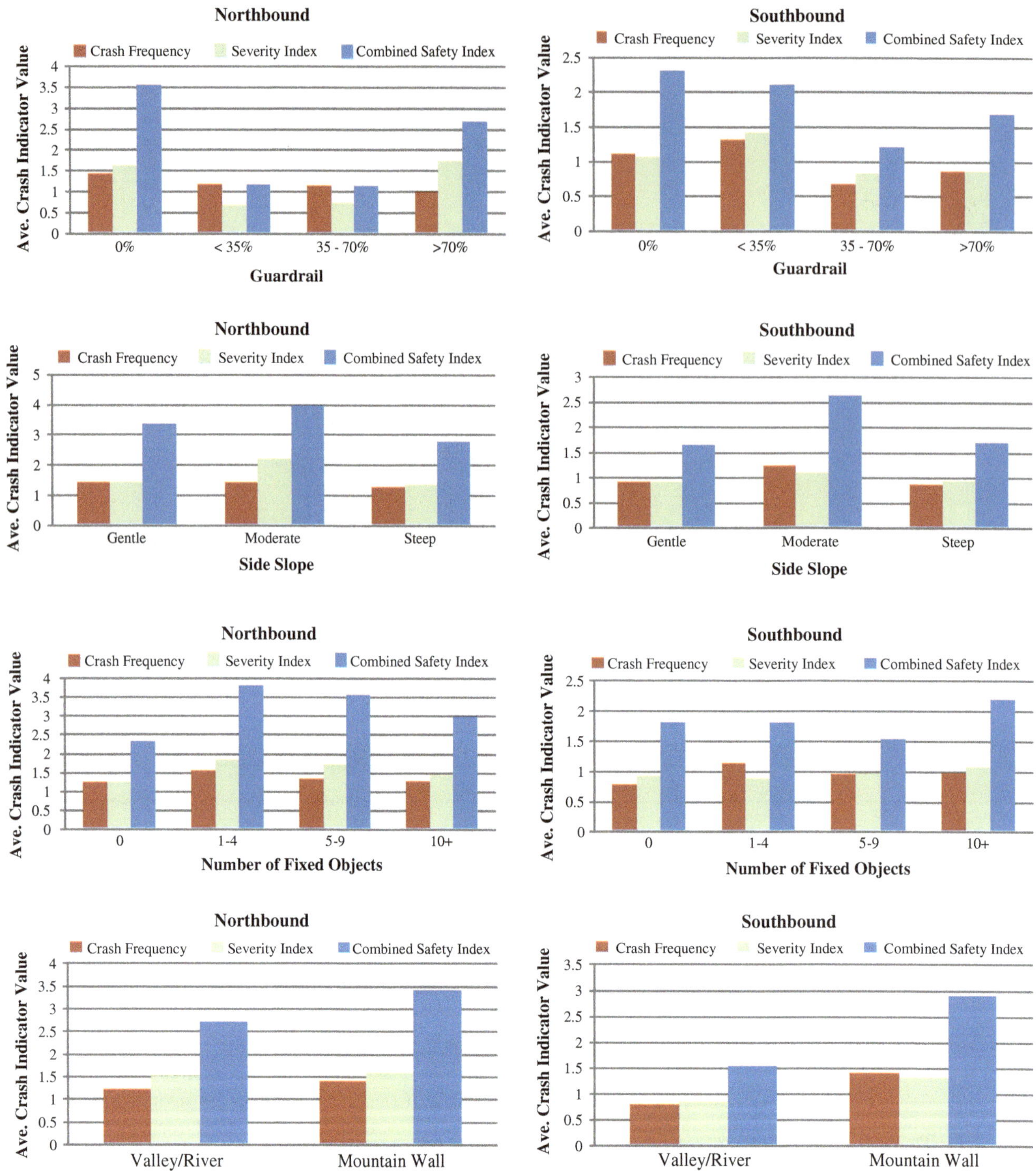

Fig. 5 Average crash frequency, severity index, and combined safety index in relation to the various roadside features

One other roadside feature that was investigated by this study is the adjacent roadside being on the side of the valley/river or on the other side, which is in most parts the mountain wall. The study corridor is a mountainous corridor with relatively winding alignment, and therefore, the valley/river and the mountain sides are often present in the roadway cross section with the exception of limited distances where the mountain is not in close proximity to the roadway. The trend shown in Fig. 5 is that the mountain wall side is associated with higher crash frequency and severity, and thus a higher combined safety index. To further understand this trend, it was necessary to examine the guardrail installation for segments along the valley side versus those that are

Table 4 Average crash frequency, severity index, and combined safety index for segments with and without guardrail installations

	No guardrail	Guardrail	% Difference	*t* test 90 % Confidence
NB—Crash Frequency	1.413	1.054	25.4	*Significant*
NB—Severity Index	1.614	1.355	16.0	Insignificant
NB—Combined Safety Index	3.545	2.135	39.8	*Significant*
SB—Crash Frequency	1.098	0.866	21.2	Insignificant
SB—Severity Index	1.052	0.900	14.4	Insignificant
SB—Combined Safety Index	2.311	1.649	28.6	Insignificant

Table 5 Information on guardrail installations for segments with different side slopes

Guardrail %	0 %	<35 %	35 %–70 %	>70 %	Pearson χ^2 test
Northbound					
Gentle slope	35 (97.2 %)	0 (0 %)	1 (2.8 %)	0 (0 %)	$\chi^2 = 26.541$
Moderate slope	34 (82.9 %)	4 (9.8 %)	1 (2.4 %)	2 (4.9 %)	$P < 0.001$
Steep slope	52 (64.2 %)	2 (2.4 %)	5 (6.2 %)	22 (27.2 %)	
Southbound					
Gentle slope	15 (68.2 %)	3 (13.6 %)	2 (9.1 %)	2 (9.1 %)	$\chi^2 = 22.934$
Moderate slope	18 (46.1 %)	4 (10.3 %)	3 (7.7 %)	14 (35.9 %)	$P = 0.001$
Steep slope	28 (28.9 %)	3 (3.1 %)	10 (10.3 %)	56 (57.7 %)	

Table 6 Average crash frequency, severity index, and combined safety index for segments with and without fixed roadside objects

	Objects	No objects	% difference	*t* test 90 % Confidence
NB—Crash frequency	1.346	1.227	8.8	Insignificant
NB—Severity index	1.603	1.242	22.5	Insignificant
NB—Combined safety index	3.360	2.318	31.0	*Significant*
SB—Crash frequency	1.017	0.769	24.3	Insignificant
SB—Severity index	0.973	0.915	6.0	Insignificant
SB—Combined safety index	1.941	1.795	7.5	Insignificant

Table 7 Information on guardrail installations for segments along the valley and the mountain sides

Guardrail %	0 %	<35 %	35 %–70 %	>70 %	Total (%)
Northbound					
Mountain wall	105 (92.1 %)	5 (4.4 %)	0 (0.0 %)	4 (3.5 %)	100
Valley/river	16 (36.4 %)	1 (2.3 %)	7 (15.9 %)	20 (45.4 %)	100
Southbound					
Mountain wall	40 (90.9 %)	3 (6.8 %)	0 (0.0 %)	1 (2.3 %)	100
Valley/river	21 (18.4 %)	7 (6.1 %)	15 (13.2 %)	71 (62.3 %)	100

along the mountain side. The results are presented in Table 7 below.

It is evident from the numbers in this table that guardrail installation is heavily overrepresented at segments along the valley side compared with those along the mountain side. This may partly explain the higher crash indicator values for those segments along the mountain side.

5.5 Multivariate analysis of crash data

After observing some of the trends between the safety indicators on one hand and speed, road geometry, and roadside features on the other hand, multivariate linear regression and correlation analyses were performed to further evaluate the relationships between these variables.

Table 8 summarizes the results of the regression analysis which show the relationship between the response variable (safety indicators) and predictor variables (speed, geometry and roadside features) in the southbound direction.

Upon examining this table, the following observations can be made:

(1) Only two models were found statistically significant at the 95 % confidence level as affirmed by the *F*-test results. Those models are for crash frequency and the combined safety index. The model for severity index was not found statistically significant.
(2) The coefficients of determination (R^2) are relatively low for the two models that were found significant with 0.333 and 0.350 for the crash frequency and the combined safety index models, respectively. This shows that little variations in the safety indicators are explained by the predictor variables.
(3) The degree of curve and the presence of valley versus river at roadside were both found significant for the crash frequency model, while only ΔV was found significant in the combined safety index model at the 95 % confidence level. The *t* test results confirmed that all other variables investigated were not found significant at the 95 % confidence level.

Correlation analysis in the southbound direction was performed using the same above variables to gain more insights into the safety effects of speed, geometry, and roadside features. Correlation results are provided in Table 9 below.

As clearly shown in Table 9, the correlation coefficients were low overall with the highest coefficient being 0.272 between ΔV and the combined safety index. However, those correlation coefficients have all exhibited logical relationships with the three safety indicators.

Similar regression and correlation analyses were performed using the data in the Northbound direction, however, the models developed and the correlations found revealed even weaker effects of the study variables on safety indicators. As such, results were not included in the current section.

6 Summary of findings

The current study presents a crash analysis along a rural corridor in southwest Montana. The study corridor is located on US-191 within the Gallatin Canyon between the cities of Bozeman and Big Sky. The approach used in this study is macroscopic in that it examined the whole corridor systematically without focusing on any specific crash or any particular site. However, the approach can also be viewed as microscopic in that detailed crash-related information at a relatively high resolution was used where every tenth of a mile was characterized as an individual segment within the corridor. The approach could be applied at higher resolution should more accurate crash location information become available. The study investigated crash experience at the study corridor along with some of the fundamental relationships between crashes on one hand and roadway alignment, roadside features, and speed on the other hand. These relationships shed light on many of the crash causal factors in a rural environment. The major findings of this study are:

- While ROR constituted more than 56 % of total crashes, crashes involving multiple vehicles tended to be higher in severity. Further, ice, snow, and slush on pavement were associated with more than 43 % of all crashes investigated by this study.
- Horizontal curvatures were found to have strong association with higher crash frequencies even on flat curves that would not restrict the traveling speed based on the highway design equations. This requires a rethinking of the speed management techniques currently in use at the location of curves.
- Higher crash frequencies were also found to be associated with the difference between prevailing free-flow speed and the speed dictated by design equations (at segments with small radii or limited sight distance) or speed limit.
- Safety benefits of guardrails were clearly reflected in the results of this investigation as evidenced by the lower crash frequency and severity despite the more

Table 8 Summary results from multivariate linear regression analysis at US 191 Corridor (Southbound)

Safety indicator	Regression model			*P* value from *t* test[b,c]					
	F test[a]	R^2	SE	DOC	ΔV	Guardrail	Sideslope	Roadside Objects	Valley/river
Crash frequency	*0.0060*	0.333	1.14	**0.026**	0.561	0.814	0.486	0.956	**0.025**
Severity index	0.2510	0.223	1.33	0.646	0.187	0.935	0.921	0.740	0.134
Combined safety index	*0.0027*	0.350	3.25	0.231	**0.013**	0.681	0.709	0.840	(0.094)

[a] Values underlined in italic refer to models that were found significant using the *F* test

[b] Values in bold are for coefficients that were found significant using the *t* test

[c] Values in brackets are those that passed significance testing at the 90 % confidence level only

Table 9 Correlation analysis results between safety indicators and study variables

	DOC	Average overall ΔV	Guardrail	Sideslope	Roadside objects	Valley/river?
Crash frequency	0.190	0.152	−0.110	−0.064	0.036	−0.225
Severity index	0.085	0.140	−0.081	−0.019	0.051	−0.158
Combined safety index	0.189	0.272	−0.091	−0.041	0.045	−0.180

hazardous roadside features at the locations of guardrail installations.

- Results suggest that crash frequency and severity were associated with the presence of fixed roadside objects and the steepness of side slope along the study corridor.
- While the deep valley may pose significant hazard for run-of-the-road vehicles, segments along the valley were associated with lower crash frequency and severity. This is believed to be related to guardrail installation that exists more on this side of the road. These results suggest the effectiveness of guardrails as a low-cost safety countermeasure in many hazardous rural locations.
- The multivariate regression analysis did not reveal further insights and evidence about the relationships between the speed, geometry, and roadside features with safety indicators. This is despite the fact that most of those variables were observed to have clear connection with the crash indicators using the preliminary analyses presented earlier in this paper. This suggests that crash occurrence is related to some other important factors that were not investigated in this study such as the driver and weather conditions.

The authors would recommend a few measures in an attempt to reduce the high crash rates at this corridor. Among these measures are speed management strategies (e.g., variable speed limit) that would control speeds at the more hazardous segments in this corridor. This is essential especially since a significant number of crashes occur during inclement weather events. The use of guardrail at other parts of the corridor, not necessarily along steep side slopes, could help further improve safety within the corridor, as suggested by study results. Another safety measure is the use of ITS warning systems alerting drivers of pavement condition during and after inclement weather events.

Acknowledgements The authors would like to acknowledge the financial support to this research by the National Science Foundation (NSF) through the Western Transportation Institute (WTI) of Montana State University. The assistance of Mr. Pierre Jomini and Mr. Scott Keller of the Montana Department of Transportation (MDT) was critical to this project and was acknowledged by the authors.

References

1. National Highway Traffic Safety Administration (2013). Early Estimate of Motor Vehicle Traffic Fatalities in 2012. Publication DOT HS 811 741. NHTSA, U.S. Department of Transportation, May 2013
2. Anderson I, Krammes R (2000) Speed reduction as a surrogate for accident experience at horizontal curves on rural two-lane highways. Transportation Research Record No. 1701, Transportation Research Board, Washington D.C., pp 86–94
3. Lamm R, Beck A, Ruscher T, Mailaender T, Cafiso S, La Cava G (2007) How to make two-lane rural roads safer: scientific background and guide for practical application. WIT Press, Massachusetts
4. Garber N, Kassebaum E (2008) Evaluation of crash rates and casual factors for high-risk locations on rural and urban two-lane highways in Virginia. Publication FHWA/VTRC 09-R1. Virginia Transportation Research Council
5. Caliendo C, Lamberti R (2001) Relationships between Accidents and Geometric Characteristics for Four Lanes Median Separated Roads. In: Proceeding of the Road Safety on Three Continents, Moscow, Russia, September 19–21, 2001
6. Cenek P, Davies R, Henderson R (2012) Crash risk relationships for improved road safety management. Publication 488. New Zealand Transport Agency
7. Hauer E (1995) Safety and the choice of degree of curve. Transportation Research Record No. 1665, Transportation Research Board, Washington, D.C, pp 22–27
8. Hadi M, Aruldhas J, Chow L, Wattleworth J (1995) Estimating safety effects of cross-section design for various highway types using negative binomial regression. Transportation Research Record No. 1500, Transportation Research Board, Washington, D.C., pp 169–177
9. Gross F, Jovanis P, Eccles K (2009) Safety effectiveness of lane and shoulder width combinations on rural, two-lane, undivided roads. Transportation Research Record No. 2103, Transportation Research Board, Washington, D.C., pp 42–49
10. Karlaftis M, Golias I (2002) Effects of road geometry and traffic volumes on rural roadway accident rates. Accid Anal Prev 34(3):357–365
11. Bester C, Makunje J (1995) The effect of rural road geometry on safety in Southern Africa. In: Proceeding of the International Symposium on Highway Geometric Design Practices, Boston, August 30–September 1, 1995
12. Council F (1998) Safety benefits of spiral transitions on horizontal curves on two-lane rural roads. Transportation Research Record No. 1635, Transportation Research Board, Washington, D.C., pp 10–17
13. Farah H, Bekhor S, Polus A (2009) Risk evaluation by modeling of passing behavior on two-lane rural highways. Accid Anal Prev 41(4):887–894
14. Zhong C, Sisiopiku V, Ksaibati K, Zhong T (2011) Crash prediction on rural roads. In: Proceeding of the 3rd International Conference on Road Safety and Simulation, Indianapolis, Indiana, September 14–16, 2011

15. Andjus V, Maletin M (1998) Speeds of cars on horizontal curves. Transportation Research Record No. 1612, Transportation Research Board, Washington, D.C., pp 42–47
16. Collins J, Fitzpatrick K, Bauer K, Harwood D (1999) Speed variability on rural two-lane highways. Transportation Research Record No. 1658, Transportation Research Board, Washington, D.C., pp 60–69
17. Tate F, Turner S (2007) Road geometry and drivers' speed choice. Road Transp Res J Aust N Z Res Pract 16(4):53–64
18. Garber N, Ehrhart A (2000) Effect of speed, flow, and geometric characteristics on crash frequency for two-lane highways. Transportation Research Record No. 1707, Transportation Research Board, Washington, D.C., pp 76–83
19. Bird R, Hashim I (2006) Exploring the relationship between safety and consistency of geometry and speed on British roads. Transportation Research Board 2006 Annual Meeting. CD-ROM. TRB, Washington, D.C., pp 1–19
20. Wang C, Quddus M, Ison S (2009) The effects of area-wide road speed and curvature on traffic casualties in England. J Transp Geogr 17(5):385–395
21. Dissanayake S, Indike R (2006) Identification of factors leading to high severity of crashes in rural areas using ordered probit modeling. J Transp Res Forum 45(2):87–101
22. Malyshkina N, Mannering F, Labi S (2007) Influence of speed limits on roadway safety in Indiana. Publication FHWA/IN/JTRP-2007/6. Joint Transportation Research Program
23. Thomas I (1996) Spatial data aggregation: exploratory analysis of road accidents. Accid Anal Prev 28(2):231–264
24. Koorey G (2009) Road data aggregation and sectioning considerations for crash analysis. Transportation Research Record No. 2103, TRB, Washington, D.C., pp 61–68
25. Mayora J, Rubio R (2003) Relevant variables for crash rate predictions in Spain's two lane rural roads. Transportation Research Board 2003 Annual Meeting. CD-ROM. TRB, Washington, D.C., pp 1–24
26. MDT State Highway Traffic Safety Bureau (2010) Traffic safety problem identification. Publication FFY 2009. Montana Department of Transportation, 2008. Accessed 30 July 2010
27. Montana Department of Transportation (2009) 2008 rural traffic flow map.

4

Air flow control based on optimal oxygen excess ratio in fuel cells for vehicles

Ai Guo · Weirong Chen · Qi Li · Zhixiang Liu · Haidan Que

Abstract Air flow control is one of the most important control methods for maintaining the stability and reliability of a fuel cell system, which can avoid oxygen starvation or oxygen saturation. The oxygen excess ratio (OER) is often used to indicate the air flow condition. Based on a fuel cell system model for vehicles, OER performance was analyzed for different stack currents and temperatures in this paper, and the results show that the optimal OER was affected weakly by the stack temperature. In order to ensure the system working in optimal OER, a control scheme that includes an optimal OER regulator and a fuzzy control was proposed. According to the stack current, a reference value of air flow rate was obtained with the optimal OER regulator and then the air compressor motor voltage was controlled with the fuzzy controller to adjust the air flow rate provided by the air compressor. Simulation results show that the control method has good dynamic and static characteristics.

Keywords Fuel cells · Oxygen excess ratio · Air flow · Fuzzy control

1 Introduction

A fuel cell, which converts chemical energy of a fuel directly to electricity continuously with a electrochemical reaction is considered as a clean and efficient power source. Among all types of fuel cells, proton exchange membrane fuel cell (PEMFC) is most promising to be used in the engine of an electric vehicle, with advantages of low operation temperature, high efficiency, low noise, and zero pollution [1–3].

A fuel cell system consists of a stack and various ancillary equipments including air compressors, cooling water circulation, humidifier, pumps, and electrical control equipment. Among all the ancillary equipments, the power consumption of air compressor is the greatest, typically accounting for up to 80 % power of the overall ancillary equipments and 15 %–20 % of the fuel cell system [2, 4]. For a fuel cell stack, air flowrate must be greater than the amount needed by the electrochemical reaction, and the oxygen excess ratio (OER), which is the ratio between the flowrate of oxygen provided and reacted, is critical to ensure the system stability and reliability. If the OER is too low, even though higher than 1, it is likely to cause "oxygen starvation" [5, 6] and results in cell reversal, which is harmful to the stack. On the other hand, a higher OER will lower the concentration difference in cathode flow field and improve the fuel cell performance, while greater air flow results in higher power consumption for the air compressor and lower system efficiency, which is called "oxygen saturation." In addition, the air pressure in cathode flow field of the fuel cell stack and the pressure difference between both sides of proton exchange membrane must be maintained in a certain range during operation. Air flow control is important to maintaining the reliability and efficiency of the system, and different air flow control techniques have been provided by researchers [7–10].

OER was first proposed by Pukrushpan et al. [11] in linear-quadratic-Gaussian (LQG) techniques to control air supply subsystem, where it was simplified to be a constant. Based on Pukrushpan's model, Chang et al. [7] analyzed the variation of a fuel cell system's net power with different stack temperatures and membrane water contents, and discovered that the optimal value of OER changes with different system parameters. They put forward a real-time extremum

A. Guo (✉) · W. Chen · Q. Li · Z. Liu · H. Que
School of Electrical Engineering, Southwest Jiaotong University, Chengdu 610031, China
e-mail: 634576728@qq.com

seeking control method. This control method is simple, and can achieve maximum power over the entire operating range. But the periodic perturbation in the method may produce unacceptable oscillations in output power, and its complicated penalty function will lead to long convergence time. Arce et al. [12] proposed two control objectives aimed at satisfying the oxygen starvation avoidance criterion and the maximum efficiency criterion, and analyzed two OER values in "oxygen starvation" state and "maximum efficiency" state on 1.2 kW fuel cells test bench, and a control architecture for a fixed OER was designed. An explicit model predictive control method combined with proportional integral (PI) control was adopted for controlling air flow to prevent oxygen starvation and guarantee the optimal transient response of the system, but the control algorithm does not maximize the net power of the fuel cell.

In this paper, based on the fuel cell system model built in [8], optimal OERs under different stack currents and temperatures were analyzed. In order to ensure the maximum net power output, a control strategy by combination of feed forward control and fuzzy feedback control was proposed. A reference value for the air flow of compressor was acquired by the optimal OER regulator, and then the compressor's air flow was adjusted by the fuzzy controller according to the reference value. The simulation results show that the control method can make the fuel cell system work with an optimal OER.

2 Fuel cell system for vehicle

A fuel cell system for vehicles is composed of four subsystems, as shown in Fig. 1: (i) air flow supply system to the cathode, (ii) hydrogen flow supply system to the anode, (iii) cooling system, and (iv) humidification system. The air flow supply subsystem includes a compressor, a cooler, a water separator, a back pressure valve, connecting pipes, etc., and is used to provide requisite air flow into the stack cathode. Hydrogen flow supply subsystem includes hydrogen tank, a solid valve, a recirculation pump, a purge valve, connecting pipes, etc. It is used to provide sufficient hydrogen flow into the stack anode.

Air temperature will elevate during compressing, so an air cooler is needed to cool the air before it enters the stack. A humidifier is used to add moisture into the air flow to hydrate the proton exchange membrane for higher proton conductivity. On the cathode side of the stack, the electrochemical reaction occurs between oxygen and protons to form water, together with heat released. In order to keep the stack temperature, coolant must be pumped into the stack and take the heat out and then released into circumstance with a radiator. The water separator is used to recover the water in the exhaust air. The back pressure valve is used to keep a suitable air flow pressure in the stack.

Up to now, a number of fuel cell models have been built for different purposes [4, 13], among which the models reported in [8, 11–14] belong to the control-oriented model. The fuel cell system model in this paper is a theoretical model developed from [8, 11], and adopts a twin screw compressor. Compared with the centrifugal compressor, a twin screw compressor has the following advantages: (1) its nominal rotation speed is almost an order of magnitude lower than a centrifugal compressor, which makes the system more reliable; (2) its flow curve in compressing map is almost a straight line at a constant speed, and is impacted by pressure far less than a centrifugal compressor, which makes the system control easier; (3) a twin screw compressor does not have "surge"

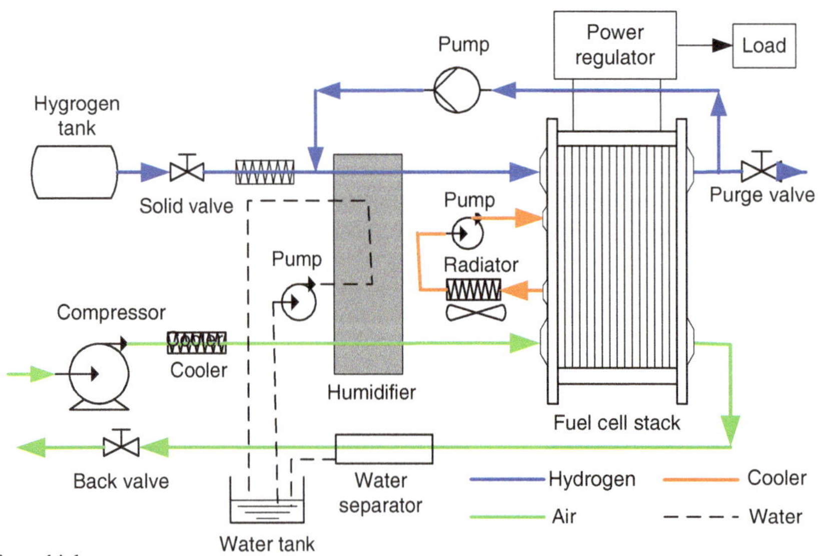

Fig. 1 Fuel cell system for vehicle

phenomenon, which makes control system simpler. The twin screw compressor modeling and the compressor map are developed from [15].

The twin screw compressor model is separated into two parts, as shown in Fig. 2. The first part is the static compressor map which determines the air flow rate through compressor and compressor power. The second part is adjustment. Typically, the operation environment of a compressor is different from the test environment, so its air flow and power must be modified according to the thermodynamic equation.

The inputs to the model include downstream pressure p_{sm}, rotation speed N_{cp}, air pressure p_{atm}, temperature T_{atm}, pressure p_0 (1.013 bar), and temperature T_0 (293.15 K) in Map test circumstances. The rotation speed is determined by compressor motor torque and torque required. The compressor is connected to supply manifold, and its downstream pressure is equal to the pressure in supply manifold, which is computed by its downstream cathode flow field.

The compressor power P_{cp} could be indexed from the compressor power map according to the rotation speed of the compressor. The air flow rate W_{cp} is determined by the compressure ratio and rotation speed from the compressor flow map. Then the revised air flow rate W_{cp_re} is calculated by

$$W_{cp\_re} = \frac{T_{amb}}{T_0} \cdot \frac{p_0}{p_{amb}} W_{cp} \tag{1}$$

The temperature of the air leaving the compressor T_{cp} is calculated with equations listed in Ref. [11].

The main parameters used in the model are given in Table 1. Fuel cell stack type is BALLARD Mark 902 (85 kW).

3 OER performance and air flow control

Since the air compressor takes majority part of the auxiliary equipment's power consumption, other auxiliary

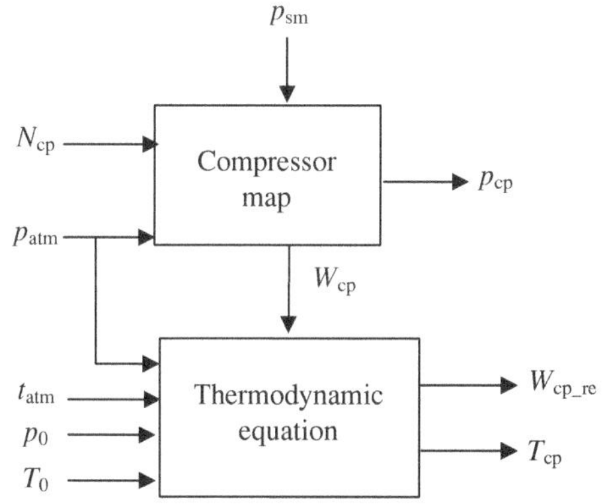

Fig. 2 Compressor block diagram

Table 1 Fuel cell system model specification [16]

Part	Parameter	Value
Fuel cell stack	Membrane type	Nafion112
	Maximum power	85 kW
	No. of cells	440
	Cell active area	300 cm^2
	Cathode volume	0.01 m^3
	Anode volume	0.01 m^3
	Nominal temperature	80 °C
	Membrane water content	14 (100 %)
Air supply system	Supply manifold volume	0.02 m^3
	Return manifold volume	0.03 m^3
Hydrogen supply system	Supply manifold volume	0.02 m^3
	Return manifold volume	0.05 m^3

equipment's power can be ignored, and the net power of the fuel cell system, P_{net}, could be calculated as

$$P_{net} = V_{st} I_{st} - P_{cp}. \tag{2}$$

where V_{st} and I_{st} are voltage and current of fuel cell stack, respectively; P_{cp} is power of the compressor.

OER λ_{O_2} is often used to indicate the air flow condition, which is defined as

$$\lambda_{O_2} = \frac{W_{O_2,ca}}{W_{rec}}, \tag{3}$$

where $W_{O_2,ca}$ is the oxygen flow rate into the stack cathode, and W_{rec} is oxygen flow rate participated in electrochemical reaction in stack cathode.

If $\lambda_{O_2} < 1$, i.e., oxygen flow rate in the stack is less than the amount required for a given current value, then the fuel cell system is in the state of oxygen starvation. If $\lambda_{O_2} \gg 1$, i.e., oxygen flow rate in the stack is much larger than the amount required for a given current value, then the fuel cell system is in the state of oxygen saturation. High λ_{O_2} corresponds to more oxygen supplied to the cathode, which improves the power generated by the stack. However, if λ_{O_2} is too large, the net power P_{net} will decrease due to the excessive power demanded by the compressor, as shown by (2), which makes the fuel cell system in the "oxygen saturation" state. As a result, there exists an optimal λ_{O_2} value that maximizes P_{net} by trading off stack power production and compressor power consumption.

3.1 OER performance

The relation between the net power and OER for various stack current is called OER performance. On a given stack current, there is an OER performance curve (Fig. 3). As shown in Fig. 3, there is a maximum net power point, where the corresponding OER value is defined as the optimal OER for each current density value. The optimal OER points are signed with asterisk * and all the points are in [1.990, 2.608]. The optimal value decreases as stack current increases in the interval [50, 80 A], then increases roughly in the interval [90, 200 A], and decreases again in the interval [210, 300 A].

The OER performance of this model is different from that of [7], where the optimal value decreases roughly as stack current increases. The reason is that different types of compressor are used in the two models. In the case of constant rotation speed, the air flow rate through the twin screw compressor is affected little by compressure ratio (in range 1.5–3) [15]. If the effect is ignored, the air flow rate is only determined by rotation speed. The higher the rotation speed, the greater the air flow rate. Yet the air flow rate through the centrifugal compressor is determined by both rotation speed and compressure ratio [9].

Stack voltage is affected by stack temperature. According to (2), the net power is also affected by stack temperature. The relationship between net power and stack temperature is shown in Fig. 4, where stack current I_{st} is equal to 180 A. Table 2 lists the optimal OER value $\lambda_{O_2\_op}$ and maximum net power P_{net_max} at different stack temperatures T_{st}.

As shown in Fig. 4, for a constant stack current, the higher the stack temperature, the greater the net power. Higher stack temperature can improve the system performance, because the electrochemical reaction is more active. Seen from Table 2, the optimal value $\lambda_{O_2\_op}$ decreases a little bit between 2.6 and 2.7 with the rising of temperature. According to further analysis, we can see that if OER changes from 2.6 to 2.7 at the same temperature, the change of net power does not exceed 30 W, which is about 0.068 % of the maximum net power. It is clear that $\lambda_{O_2\_op}$ is affected weakly by the stack temperature.

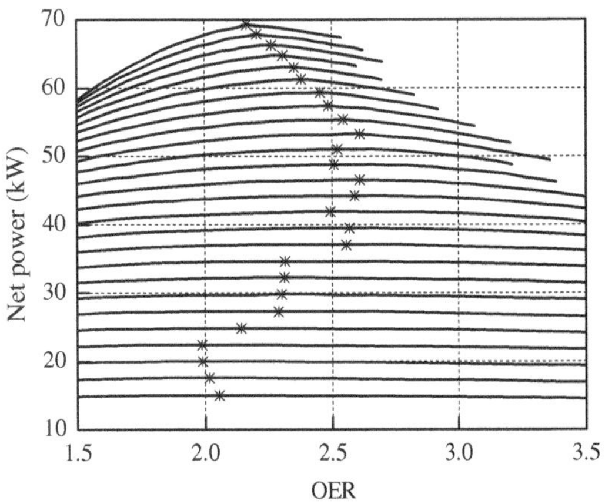

Fig. 3 OER performance of the fuel cell system for vehicle. Stack current changes from 50 to 300 A with a 10 A step length

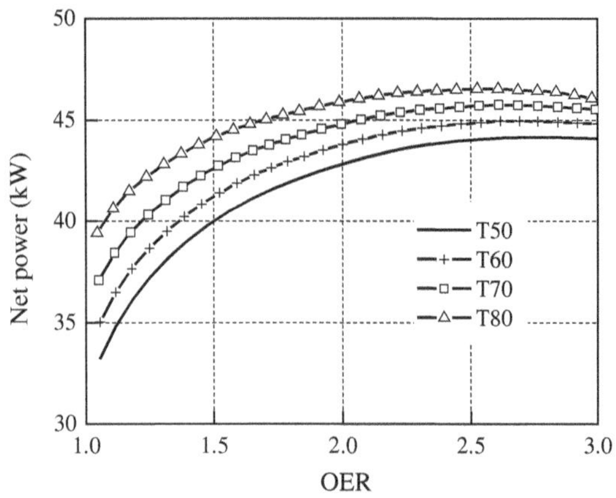

Fig. 4 Net power at different stack temperatures

Table 2 Optimal OER and maximum net power at different temperatures

T_{st} (°C)	$\lambda_{O_2\_op}$	P_{net_max} (kW)
50	2.6960	44.153
60	2.6911	44.943
70	2.6101	45.754
80	2.6081	46.543

3.2 Air flow control

The primary objective of air flow control is to avoid "oxygen starvation" and "oxygen saturation" and assure safe OER and rapid transient response. In order to realize the maximum net power, OER should be adjusted near the optimal value by changing the air flow into cathode. Figure 3 demonstrates that once OER exceeds the optimal value, the parasitic power consumption increases greatly, which deteriorates the fuel cell system. The greater the stack current, the more prominent this feature. The fuel cell system for vehicle should work near the optimal OER to obtain good performance.

The above analysis shows that the stack temperature has a weak effect on the optimal OER. For the sake of control convenience, temperature influence is ignored so that in the case of the constant temperature, we can obtain the corresponding relation between the optimal OER and the stack current. The control method is shown in Fig. 5. The optimal OER regulator includes a lookup model and a computation model. The lookup model is a relationship table between optimal OER and stack current. It is used to obtain the optimal OER $\lambda_{O_2\_op}$ according to the stack current I_{st}. The computation model computes the reference value of the air flow of the compressor W^*_{cp} with the known stack current I_{st} and optimal OER $\lambda_{O_2\_op}$. The fuzzy controller (FC) controls the compressor motor voltage V_{cm} to adjust the air flow through the compressor.

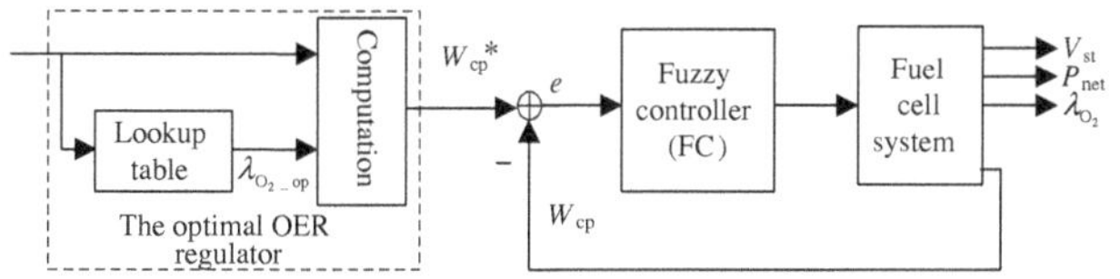

Fig. 5 Control block diagram

In the computation model, the reference value of the air flow W^*_{cp} is calculated as in [6]:

$$W^*_{cp} = \left[1 + \frac{M_v}{M_a^{atm}} \frac{\phi_{atm} p_{sat}^{atm}}{(p_{atm} - \phi_{atm} p_{sat}^{atm})}\right] \frac{1}{y_{O_2}} \lambda_{best} M_{O_2} \frac{n I_{st}}{4F}, \quad (4)$$

where ϕ_{atm} is air relative humidity; p_{sat}^{atm} is the vapor saturation pressure with relative humidity ϕ_{atm}; p_{atm} is atmospheric pressure; M_v is the molar mass of vapor; M_a^{atm} is the air molar mass with relative humidity ϕ_{atm}; M_{O_2} is oxygen molar mass; n is the number of single fuel cell; F is Faraday's constant; and, y_{O_2} is oxygen mass fraction in the air and is calculated by

$$y_{O_2} = 0.21 M_{O_2} / M_a^{atm}. \quad (5)$$

A typical fuzzy controller is selected, whose block diagram is shown in Fig. 6, where error signal e is input, compressor motor voltage v_{cm} is output, k_e and k_{ec} are quantity coefficients of error and error change rate, respectively, and k_u is proportion coefficient. Different parameters and their relations will affect the control characters of the fuzzy controller, especially dynamic characters. In order to attenuate the steady-state error, the incremental mode of output is adopted in the fuzzy controller.

4 Simulation results and analysis

Simulations were performed to analyze the proposed OER control method for the fuel cell system. The stack temperature was set to 80 °C, and membrane water content to 14. Three parameters of the fuzzy controller are fixed using the optimization method, namely:

$$k_e = 14.9764, \quad k_{ec} = 0.1972, \quad \text{and} \quad k_u = 0.5532.$$

When stack current I_{st} is 180 A, the optimal value $\lambda_{O_2\_op}$ is obtained to be 2.608 in the lookup table. According to Formula (4), the reference value of air flow rate through the compressor is calculated to be 74.27 g/s. Setting the stack current to be a step function from 30 to 180 A at $t = 0$, the simulation result is shown in Fig. 7. It is seen that the rise time of the air flow rate is less than 0.8 s with no overshoot, and the stable air flow rate W_{cp} (Fig. 7a) is 74.24 g/s with a small error 0.02 g/s. The air pressure (Fig. 7b) and oxygen partial pressure (Fig. 7c) in the stack gradually increase to 1.6821 and 0.17062 bar, respectively, at about $t = 0.9$ s. In the stack, oxygen flow rate increases with air flow rate gradually, and the required oxygen flow rate does not change under constant stack current, so the OER value λ_{O_2} (Fig. 7d) increases gradually. The stable value of λ_{O_2} is 2.617, the optimal value is 2.608, and the error is −0.009. One can see that the OER value λ_{O_2} promptly changes from oxygen starvation state to hyperoxic state in the initial stage, and it takes only 0.8 s for OER to reach its stability.

The stack current is a series of step functions (Fig. 8a). When the stack current varies, the optimal OER (the dotted line in Fig. 8c) and the reference value of the air flow through compressor (Fig. 8b) change, which are counted by the

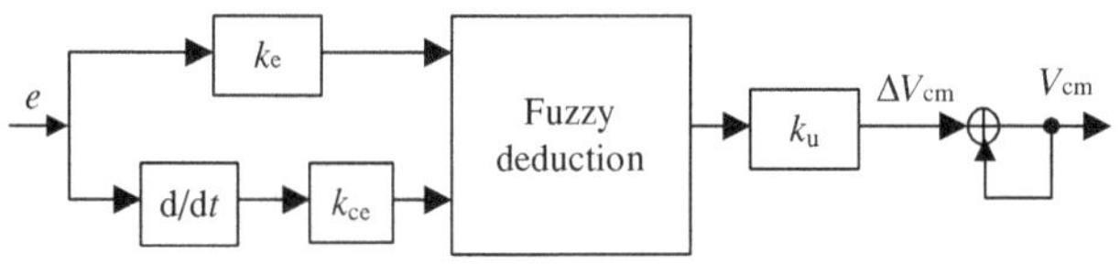

Fig. 6 Fuzzy controller block diagram

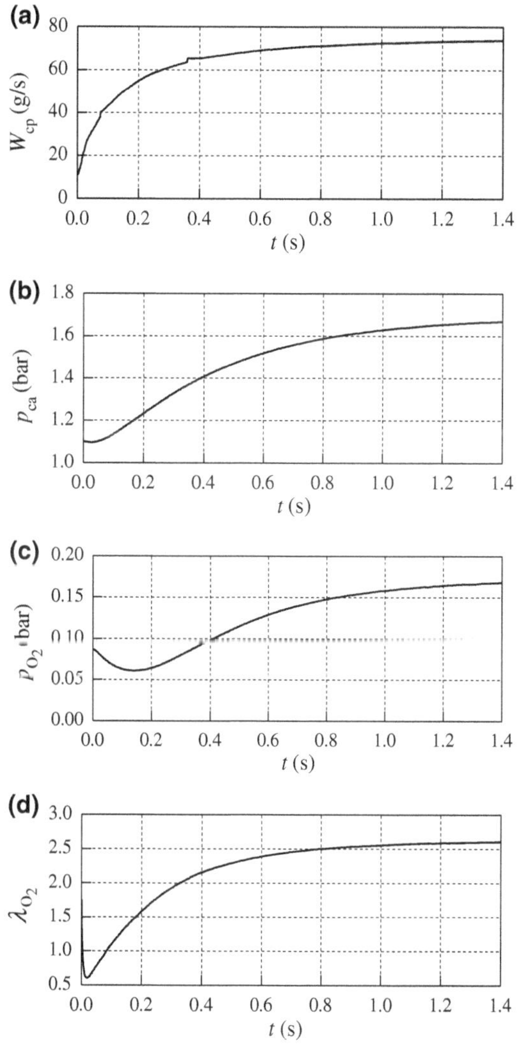

Fig. 7 The step response of stack current. **a** air flow rate. **b** cathode pressure. **c** partial oxygen pressure. **d** OER

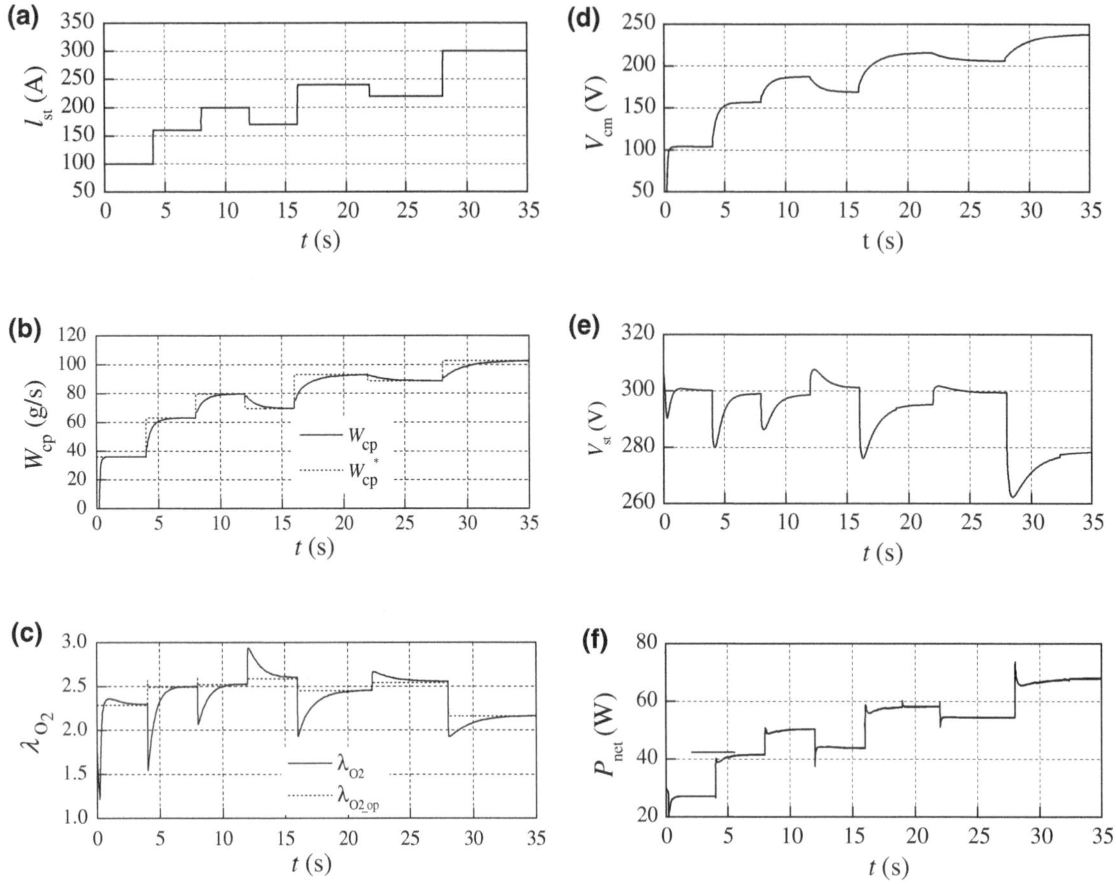

Fig. 8 The response for a series of step stack current. **a** stack current. **b** air flow rate. **c** OER. **d** motor voltage. **e** stack voltage. **f** net power

optimal OER regulator. On the basis of the reference value (the dashed line in Fig. 8b), the FC adjusts the motor voltage (Fig. 8d) to regulate the air flow through the compressor (the solid line in Fig. 8b). When current steps up (at 4, 8, 16, and 28 s), the oxygen amount required suddenly rise, and the FC adjusts the motor voltage (Fig. 8d) according to the air flow rate and its change rate, to change the air flow through the compressor. Different from the quick response of the stack current, the air flow rate increases laggardly (Fig. 8b) because of the hysteresis effect of the compressor response, causing instantaneous drops of OER (solid line in Fig. 8c). Simultaneously, a sudden decrease of oxygen partial pressure in the stack cathode causes a sag of the stack voltage (Fig. 8e). When current steps down (at 12 and 22 s), the situation is just the opposite. The net power of system (Fig. 8f) changes according to the transient responses of the stack current and voltage.

5 Conclusions

Air flow control is very important to a fuel cell system for vehicles. Based on a fuel cell model for vehicles, the relation between OER and net power under various stack currents, and the relation between OER and stack temperature were analyzed. The results show that for a fixed stack current, stack temperature has weak influence on the optimal OER, which could be ignored for the simplification of controller design. In order to ensure the fuel cell system to work at the optimal OER, a control strategy with an optimal OER regulator and air flow feedback control (fuzzy control) was proposed. When stack current changes, the optimal OER regulator can promptly compute the reference value of air flow rate corresponding to the optimal OER, and the FC controller controls the compressor motor voltage to change air flow in time. Simulation results show that the control strategy could make fuel cell system work at the optimal OER whether for a given load or for a variable load. In addition, the designed controller for fuel cell system has good dynamic and steady-state characteristics.

Acknowledgments The paper was supported by the National Natural Science Foundation of China (No. 51177138), the Research Fund for the Doctoral Program of High Education of China (No. 20100184110015), and Sichuan Province International Technology Cooperation and Exchange Program (No. 2012HH0007).

References

1. Lukic S, Cao J, Bansal R et al (2008) Energy storage systems for automotive applications. IEEE Trans Ind Electron 55(6): 2258–2267
2. Ehsani M, Gao Y, Emadi A (2010) Modern electric, hybrid electric, and fuel cell vehicles—fundamentals, theory, and design. China Machine Press, Beijing
3. Zhang LY, Quan SH (2011) Modeling fuel cell and optimal control. Publishing House of Electronics Industry, Beijing
4. Gou B, Na WK, Diong B (2011) Fuel cells modeling, control and application. China Machine Press, Beijing
5. Gerard M, Cuvezier JPP, Hissel D et al (2010) Oxygen starvation analysis during air feeding faults in PEMFC. Int J Hydrogen Energy 22(35):12295–12307
6. Thounthong P, Sethakul P (2007) Analysis of a fuel starvation phenomenon of a PEM fuel cell. In: Proc. PCC, Nagoya, 2007: 731–738
7. Chang YA, Moura SJ (2009) Air flow control in fuel cell systems: an extremum seeking approach. In: Proc. IEEE Amer. Control Conf., Hyatt Regency Riverfront, St. Louis, 10–12 June 2009: 1052–1059
8. Pukrushpan JT, Stefanopoulou AG, Peng H (2004) Control of fuel cell breathing. IEEE Trans Control Systems 24(2):30–46
9. Danzer M, Wilhelm J, Ascheman H et al (2008) Model-based control of cathode pressure and oxygen excess ratio of a PEM fuel cell system. J Power Sources 176(2):515–522
10. Talj R, Ortega R, Astolfi A (2011) Passivity and robust PI control of the air supply system of a PEM fuel cell model. Automatica 47(12):2554–2561
11. Pukrushpan JT, Peng H, Stefanopoulou AG, Simulation and analysis of transient fuel cell system performance based on a dynamic reactant flow model (2002) In: Proceedings of ASME MECE. New Orleans: Louisiana, 2002: 17–22
12. Arce A, del Real AJ, Bordons C et al (2010) Real-time implementation of a constrained MPC for efficient airflow control in a PEM fuel cell. IEEE Trans Ind Electron 57(6):1892–1905
13. Yang Y, Wang FC, Chang HP et al (2007) Low power proton exchange membrane fuel cell system identification and adaptive control. Power Sources 164(2):761–771
14. Gabin WG, Dorado F, Bordons C (2010) Real-time implementation of a sliding mode controller for air supply on a PEM fuel cell. J Process Control 20:325–336
15. Ma TC (2007) Research on the control of fuel cells engine—system analysing and modeling [Dissertation], Shanghai, Tongji University

5

A construction strategy for a tunnel with big deformation

Liang Chen · Shougen Chen · Xinrong Tan

Abstract By integrating literature reviews, site observation, field monitoring, theoretical analysis, summarization, etc., a construction strategy was proposed and verified for tunneling with big deformation in this paper. The tunnel was in phyllite, shotcrete cracks and steel arch distortion were observed, and a big deformation with a maximum of 2.0 m was monitored during the initial stage of the construction. Through carefully examining the site observation and laboratory test results, a construction principle was established for the tunneling on the basic concept of maintaining the rock strength/stiffness and keeping the rock dry, by providing confinement pressure to the rock, reducing the rock exposure time, keeping water out of the tunnel, etc. To achieve the construction principle, a set of specific construction measures with 11 items was further proposed and applied to the construction. To check the effectiveness of the construction measures, field monitoring was carried out, which showed that the rock deformation was well controlled and the tunnel became stable. An allowable deformation was then determined using the Fenner formulae and the monitored data in order to guide further construction, which received a good result. From this study, it can be concluded that providing quick strong initial support and reserving core soil at the working face are extremely important to control the rock deformation and keep the tunnel stable.

Keywords Tunnel engineering · Big deformation · Construction strategy

L. Chen · S. Chen (✉) · X. Tan
Key Laboratory of Transportation Tunnel Engineering, Ministry of Education, Southwest Jiaotong University, Chengdu 610031, China
e-mail: csgchen2006@163.com

L. Chen
e-mail: clng9716@163.com

X. Tan
e-mail: tanxr2002@126.com

1 Introduction

The occurrence of big deformation in tunneling had become a serious problem since 1906 when an extremely big deformation was found in the Simplon Tunnel that is located between Italy and Switzerland [1, 2]. Big deformation could cause a lot of difficulty in several aspects [3–5]: (1) rock support failure, which induces tunnel instability such as working face collapsing and roof fall; (2) tunnel section reduction, in which initial support invades the tunnel and occupies the space of concrete lining, and an expansion excavation has to be done again; (3) lining crack, which causes the tunnel's instability and water leakage during the operation.

Based on relevant studies worldwide, big deformation in tunneling can be classified into four categories [6–8]: (1) squeezing deformation; (2) swelling deformation; (3) loosing deformation; and (4) soft rock deformation under high in situ stress, which frequently causes big deformation. There could be many factors influencing rock deformation [9–11]. Firstly, the in situ stress and rock strength would be the most important factors. Obviously, a higher in situ stress generates a higher stress in the rock and a bigger load onto the tunnel structure. Under a high in situ stress, tunneling in soft rock is prone to induce big deformation. In other words, big deformation is believed to easily occur when tunneling in a soft rock with a high in situ stress [12]. Secondly, the ground water would be another factor as it could significantly reduce the strength and stiffness of the soft rock. Laboratory tests show that the water existence can reduce the rock strength up to 90 % depending on the water flow rate [13]. Thirdly, the construction method, especially the support time, would be another key factor as it may boost rock overrelaxation and strength reduction if the support was delayed [14]. In addition, the intersection angle between the tunnel axis and the rock structures may affect the rock failure mode and the rock deformation [15].

The construction experience confirmed the effectiveness of some measures in avoiding big deformation [16, 17]. The most effective measure would be to strengthen the working face by applying rock blots and shotcrete, or reserving core soil at the working face. Once the working face becomes stable, big deformation can be minimized. Controlling floor heave and strengthening sidewall base are other effective measures, and an invert arch could be considered if required [18]. Providing stronger initial support and earlier lining can effectively control the rock deformation. Furthermore, rock deformation monitoring and advanced geological prediction during the tunneling are necessary to provide adequate informational data to the measure decision [19, 20].

However, more and more big deformation cases in tunneling are still occurring [21]. The problem is that in most cases, no attention was paid at the beginning and no effective measures were adopted before the big deformation was found, which missed the best time to control rock deformation [4]. In addition, there is a lack of a comprehensive summary from previous similar projects in providing references to guide further construction [22].

The Dongsong Hydropower Station is located on the Shuoqu River in Nishi Town and Dongsong Town of Xiangcheng County, Sichuan Province, China [23]. It is the fourth cascade hydropower station with a capacity of 150 MW. The diversion tunnel of the station was laid at the right side of the river with a total length of 17.862 km and an overburden of 150–500 m. The tunnel section is in a horseshoe shape with a width of 7.88 m and a height of 7.96 m.

The major rock encountered is phyllite with a uniaxial compressive strength (UCS) of less than 4.0 MPa, which belongs to extremely soft rock according to the Chinese specifications. The water flow rate of the tunnel is about 103.5 m^3/s. A big deformation was observed during the tunneling with a maximum deformation of 2.0 m, causing great difficulty and severe delay.

This paper is to investigate the strategy for the tunnel construction. The site observation was firstly briefed. A tunneling strategy and a set of measures were then proposed and applied to the construction. Then, an allowable deformation was determined to guide further construction of the tunnel.

2 Site observation

As the tunnel is very long, the construction was divided in nine segments. Eight horizontal branch tunnels were used in between for achieving earlier completion of the project. In other words, the tunnel has 18 working faces in order to speed up the advance. The construction was started in February 2009, but a big deformation was observed in branch tunnels 2, 4, and 6. The construction became very difficult and the advance was much delayed. As of May 2010, only a total excavation length of 6.8 km was completed in 16 months.

2.1 Site observation from branch tunnel 2

Branch tunnel 2 is 357 m long toward the intersection with the diversion tunnel. The excavation of the tunnel was started in May 2008 and completed in September 2009. A big deformation occurred and the advance was very slow with an average monthly advance of 21 m.

The excavation of the diversion tunnel through the branch tunnel 2 was then started from two working faces, one going upstream and the other downstream. As of October 2010, the excavation of only 106 m was completed within 14 months, equivalent to an average monthly advance of <8.0 m. During that period, the tunneling was extremely difficult due to big deformation occurrence at the crown and sidewall. Generally, the deformation ranged from 10–50 cm, while the mountain side deformed more than the river side. A deformation of 2.0 m occurred at the intersection of the branch tunnel and the diversion tunnel. Due to the big deformation, the shotcrete cracked and steel arch distorted as shown in Figs. 1 and 2. The initial support invaded the tunnel and occupied the lining space, and a second expansion excavation had to be done. Site observation showed that the rock was wet, very soft, and extremely fractured; once water was present, the rock could not be stabilized and advanced support had to be done.

2.2 Site observation from the downstream of branch tunnel 4

Branch tunnel 4 is 436 m long. Excavation of the tunnel was started in June 2008 and completed in March 2009 with an average monthly advance of about 44 m.

The excavation of the diversion tunnel through branch tunnel 4 was then started from two working faces, one going upstream and the other downstream. As of April 2011, the excavation of 156 m downstream was completed,

Fig. 1 Steel arch distortion

Fig. 2 Shotcrete cracks

Fig. 3 Shotcrete cracks

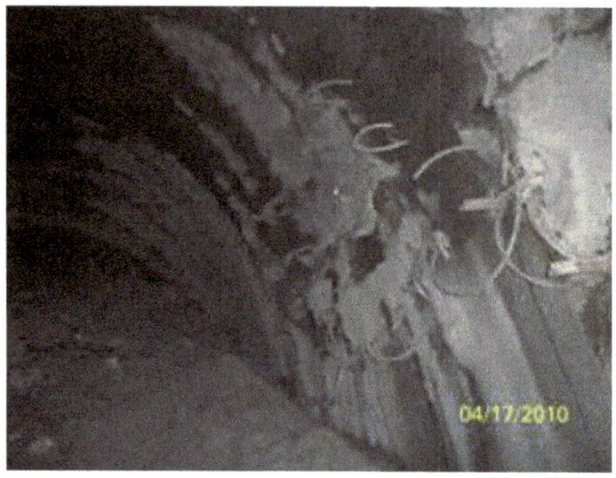

Fig. 4 Working face collapsed

equivalent to an average monthly advance of only 6.0 m, which was the worst case in the tunneling.

During the tunneling, a big deformation occurred downstream (S8 + 267.92–S8 + 409.46) as well as at the intersection of the branch tunnel and the diversion tunnel (K8 + 237.92–K8 + 267.92). The deformation ranged from 6–10 cm at the crown and 10–30 cm at the sidewalls, and the second expansion excavation had to be carried out. The maximum deformation velocity was 5.5 cm/day. The shotcrete cracked and the steel support distorted very severely (Fig. 3). In particular, the working face at the downstream area collapsed in April 2010 (Fig. 4), and its treatment was not completed until April 2011, taking more than 1 year. Some water was found in the site, which made the rock wet, soft, and extremely fractured. The in situ stress is high as the overburden is around 440 m.

2.3 Site observation from the upstream of branch tunnel 6

Branch tunnel 6 is 360 m long. The excavation of the tunnel was started in July 2008 and completed in July 2009, taking 13 months with an average monthly advance of about 33 m.

The excavation of the diversion tunnel from branch tunnel 6 was then started from two working faces: one going upstream and the other downstream. As of March 2011, an excavation of only 281 m in the upstream direction was completed, equivalent to an average monthly advance of only 13.0 m.

During the construction, a big deformation occurred at the upstream area (S13 + 493.0–S13 + 630.0). The deformation ranged from 9–22 cm at the crown and 10–25 cm at the sidewalls. A second expansion excavation had to be carried out. The maximum deformation was 37 cm. The shotcrete bulged and the steel support distorted very severely. To control the big deformation, steel support was applied as shown in Fig. 5. The steel support was then cast into the lining as a permanent support (Fig. 6). The reinforcement anchor was also applied. Fortunately, no water was observed, but the in situ stress was very unfavorable with an overburden of 550 m and big eccentric compression from the mountain side.

In summary, the big deformation occurred in the tunneling because of three reasons. Firstly, the rock is phyllite, which is very soft with a low strength. Particularly when water is present, the rock will lose its strength and the tunnel becomes unstable. Secondly, the in situ stress is very

Fig. 5 Steel support

Fig. 6 Steel support casted in lining

high with a serious eccentric compression. The rock bedding is oblique to the tunneling direction with a small angle. The excavation released the normal stress on the bedding, which caused a big deformation. Finally, the initial support (shotcrete and rock bolt) was not strong enough and could not provide enough normal stress to the bedding. In addition, the lining was installed too late to allow more stress release.

The site observation also indicated that the rock weathered very fast after the excavation, and a quicker application of shotcrete was effective to prevent the rock from weathering. Laboratory tests found that the rock properties were closely related to the confinement pressure. With the decrease of confinement pressure, both rock strength and Young's modulus decrease sharply. Particularly when the rock contained water, the rock loses a lot of strength.

3 Tunneling strategy

Among the reasons causing big deformation, compared with high in situ stress and soft rock that cannot be changed during the construction, the rock strength loss and rock stiffness decrease can be improved by providing quick and strong support pressure to the rock and keeping the rock dry. Based on the site observation and laboratory test results, a set of tunneling strategies was proposed to overcome the big deformation problem.

3.1 Construction principle

To avoid big deformation, a construction principle was established for the tunneling including the following:

(1) Effort should be made to provide confinement pressure to the rock.
(2) Exposure time of rock at/near the working face should be shortened to avoid rock weathering and loosing.
(3) The water should be kept out of the tunnel by reducing the water stay time and strengthening water drainage.
(4) The working face would be kept stable to avoid it collapsing, which will seriously delay the tunnel advancing and increase treatment cost.
(5) Field data and information should be collected, back analyzed, and returned to guide the construction and rock support design.
(6) The tunnel advance should be improved if the safety can be assured.
(7) An allowable rock deformation should be given in the design to prevent the second expansion excavation.

3.2 Specific measures for the construction

Based on the construction principle above, a set of specific construction measures including 11 items was proposed:

- An advanced geological prediction needs to be done. The geology in front of the working face needs to be explored before the excavation to predict water condition, fractured zone, etc.
- The rock deformation monitoring needs to be done during the construction. Information such as crown subsidence and convergence needs to be collected and back analyzed to judge the tunnel stability, optimize the rock support parameters and construction scheme, and determine the allowable deformation, initial support parameters, and lining installation time.
- An allowable rock deformation is needed to avoid second expansion excavation. The determination of the allowable deformation would be based on the achievement from the advanced geological prediction and rock deformation monitoring results. A method to predict the

rock deformation would be established to support the determination of allowable rock deformation.

- After the excavation, shotcrete would be applied immediately to the exposed rock at/near the working face in order to reduce the rock exposing time, avoid rock weathering, and prevent rock strength and stiffness from decreasing.
- Water would be drained out of the tunnel to avoid water pond in the tunnel to prevent the rock from soaking and softening.
- If the rock is heavily fractured, keeping the working face stable is most important. This can be assured by controlling the step advance, strengthening support, reserving core soil to support the working face, and preventing the working face from collapsing.
- Integrated support of the steel arch, rock bolt, reinforcement mesh, and shotcrete would be applied immediately after the excavation. Keeping the support as a whole with a strong connection and improving the support stiffness are very important.
- The steel arch is a major load-bearing support. Except the steel arch stiffness itself, the longitudinal steel arch stiffness is also very effective to resist the big deformation. This can be achieved by connecting steel arches using longitudinal steel beams. In addition, keeping the steel arch close at bottom as a loop is also effective.
- Although it is verified that long rock bolts (9 m or longer) are workable to resist big deformation, long rock bolts are not applicable in this tunnel as the tunnel diameter is small. Short rock bolts (4.5 m) are suggested to connect the steel arches to the rock.
- Shotcrete is useful to prevent the rock from weathering and to provide an integrated support pressure to the rock. Shotcrete would be applied again in time if it was cracked.
- Lining can provide a strong support; earlier application of the lining can effectively resist big deformation. It is suggested to apply lining within a distance of twice the tunnel diameter from the working face.

3.3 Measures to speed up the tunneling

The specific measures above were preliminarily applied to the tunneling in the branch tunnel 2 and a good result was obtained, but with a very slow advance. The main reason causing slow advance is that the lining is applied using a full frame that takes a long time (9–12 m per half a month) and the working face must be stopped during the lining installation. To solve this problem, the following measures were taken:

(1) Instead of the full frame, a formwork jumbo was used to allow the transportation passing through.
(2) The upper part of the lining was applied first to provide fast support to the rock, and the invert part of the lining was then followed.
(3) A good road condition was kept to allow easy entrance of transportation and manpower to improve the construction effectiveness.
(4) The ventilation is strengthened to improve the air condition to enhance the construction effectiveness.
(5) The full face excavation was suggested, but core soil must be reserved to assure the working face stability.

After adopting the measures above, the tunneling was speeded up with an average advance of more than 60 m/month.

4 Field monitoring

In order to check the effectiveness of the tunneling strategy above, a series of field monitoring procedures was carried out during the construction at branch tunnel 2. A total of 18 monitoring sections was carried out, 5 of them being at the upstream area and the others at the downstream. The monitoring items include (1) convergence, (2) rock support (rock bolt, steel arch, and lining) stress, and (3) contact pressure between rock and lining.

4.1 Convergence

The convergence was conducted at three sections, and monitored data are listed in Table 1 (the third column). In fact, the monitored data are not the total convergence, but are only partial as the monitoring started at a distance behind the working face. To evaluate the convergence developing with the advancing, numerical modeling was carried out and a relation of displacement percentage versus distance coefficient was obtained as shown in Fig. 7. From the relation, the prior convergence occurring before the monitoring as listed in Table 1 (the fifth column) could be calculated according to the distance behind the face (the forth column) when the first monitoring data were read. The total deformation was then calculated from the sum of the monitored convergence and the prior one as listed in Table 1(the last column).

As an example, Fig. 8 shows the convergence monitoring data at Section S3 + 770.0. It can be seen that the monitored convergence is small with a maximum of only 18 mm. More importantly, the convergence initially increases with the tunnel advancing, but gradually tends to be a certain value, indicating that the tunnel is stable.

Table 1 Monitored and total convergence

No.	Chainage	Monitored convergence (mm)	Distance behind face (m)	Prior convergence (mm)/(%)	Total convergence (mm)
1	S3 + 634.4	160	5.0	98/(38)	258
2	S3 + 762.8	30	2.4	5/(13)	35
3	S3 + 770.0	18	2.5	3/(14)	21

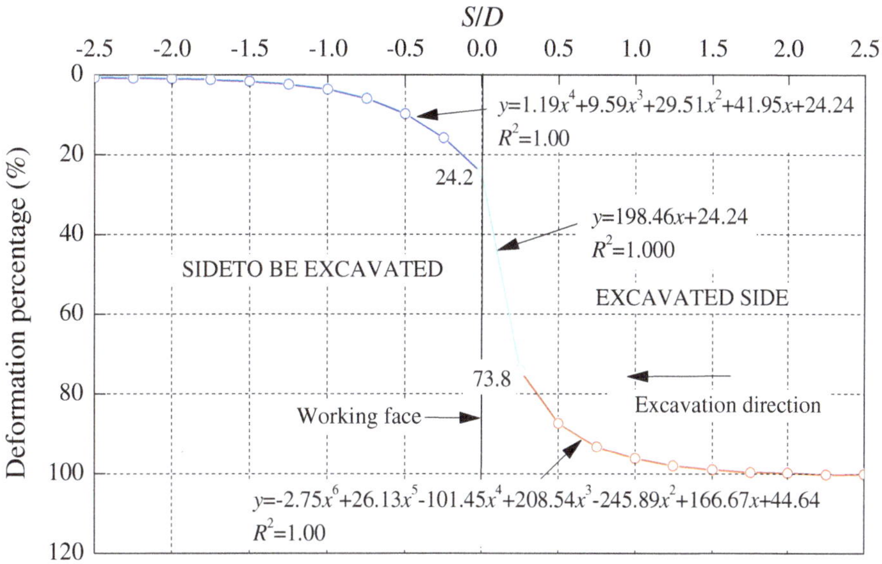

Fig. 7 Relation of displacement percentage versus distance coefficient. *S* is distance from the working face and *D* is the tunnel diameter

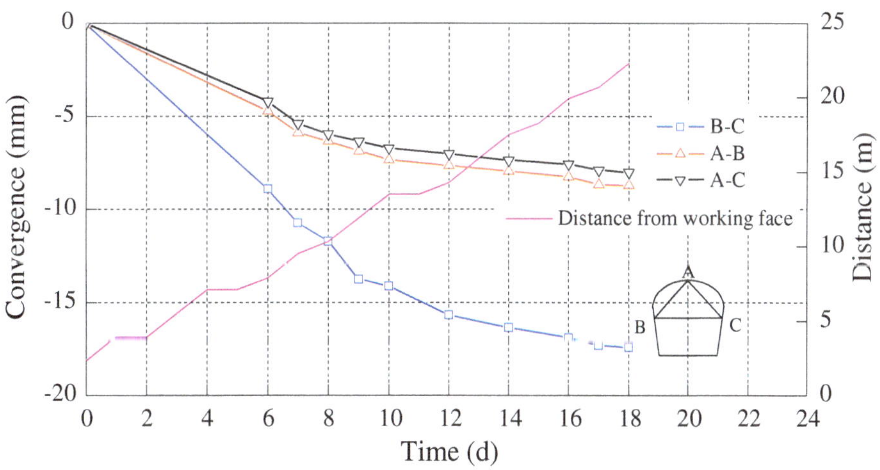

Fig. 8 Convergence monitoring data at Section S3 + 770.0

4.2 Contact pressure between rock and shotcrete

Figure 9 shows the monitoring data of the contact pressure between the rock and shotcrete at Section S3 + 762.8. As shown in the figure, the contact pressure is very small with a maximum value of 0.47 MPa, and the contact pressure tends to be stable.

4.3 Steel arch stress

Figure 10 shows the monitoring data of steel arch stress at D of Section S3 + 766.8. As shown in the figure, the steel arch stress with tunneling converges very fast, and the stress is small with a maximum stress of less than 100 MPa.

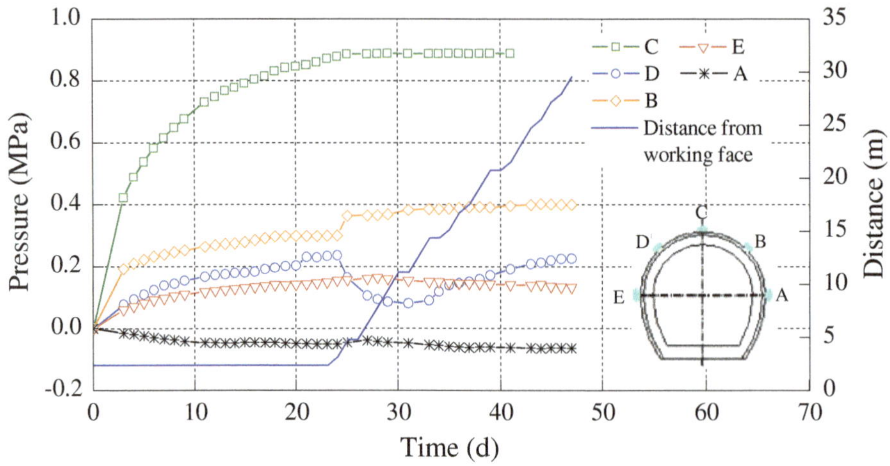

Fig. 9 Monitoring data of contact pressure at Section S3 + 762.8

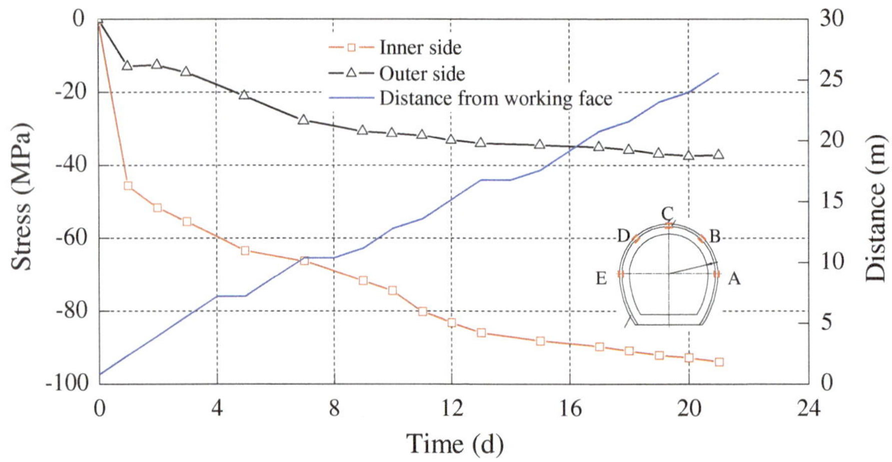

Fig. 10 Monitoring data of steel arch stress at D of Section S3 + 766.8

4.4 Lining strain

Figure 11 shows the monitoring data of lining strain at Point A of Section S3 + 657.6. As shown in the figure, the concrete strain is small with a maximum of <180 με, and the strain tends to be stable.

The monitoring results above indicate that the rock deformation is effectively controlled by applying the proposed tunneling strategy. The stress on the support is small and the contact pressure between the rock and shotcrete is within the allowable value, which suggests that the tunnel is stable and the tunneling strategy is applicable.

4.5 Site observation

From the site observation, no shotcrete failure was found near Sections S3 + 762.8 and S3 + 770.0, but shotcrete cracking and steel arch distortion were observed near Section S3 + 634.4. This agrees well with the monitored deformation. However, the support failed though the tunnel became stable, which indicates that an allowable deformation would be given to the support design to prevent the support from failing.

5 Determination of allowable deformation

To determine the allowable deformation, a maximum deformation would be obtained. According to the Fenner formulae, the yielding radium, normal stress at the border between the elastic zone and the yielding zone, and tunnel deformation are related to rock properties, overburden, tunnel radium, and support pressure as

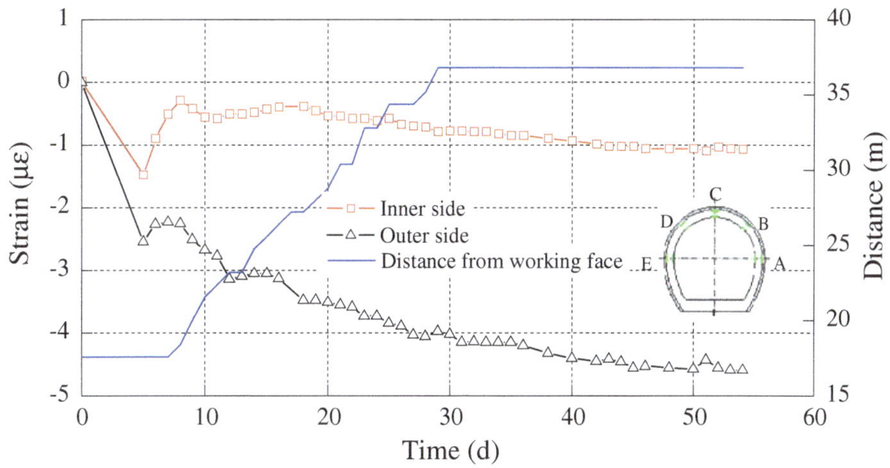

Fig. 11 Concrete strain of lining at A of Section S3 + 657.6

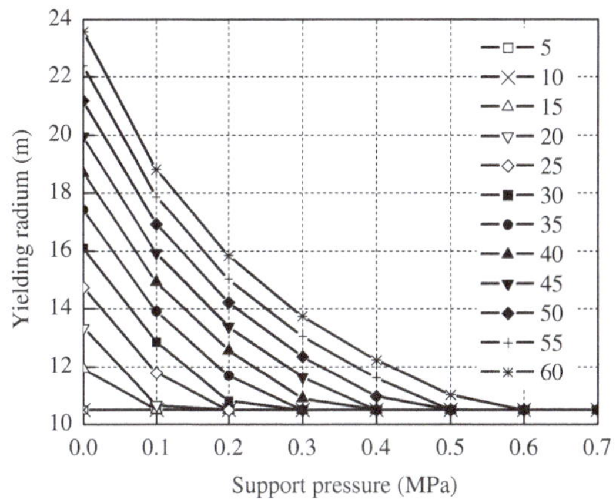

Fig. 12 Yielding radium versus support pressure for different overburden values

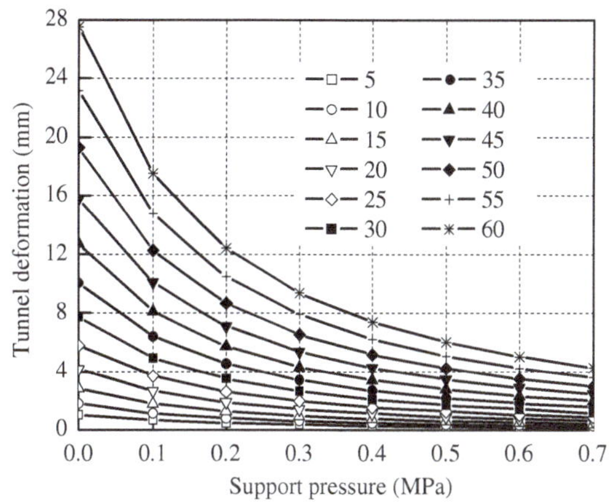

Fig. 13 Tunnel deformation versus support pressure for different overburden values

$$\left.\begin{aligned} r_0 &= a\left[\frac{2}{\xi+1}\cdot\frac{\sigma_y(\xi-1)+R_b}{p_a(\xi-1)+R_b}\right]^{\frac{1}{\xi-1}}, \\ \sigma_{r0} &= \frac{R_b}{\xi-1}\left[\left(\frac{r_0}{a}\right)^{\xi-1}-1\right]+\left(\frac{r_0}{a}\right)^{\xi-1}\cdot p_a, \\ u_a &= \frac{1}{2K}(\sigma_y\sin\phi+c\cos\phi)\frac{r_0^2}{a}, \end{aligned}\right\} \quad (1)$$

where

$$\xi = \frac{1+\sin\phi}{1-\cos\phi},$$

$$R_b = \frac{2c\cos\phi}{1-\sin\phi},$$

c is cohesion, ϕ is the friction angle, a is the tunnel radium, r_0 is the yielding radium, σ_{r0} is the normal stress at the border between the elastic zone and the yielding zone, u_a is the tunnel deformation, K is the bulk modulus of rock, P_a is the support pressure, andσ_y is the vertical in situ stress

Figures 12, 13 show the yielding radium and tunnel deformation versus support pressure for overburden of different depths. From the figures, it can be found that a higher overburden causes a bigger yielding radium and tunnel deformation, while a bigger support pressure could decrease the yielding radium and tunnel deformation.

Based on Eq. (1) and taking account of the respective installation time of the initial support and lining, the tunnel deformation was estimated as 10–30 cm for evaluated cohesion of 0.05–0.20 MPa. Compared to the monitoring results as listed in Table 1, it can be seen that the deformation estimation agrees well with the monitored result. Therefore, the allowable deformation for this project was determined as 30 cm, which was verified by the further construction.

6 Conclusions

Big deformation with a maximum up to 2.0 m was observed in the tunneling of the diversion tunnel of Dongsong Hydropower Station. The shotcrete was cracked and the steel arch was distorted due to the big deformation. In order to solve this problem, a construction strategy was proposed and an allowable deformation was determined for further construction of the tunnel.

- A construction principle was established based on a concept of mainly maintaining the rock strength/stiffness and keeping the rock dry, by providing a confinement pressure to the rock, reducing the rock exposure time, keeping water out of the tunnel, etc. A set of construction measures with 11 items was then proposed and applied to the construction, and a good result was achieved. It was found that among them, reserving core soil at the working face and applying immediate initial support to the rock after excavation were most effective to control the rock deformation.
- Presetting a bigger allowable deformation in the support design is necessary as the soft rock needs an obvious deformation before becoming stable; otherwise, the support may fail. The allowable deformation can be determined through an integrated manner of theoretical prediction and monitoring verification. In this study, an allowable deformation of 30 cm was obtained and applied to the support design, which had a good achievement in further construction of the tunnel.

References

1. Guan BS (2008) Key issues on tunnel construction. People's Transportation Publisher, Beijing
2. Chapman D, Metje N, Staark A (2010) Tunnel construction. Spon Press, London
3. Guan BS, Zhao Y (2011) The construction technology of soft rock tunnels. People's Transportation Publisher, Beijing
4. Brady BHG, Brown ET (1993) Rock mechanics for underground mining. Chapman & Hall, London
5. Zhang Z, Guan BS (2000) A study on deformation patterns of soft rock tunnels under high in situ stress. Chin J Geotech Eng 22(6):696–700
6. Cheng F, Chen SG, Tan X et al. (2009) Wang, Deformation control for tunneling with a small separation in soft strata. Paper presented at the ISRM regional symposium Eurock 2009, Cavtat, Dubrovnik, 29–31 Oct 2009
7. Chen SG, Zhao YB, Zhang H (2009) Analysis of large rock deformation under high in situ stress. Paper presented at the 9th international conference on analysis of discontinuous deformation, Singapore, 25–27 Nov 2009
8. Bieniawski ZT (1984) Rock mechanics design in mining and tunneling. A. A. Balkema, Boston
9. Kolymbas D (2005) Tunneling and tunnel mechanics. Springer, Berlin
10. Singh B, Goel RK (1999) Rock mass classification. Elsevier, Amsterdam
11. Wang M, Zhang J (1998) A study on the mechanical effect of measures to control tunnel deformation. Chin J Geotech Eng 20(5):27–30
12. Chen SG, Zhang H, Tan X et al (2011) Key technologies for construction of Jinping traffic tunnel with an extremely deep overburden and a high water pressure. J Mod Transp 19(2):94–103
13. Zhang L, Zhou X, Zhao C (2006) The design and construction technologies of soft rock tunnel. China Water Power Press, Beijing
14. Chen SG, Hu W (2009) A comprehensive study on subsidence control using COSFLOW. Int J Geotech Geolog Eng 27(3):305–314
15. Ma H, Chen SG, Hu C, et al. (2011) A study on the stability of a big-section tunnel in karst area. Paper presented at the 10th international conference of discontinous deformation, Hawaii, USA, 6–9 Dec 2011
16. Gui R, Liu Y (2011) An analysis on time and space effect during tunneling in soft rock. J Nanhua Univ 25(1):28–32
17. Haruyama K, Teramoto S, Taira K (2005) Construction of large cross-section double-tier metropolitan inter-city highway tunnel by NATM. Tunn Undergr Space Technol 20:111–119
18. Chou WI, Bovet A (2002) Predictions of ground deformations in shallow tunnels in clay. Tunn Undergr Space Technol 17(1):3–19
19. Peck RB (1969) Advantages and limitations of the observational method in applied soil mechanics. Geotechniques 19:171–187
20. Ma H, Chen SG, Tan X et al. (2012) Advanced geological detection for tunneling in karst area, ASCE Geotechnical Engineering State of the Art and Practice, San Francisco, 25–29 March 2012
21. Li R (2005) The processing for large-scale deformation of early support in road tunnel. J West Mine Explor 7:118–119
22. Li YL, Feng XG, Jiang Y et al (2005) Large deformations encountered in the surrounding rocks of tunnels and their prediction. Mod Tunn Technol 42(5):46–51
23. Chen SG (2011) A development on technologies of tunnel construction with big deformation. A research report, Southwest Jiaotong University

6

Dynamic airspace sectorization via improved genetic algorithm

**Yangzhou Chen · Hong Bi · Defu Zhang ·
Zhuoxi Song**

Abstract This paper deals with dynamic airspace sectorization (DAS) problem by an improved genetic algorithm (iGA). A graph model is first constructed that represents the airspace static structure. Then the DAS problem is formulated as a graph-partitioning problem to balance the sector workload under the premise of ensuring safety. In the iGA, multiple populations and hybrid coding are applied to determine the optimal sector number and airspace sectorization. The sector constraints are well satisfied by the improved genetic operators and protect zones. This method is validated by being applied to the airspace of North China in terms of three indexes, which are sector balancing index, coordination workload index and sector average flight time index. The improvement is obvious, as the sector balancing index is reduced by 16.5 %, the coordination workload index is reduced by 11.2 %, and the sector average flight time index is increased by 11.4 % during the peak-hour traffic.

Keywords Dynamic airspace sectorization (DAS) · Improved genetic algorithm (iGA) · Graph model · Multiple populations · Hybrid coding · Sector constraints

1 Introduction

In air traffic management, airspace is often partitioned into sectors, and each of the sectors is controlled by one or several controllers. Airspace sectorization is to determine reasonable sector number and sector structure such that air traffic safety and efficiency are ensured. Dynamic airspace sectorization (DAS) is to divide the airspace into several sectors according to the traffic situation. Recently, DAS becomes an important issue, and many approaches have been proposed to solve the problem.

Traffic safety and efficiency are ensured by balancing the controller's workloads between sectors and making them within a reasonable threshold on the basis of traffic situation. Apart from this, the sectors are required to meet the convexity constraint, connectivity constraint, minimum distance constraint, and minimum sector crossing time constraint [1, 2].

Up to now, genetic algorithm (GA) has been applied in DAS approaches. It was first attempted by Delahaye et al. [3, 4], who proposed two approaches, one based on weighted graph and the other using Voronoi diagram model of the airspace. Xue [5] further improved the GA efficiency by combining the algorithm with an iterative deepening algorithm, and then directly applied it to a real flight track data. In recent years, the study of DAS has been extended from 2D airspace to 3D airspace [6–8]. They used Voronoi diagram, cells and agent-based models to establish the airspace model. These models are based on the generating points or seed points, and point locations are optimized by applying GAs to realize optimal sectors. As to 3D airspace. Tang et al. [9, 10] proposed an improved agent-based model (iABM), and combined it with GA to achieve the optimized sectorization. He also identified the gaps in the iABM and three additional models such as KD-tree, graph bisection, and Voronoi diagram in 3D, and evaluated their constraints and objective indices.

The aforementioned researches used GA in a way similar to Ref. [4] and made improvements in different aspects. However, one limitation of these works is that in each generation the fitness of every individual must be calculated on

Y. Chen · H. Bi (✉) · D. Zhang · Z. Song
College of Electronic Information and Control Engineering,
Beijing University of Technology, Beijing 100124, China
e-mail: 499053243@qq.com

the basis of the new partitions. These new partitions are produced by performing the model algorithm time and again after accomplishing GA operators, resulting in lower algorithm efficiency. Another limitation is that in the design of GA the emphasis was to establish a relation between the multi-objectives of DAS and the fitness function, while the sector constraints were not considered. In fact, the sector constraints cannot be completely satisfied because of the limitation of model algorithm (see [9] for the details). Figure 1 shows an example of Voronoi diagram. Although the safety distance and residence time are taken into account in the fitness function, boundaries may superimpose on the air-routes or key-points, and it is hard to satisfy the minimum distance constraint. This is because the airspace structure is taking into account when establishing model.

In this paper, we discuss the feasibility of full application of GA in DAS by improving the method in Ref. [3]. The improved GA (iGA) can overcome the aforementioned shortages in the previous works. By using several populations competed genetic algorithm (SPCGA) [11], the optimal sector number and airspace sectorization can be determined. We also studied the search capability by adding a crossover operator and designed a repair strategy to meet the convexity constraint.

2 Weighed graph model of airspace

In this section, we will first set up an airspace model by using a weighed graph. The model is limited to 2D airspace, because the DAS in 2D airspace is easily extended to the one in 3D airspace. The vertices of the weighed graph consist of key-points of the airspace such as airports, waypoints, conflict points, etc., and its edges describe the air-routes between these key-points. Both of the vertices and edges are endowed with weighed values representing workloads of controllers in the airspace.

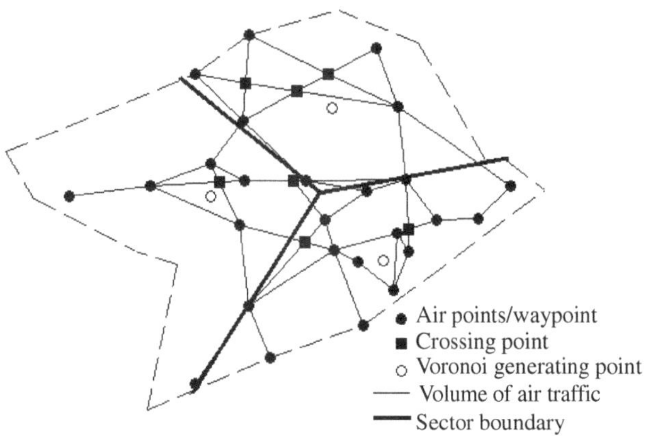

Fig. 1 DAS by Voronoi diagram method

2.1 Establishing the weighed graph

Static structural information in airspace includes air-routes, locations of key-points (airports, waypoints, conflict points), etc. The key-points are described as the vertices of an undirected graph. The air-routes between airports and waypoints are described as the edges of the undirected graph. That is, we set up a graph $G = \{V, E\}$ with the set of vertices V and the set of edges E to describe the airspace structure.

Most of the air-routes in China airspace are fixed. Figure 2 shows the distribution of air-routes and key-points of North China airspace. We choose the main airspace of North China based on Fig. 2 to establish the undirected graph as shown in Fig. 3.

Next, we endow the vertices and edges of the graph with workloads of controllers. Workloads consist of the three styles, i.e., monitoring workload, coordination workload, and conflict avoidance workload [12]. Each style is calculated on the basis of airspace complexity and then added to the undirected graph. Monitoring workload and conflict avoidance workload are endowed to vertices, and coordination workload is to the edges. Thus, we have the final weighted graph that describes both of static structural information and dynamic workloads.

As illustrated above, sectors are required to meet constraints such as convexity, connectivity, minimum distance, and minimum sector crossing time. Some constrains are imposed by adding a square protection zone around two vertices with a distance between them less than $2d_{\min}$, where $d_{\min}$ is the required minimum distance (see Fig. 4). The width of the square protection zone is $2d_{\min}$ and the

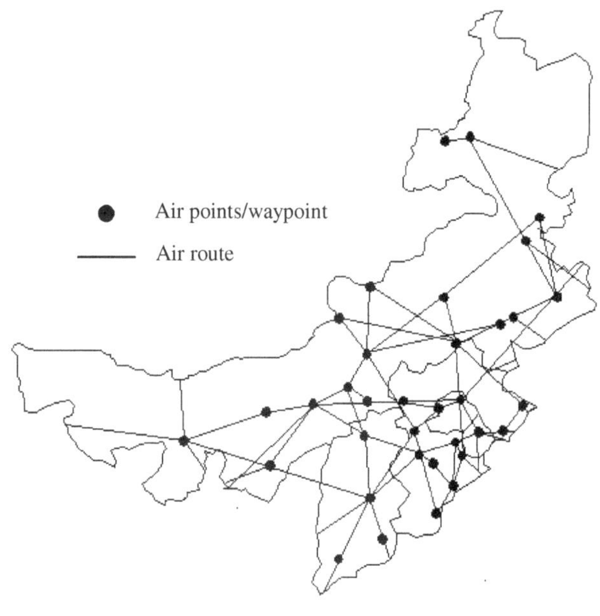

Fig. 2 Airspace structure of North China

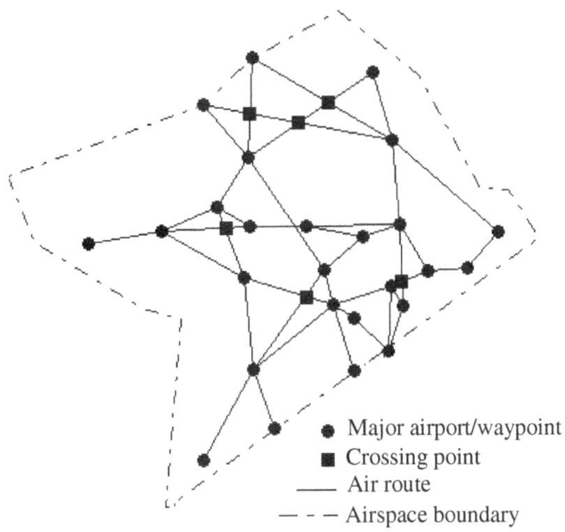

Fig. 3 Undirected graph of main airspace of North China

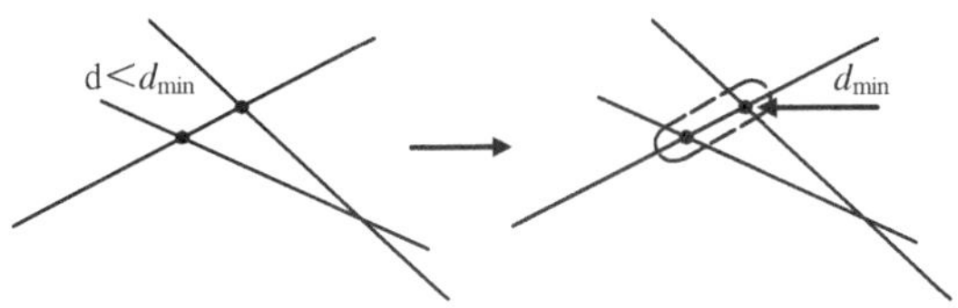

Fig. 4 The protection zone

length is $2d_{\min} + d$ so that a sector boundary cannot pass through the boundary of protection zone, where d is the distance between the two vertices.

2.2 Optimization problem of DAS

Now we calculate a reasonable sector number and describe an optimal objective for the optimization problem of the DAS.

The reasonable sector number is one of the important factors to ensure the controllers' workloads in safety limit range. Sector number depends on traffic conditions and it is generally not unique. So we assume that a reasonable threshold of sector number is $[K_{\min}, K_{\max}]$, where the integers $K_{\min}$ and $K_{\max}$ are the possible minimal and maximal sector numbers.

According to the Doratask method, the average workload of every controller must be less than 80 % of statistics time. By considering the airspace capacity and flight safety, we determine the minimum sector number $K_{\min}$ as [7]

$$K_{\min} = \left\lceil \frac{W_{\mathrm{T}}}{80\,\% \cdot T} \right\rceil, \tag{1}$$

where T is the statistic time, W_{T} is the total workload of a given airspace, and notation $\lceil\cdot\rceil$ means the greatest integer function.

A larger sector number is desirable for the traffic safety, but may cause a waste of control resources. There must be a balance between the traffic safety and control resources, which relates to staffing, control efficiency, control cost, etc. So the maximum sector number should be determined on basis of the airspace situation. In this paper, we do not focus on airspace resource factors and assume that the average workload of every controller must be more than $E\%$ of statistic time to meet the work efficiency and save the control cost, i.e.,

$$K_{\max} = \left\lceil \frac{W_{\mathrm{T}}}{E\% \cdot T} \right\rceil. \tag{2}$$

As for setting up the optimization problem of DAS, in order to improve the airspace capacity and control efficiency, a multi-objective sectorization is required to balance the controller's workloads between sectors and maximize both the average sector flight time and the distance between sector boundaries and the key-points.

Here, we take into account of the following three objectives as the multi-objective function of DAS, i.e., balancing sectors' workloads f_{b}, minimizing coordination workloads f_{c}, and maximizing the average sector flight time f_{t}, which are defined as follows, respectively:

$$f_{\mathrm{b}} = \sum_{k=1}^{k=K} \frac{\left|w(k) - W_{\mathrm{T}}K^{-1}\right|}{W_{\mathrm{T}}K^{-1}}, \tag{3}$$

$$f_{\mathrm{c}} = \frac{1}{W_{\mathrm{T}}} \sum_{k=1}^{k=K} w_{\mathrm{c}}(k), \tag{4}$$

$$f_{\mathrm{t}} = \min_{k} \left\{ \frac{S_{\mathrm{T}}(k)}{N(k)} \right\}, \tag{5}$$

where $w(k)$ is the controller's workload including monitoring and conflict avoidance workload in the sector k; $W_{\mathrm{T}} = \sum_{k=1}^{K} w(k)$ is the total workload in all the sectors; $w_{\mathrm{c}}(k)$ is the coordination workload in the sector k; $S_{\mathrm{T}}(k)$ is the flight time in sector k; and $N(k)$ is the flight number in sector k. Finally, the multi-objectives are integrated into a single objective function:

$$f = a_1 f_{\mathrm{b}} + a_2 f_{\mathrm{c}} - a_3 f_{\mathrm{t}}, \tag{6}$$

where positive numbers a_1, a_2, a_3 are appropriately selected weight coefficients. Since there may be implicit conflict between these three objectives, we use weighted method to form a single objective function.

As the security must be ensured after sectorization, some constraints of DAS are considered:

(1) Convexity constraint. An aircraft cannot enter the same sector twice.
(2) Minimum distance constraint. The distance between a sector boundary and a key-point must be not less than $d_{\min}$.
(3) Connectivity constraint. Every sector should be geometrically connected.

3 DAS via an improved GA

Solving DAS problem includes two steps. The first is to determine sector number and accomplish airspace partition, and the second is sector boundary optimization. Both of the two steps will be done using a hybrid encoding multi-objective GA on the basis of the weighted graph model. Precisely, the sector number determination and the airspace partition problem will be solved by using a SPCGA. During this process, the midpoint on air-routes between two key-points is selected as temporary boundary points in the calculation of the workloads. After sector partition is accomplished, we optimize the sector boundary by changing the locations of boundary points with a float code GA [13].

In the process of using GAs, the sector constraints are realized in the following ways. Convex constraints are naturally satisfied since the weighted graph contains air-route information. Connectivity constraints will be realized by a proper way of individual initialization and repair strategy of GA operator. The minimum distance constraint and the shortest residence time constraint are accomplished by adding the protection zones around key-points so that sector boundary cannot extend beyond these protection zones.

3.1 Sector number determination and sector partition

Now we begin to determine the optimal sector number and accomplish airspace partition via using a GA based on integer code. Arbitrarily given sector number $K \in [K_{\min}, K_{\max}]$, K sub-populations are initialized and each of them has n individuals (chromosomes). Each of the chromosomes contains the information of sector partition and the corresponding sector number K. We number the vertices of the graph randomly and the coding of the chromosome is made according to the connected components. Figure 5 shows an example of these chromosomes.

The sector number within each sub-population should be same, i.e., the gene indicating sector number must be same. This will make it easy to design GA operators and to satisfy the constraints. The initialization of the sub-populations is as follows:

Step 1 K different vertices are randomly selected from the graph and then labeled with different symbols according to the order of the point numbers. Those vertices consist of initial K connected components.

Step 2 The neighbors of a component are checked. A vertex is viewed as a neighbor of a connected component if there is a link between this vertex and a vertex belonging to the connected component. If this vertex is free, then it is associated to the component, or else the vertex has been already associated to another component.

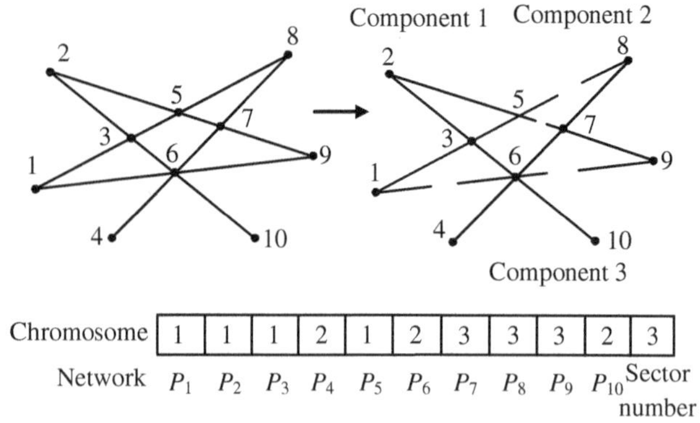

Fig. 5 Integer code of the chromosome

Step 3 Step 2 is repeated till all the vertices of the graph are labeled.

Step 4 Add the gene indicating sector number.

Figure 6 shows the procedure of the first three steps of the individual initialization.

Fitness function is the objective function represented in (6).

GA operators consist of reproduction, crossover, and mutation. As to these operators, in order to satisfy the connectivity constraint when producing a generation, a repair strategy is needed. Some common repair strategies are to refuse infeasible solution, improve GA operation, or repair infeasible solution and penalty function. In this paper, the first and second ones are applied to repair the mutation operator, and the third one is applied to repair the crossover operator. Now we expound the repaired operators as follows.

Reproduction operator Tournament selection is adopted as the reproduction operator in order to ensure not only the reproduction of better individuals but also the population diversity to a certain extent. The other reproduction operator such as the proportional fitness assignment (e.g.,

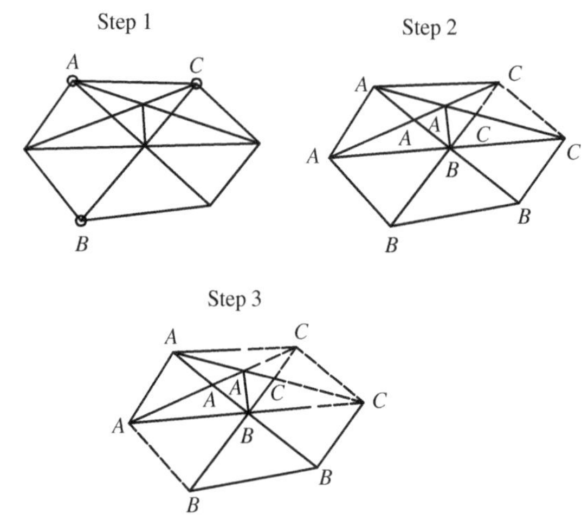

Fig. 6 Individual initialization

roulette wheel selection) is not so effective since for the individuals who have the similar fitness may have the similar offspring number and thus better individuals do not have enough chance to be reproduced.

Crossover operator Crossover operator plays an important role in GA since it helps to expand the solution space and get the global optimal. We choose the consistent crossover because of its strong search capacity [14, 15]. However, the individual produced by the crossover operator may not satisfy the connectivity constraint, so it needs to repair (see Fig. 7):

(1) The connectivity of K components is first checked. If it is not satisfied, one gets new $\bar{K}$ connected components with connectivity, where $\bar{K} > K$.
(2) The connected components are sorted according to the number of vertices. The former K connected components are chosen as the main part of the repaired strategy.
(3) First, the number of lines between the residual $\bar{K} - K$ connected components and the main part is checked. Then find the main part with which the residual connected component has the maximum line number. Last, the residual connected component is merged with this main part. This method can reduce the number of lines between components after performing the repair strategy. Then we get K repaired connected components.

Mutation operator The strong mutation and weak mutation mentioned in Delahaye's work are used (see [3, 4] for the details).

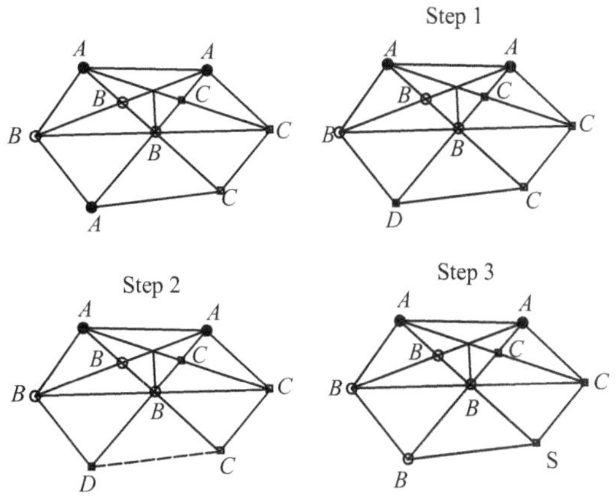

Fig. 7 Repair strategy

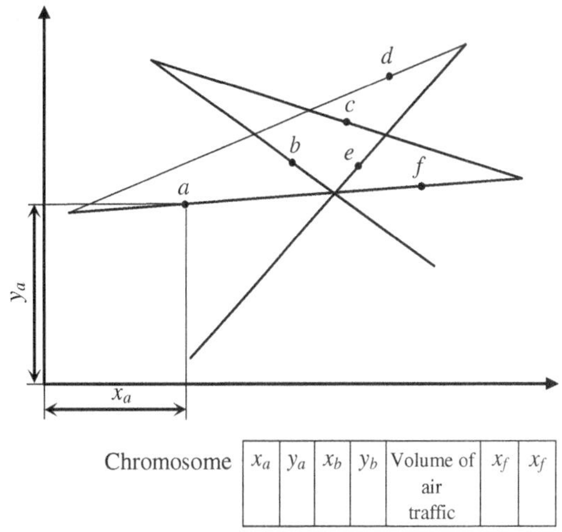

Fig. 8 Float code of the chromosome

3.2 Boundary optimization

The sector boundaries are created roughly by the airspace partition and then the positions of boundary points are adjusted via boundary optimization. We optimize the sector boundary by GA based on float code. The best sub-population is chosen as an input of the GA. The fitness function and the reproduction operator are the same as in Sect. 3.1.

The sector boundary can be described by the positions of boundary point. So, the individuals must contain the coordinate information of all the boundary points. An individual is generated by float code (see Fig. 8).

As the boundaries are optimized based on the airspace partition, the coordinates of boundary points in initial population are the coordinates of midpoints on the air-routes between the different components.

The other two operators in the GA are crossover and mutation. Consistent crossover is adopted as the crossover operator. Instead of exchanging the genes directly, the midpoint of allele in parent is used as the allele after crossing. This can improve the algorithm's convergence efficiency.

The mutation operator can move a boundary point on its edge within an effective range, and change the allele which contains the coordinate information of the boundary point (see Fig. 9).

The effective range of boundary point is constrained by the protection zones around the vertices beside the boundaries. The protection zones are centered at the vertices with radius $d_{\min}$. Neither the boundaries nor the boundary points are in the protection zones, which ensures that the distance between vertex and boundary is larger than the minimum distance $d_{\min}$. So, the boundary points are located on edge rather than in the protection zones.

4 Application to North China airspace

4.1 Algorithm performance

First, we assumed the reasonable threshold of sector number is $K \in \{3, 4, 5\}$. Figure 10 shows the variation of

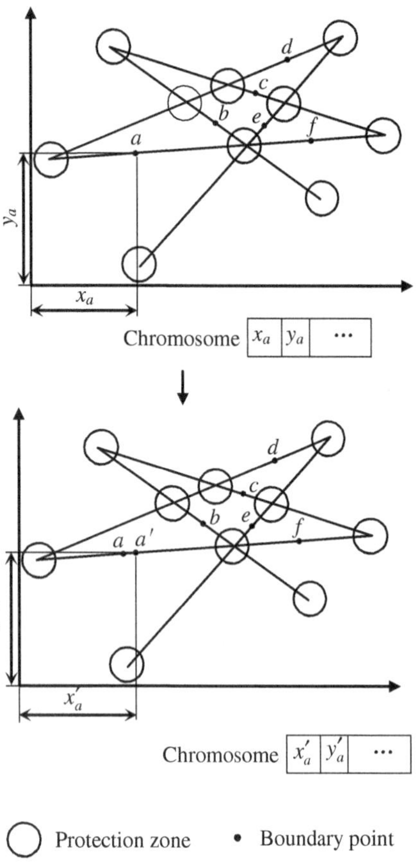

Fig. 9 Mutation operators of float code

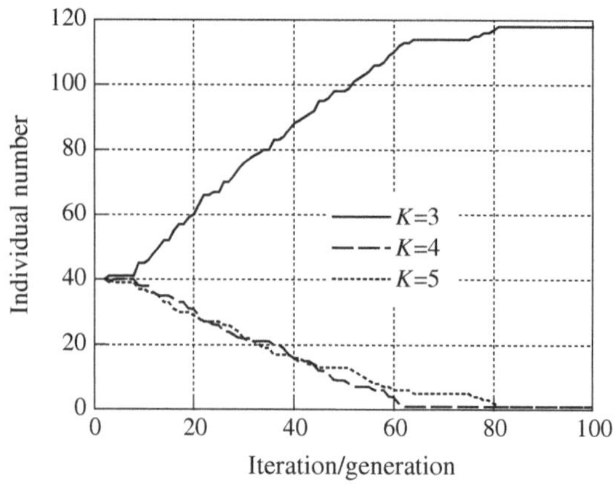

Fig. 10 Change of individual number

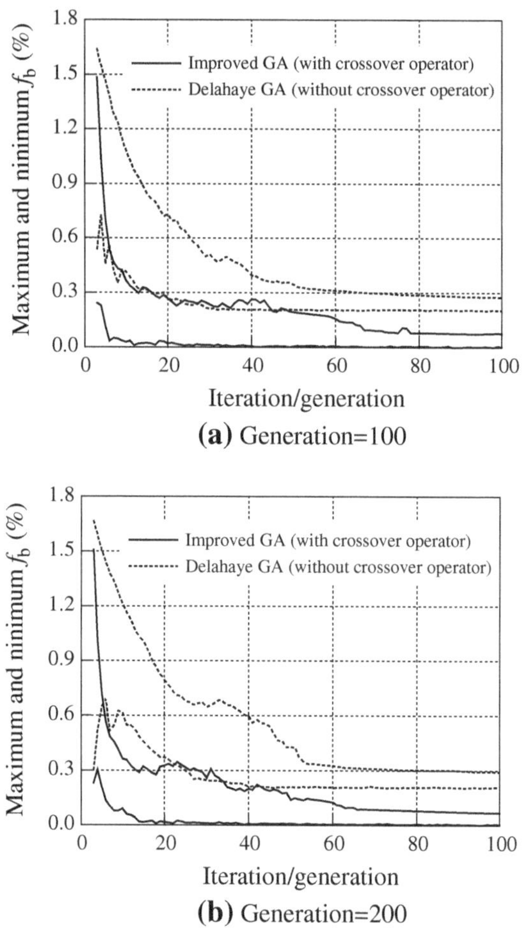

Fig. 11 Comparison of two GAs

individual number in each sub-population when using the SPCGA for the sector partition. One can see that the larger the sector number is, the smaller the individual number becomes. Thus, the sub-population with the best sector number is remained and evolved.

The GA used by Delahaye et al. abandoned the crossover operator, which influences the search capability of GA. Figure 11 compares the search capability of the iGA with the GA used by Delahaye et al. [3], where the variation of the maximum and minimum f_b during the evolvement for the sector partition is shown. One can see that with iGA, a better result is achieved (the mutation operator here does not respect connectivity constraint to make f_b approach 0). This is the reason that the added crossover operator can improve the search capability of GA).

4.2 Result of DAS

In this section, the iGA is validated with real air traffic data of North China airspace. Figure 12 shows the change of traffic flow in a whole day. As the volume of air traffic varies with time, so does the range of sector number (see Table 1). The iGA is applied to each time interval, and the final number of sectors demonstrates that the new sectorization computed by the proposed DAC algorithm is adaptive to the time-varying air traffic with a varying number of sectors.

Furthermore, the result of DAS during 19:00–21:00 configured by the iGA is shown in Fig. 13. It is obvious that all of the sectors have smooth boundaries and good geometrical shapes. Figure 14 shows the maximum

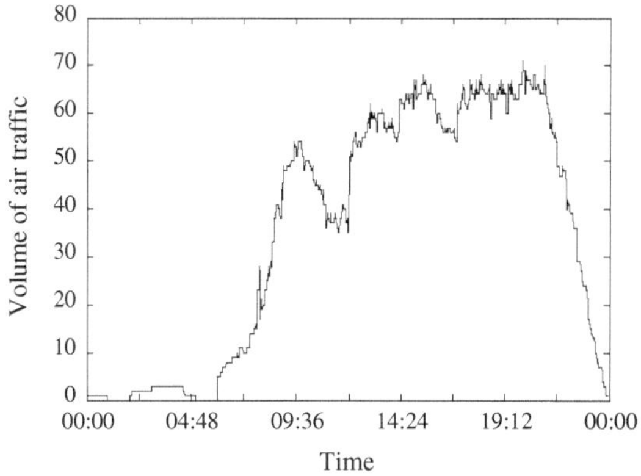

Fig. 12 Change of traffic flow

Table 1 Number of sectors in different time intervals

Time	Range of sector number	Final sector number
1:00–3:00	3–4	3
10:00–12:00	4–6	5
19:00–21:00	5–7	6

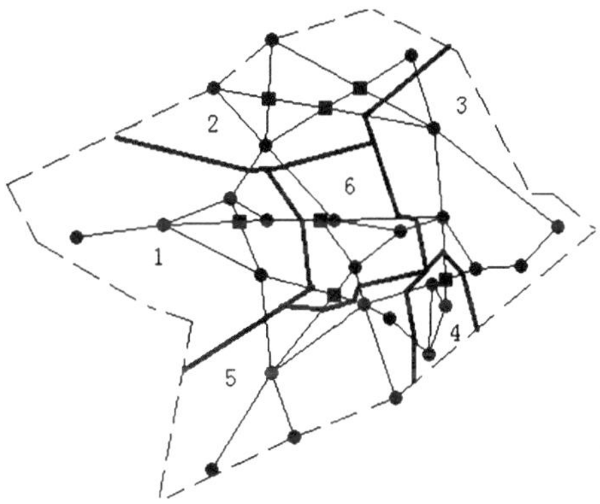

Fig. 13 Result of DAS (19:00–21:00)

workload and minimum workload in each sector during 19:00–21:00. The maximum workload in one sector should be less than or equal to 80 % of T for the majority of the time for safe operation. The minimum workload in one sector should be more than or equal to E% of T for the majority of the time for efficient operation. The workload threshold is satisfied for the majority of the time.

Then, the performance of sectorization via iGA is analyzed and compared with the sectorization using Delahaye GA (see Table 2).

In most cases, the iGA can result better sectors balancing f_b, a smaller minimization of coordination workload f_c, and a larger maximization of sector average flight time f_t, because of the boundary optimization part and the crossover operator added.

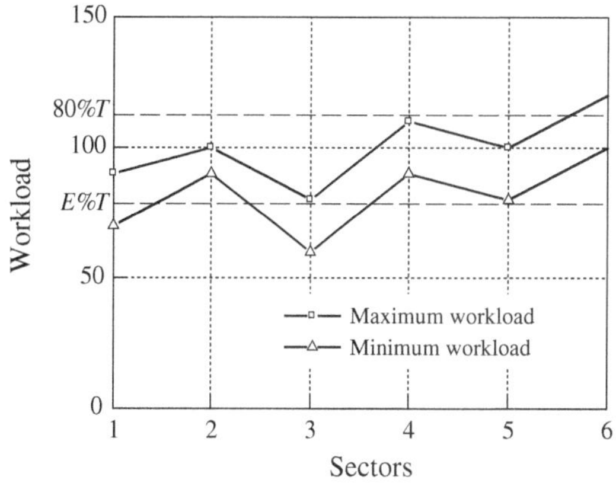

Fig. 14 The maximum workload and minimum workload in each sector

Table 2 Performance comparison between the results of the two GAs

Time	Improved GA			Delahaye GA		
	f_b (%)	f_c (%)	f_t (%)	f_b (%)	f_c (%)	f_t (%)
1:00–3:00	2.11	26.18	8.95	2.98	34.33	8.25
10:00–12:00	2.71	31.22	9.21	3.01	33.21	8.91
19:00–21:00	2.63	30.12	10.5	3.15	35.12	9.6

Then, we use 30-min interval sectors between 19:00 and 21:00 to explain the sector balancing. Figure 15 shows the workload in sectors (the maximum workload $W_{\max}$ and the minimum workload $W_{\min}$) in each sample interval.

Here, we define a coefficient of sector workload balancing (C_b) as

$$C_b = (W_{\max} - W_{\min})/W_{\max} \times 100\,\%. \quad (7)$$

Figure 16 shows the change of C_b in each 30-min interval. The less C_b, the more balanced the sector workload. The workload balancing in sectorization via iGA is better than the one via Delahaye GA.

5 Conclusions

In this paper, a new DAS method was proposed on the basis of a weighted graph model. We analyze indexes including sector balancing indexes, coordination workload index, and sector average flight time index to evaluate the results of DAS. The results show that, by applying the proposed method, both an optimal sector number and a better DAS have been achieved under the sector constraints. The practical application has shown that more accurate boundary and better flight safety can be ensured and the sector constraints are satisfied as well. One of the further improvements in the DAS method is that this

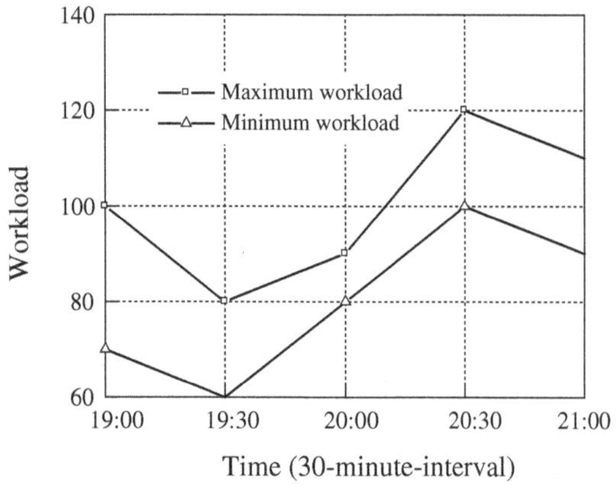

(a) Workload in sectors (iGA)

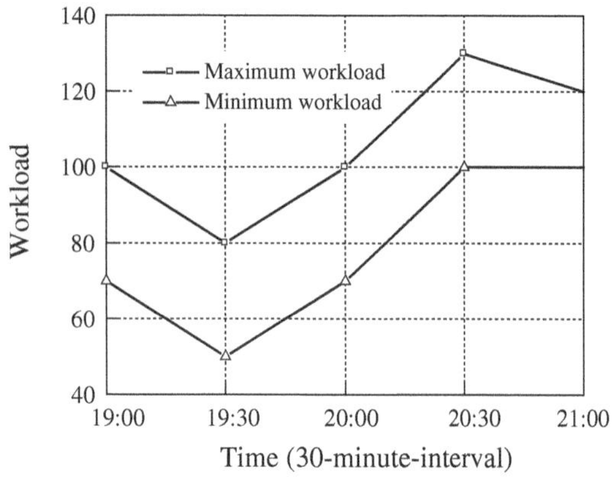

(b) Workload in sectors (Delahaye GA)

Fig. 15 Workload in sectors (30-min interval)

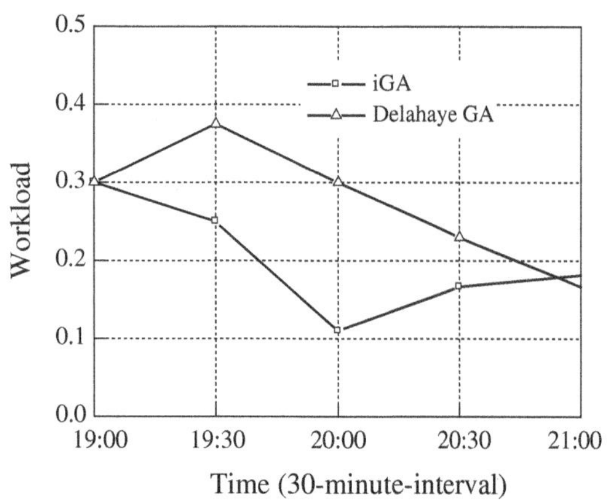

Fig. 16 Sector workload balancing (30-min interval)

method can avoid producing jagged edges in an area with more intensive routes.

Acknowledgments This study was funded by the Joint Funds of the National Natural Science Foundation of China (61079001).

References

1. Trandac H, Baptiste P, Duong V (2003) Optimized sectorization of airspace with constraints. In: The 5th USA/Europe air traffic management research and development seminar, Budapest, Hungary
2. Tien SL, Offman RH (2009) Optimizing airspace sectors for varying demand patterns using multi-controller staffing. In: The 8th USA/Europe air traffic management research and development seminar, Napa
3. Delahaye D, Schoenauer M, Alliot JM (1998) Airspace sectoring by evolutionary computation. In: Proceedings of the IEEE international conference on evolutionary computation, Piscataway, NJ, pp 218–223
4. Delahaye D, Alliot JM, Schoenauer M, Farges JL (1994) Genetic algorithms for partitioning air space. In: Proceedings of the tenth conference on artificial intelligence for applications, Washington, DC, pp 291–297
5. Xue M (2009) Airspace sector redesign based on Voronoi diagrams. Aerosp Comput Inf Commun 6:624–634
6. Delahaye D, Puechmorel S (2008) 3D airspace design by evolutionary computation. In: Proceedings of 27th digital avionics systems conference, St. Paul, USA, pp 1–13
7. Kicinger R, Yousefi A (2009) Heuristic method for 3D airspace partitioning: genetic algorithm and agent-based approach. In: Proceedings of 9th AIAA Aviation Technology, Integration and Operations (ATIO) conference, Hilton Head, South Carolina, pp 1–15
8. Han SH, Zhang M (2004) The optimization method of the sector partition based on metamorphic Voronoi polygon. Chin J Aeronaut 17(1):7–12
9. Tang JJ, Alam S, Lokan C, Abbass HA (2012) A multi-objective approach for dynamic airspace sectorization using agent based and geometric models. Transp Res C Emerg Technol 21(1):89–121
10. Tang JJ, Alam S, Lokan C, Abbass HA (2012) A multi-objective evolutionary method for dynamic airspace re-sectorization using sectors clipping and similarities. In: Proceedings of 2012 IEEE congress on evolutionary computation, Brisbane, Australia
11. Wu YH, Wang NX, Liu YZ (2005) Several populations competed genetic algorithm and its property analysis. J Northwest A&F Univ (Nat Sci Ed) 33(4):154–156 (In Chinese)
12. Li Jh, Wang T, Savai M, Hwang I (2010) Graph-based algorithm for dynamic airspace configuration. Guidance Control Dyn 33(4):1082–1095
13. Shuai XB, Zhou XG (2011) A genetic algorithm based on combination operators. Procedia Environ Sci 11(A):346–350
14. Ko CH, Wang SF (2011) The effects of two new crossover operators on genetic algorithm performance. Expert Syst Appl 38(7):8293–8302
15. Chen ZQ, Xu L, Hu WB (2012) The effect of crossover and mutation operators on genetic algorithm for job shop scheduling problem. Adv Mater Res 542–543:1251–1259

7

Extreme value of wind-excited response considering influence of bandwidth

Guoqing Huang · Xinzhong Chen · Mingshui Li · Liuliu Peng

Abstract This paper addresses the peak factors of wind-excited responses including alongwind, acrosswind tall building responses and vortex-induced vibration considering the bandwidth parameter. The influence of bandwidth parameter on the peak factor is investigated using advanced upcrossing theory taking the bandwidth influence into account. Results show that Davenport's formula without consideration of bandwidth parameter servers well in general. However, the advanced upcrossing theory leads to a better prediction of the peak factor of wind-induced response of very lightly damped buildings.

Keywords Peak factors · Wind-excited responses · Upcrossing theory · Bandwidth

1 Introduction

The closed-form formula introduced by Davenport [1] is widely used for estimating the mean extreme value and peak factor of a Gaussian stationary wind-excited structural response. This formula was developed based on the well-known "Poisson approximation" that the threshold level crossing occurs independently according to a Poisson process. While the Poisson approximation is generally valid for broad band processes, it may fall short for very narrow band processes, and lower threshold levels of practical interest, due to the tendency of crossings to occur in cluster [2, 3]. Cartwright and Longuet-Higgins [4] introduced a bandwidth parameter to account for its effect on the extreme value. Furthermore, Vanmarcke [2] proposed an improved formula with a different bandwidth parameter, which explicitly described the dependence between the crossing events and the time that the process spends above the threshold. Recently, Huang et al. [5] have presented the peak factors considering the bandwidth parameters for the non-Gaussian resultant response.

This study presents a comprehensive study concerning the peak factors of wind-excited responses considering the influence of bandwidth parameter. The responses considered include the displacement and acceleration of tall buildings in both alongwind and acrosswind directions, which are characterized as Gaussian processes, and vortex-induced vibration which is often described by non-Gaussian process due to the existence of nonlinear damping. These responses are calculated in the frequency domain based on the spectral analysis. The bandwidth parameters of these typical wind-excited responses are determined from their power spectra. The peak factors are then quantified considering the bandwidth parameters using Vanmarcke's formulation and compared with Davenport's formulation where the bandwidth parameter is neglected. These peak factors are also compared with those directly determined from time history simulations.

2 Theoretical background

The cumulative distribution function (CDF) of the extreme value of a stationary Gaussian process $X(t)$ over time duration T can be expressed as follows

G. Huang (✉) · M. Li · L. Peng
Research Center for Wind Engineering, School of Civil Engineering, Southwest Jiaotong University, Chengdu 610031, Sichuan, China
e-mail: ghuang1001@gmail.com

X. Chen
Wind Science and Engineering Research Center, Department of Civil and Environmental Engineering, Texas Tech University, Lubbock, TX 79409, USA

$$Q_{\max}(x) = \exp[-\nu(x)T], \tag{1}$$

where $\nu(x)$ is the upcrossing rate at level x. For a zero-mean stationary Gaussian process, the upcrossing rate can be estimated based on "Poisson assumption"

$$\nu(x) = \nu_0 \exp\left(-\frac{x^2}{2\sigma_x^2}\right), \tag{2}$$

where $\nu_0 = \sigma_{\dot{x}}/(2\pi\sigma_x)$ is the upcrossing rate across zero mean level; σ_x and $\sigma_{\dot{x}}$ are the standard deviations (STDs) of $X(t)$ and its derivative $\dot{X}(t)$. The Poisson approximation is named based on the fact that if the crossing rate is independent of the past history of the process, then the time lengths between the upcrossings will be independent; this makes the integer-valued process that counts the number of upcrossing in a Poisson process [3].

Davenport [1] derived the following closed-form formula for the mean and STD of the extreme value

$$\mu_{x\max} = g\sigma_x = (\beta + \gamma/\beta)\sigma_x, \tag{3}$$

$$\sigma_{x\max} = \pi/(\sqrt{6}\beta)\sigma_x, \tag{4}$$

where g is the peak factor and $\gamma = 0.5772$ is the Euler's constant; $\beta = \sqrt{2\ln(\nu_0 T)}$. The Poisson approximation may result in serious errors for very narrow band processes [3]. In such a situation, the upcrossing of level x at time t is very likely to be associated with another upcrossing approximately one period later, due to the slowly varying amplitude of the process. Such a dependence of the upcrossing time is inconsistent with the Poisson approximation that the time between the two upcrossings is independent [3]. The Poisson approximation is also less accurate for the lower threshold level of practical interest.

To avoid the limitation of Poisson approximation, the Poisson amplitude-crossing model can be used [6]. The Poisson approximation is more suitable for amplitude of the process than for the process itself. In addition, better results for small x values can be made to include the initial conditions. The modified version of amplitude-crossing rate, $\eta(x)$, is estimated as [3]

$$\eta(x) = \nu(x)\left\{1 - \exp\left[-\frac{\nu_A(x)}{\nu(x)}\right]\right\}\left[1 - \frac{\nu(x)}{\nu_{0x}}\right]^{-1}, \tag{5}$$

where $\nu_A(x) = \sqrt{2\pi(1-\alpha_1^2)}x/\sigma_X\nu(x)$ is the upcrossing rate of amplitude process $A(t)$; and α_1 is bandwidth parameter and defined as

$$\alpha_1 = m_1/(m_0 m_2)^{1/2}, \tag{6}$$

$$m_i = \int_{-\infty}^{\infty} |\omega|^i S_X(\omega)\mathrm{d}\omega = \int_0^{\infty} (2\pi f)^i S_X(f)\mathrm{d}f, \tag{7}$$

where m_i is the ith spectral moment. As α_1 tends to unity, $X(t)$ becomes a very narrow band process. Vanmarcke [2] proposed an empirical correction for $\nu_A(x)$ by fitting the simulation data and accordingly, the upcrossing rate is given as

$$\eta(x) = \nu_0 \exp\left(-\frac{x^2}{2\sigma_x^2}\right) \times \left\{1 - \exp(-\sqrt{2\pi}(1-\alpha_1^2)^{0.6}\frac{x}{\sigma_x})\right\} \times \left[1 - \exp\left(-\frac{x^2}{2\sigma_x^2}\right)\right]^{-1}. \tag{8}$$

The improved extreme value distribution is then obtained by replacing $\nu(x)$ with $\eta(x)$ in Eq. (1). The mean and STD of the extreme value can be estimated numerically.

It is noted that Cartwright and Longuet-Higgins [4] introduced a different bandwidth parameter and derived the peak factor g as

$$\varepsilon = \sqrt{1 - m_2^2/(m_0 m_4)}, \tag{9}$$

$$g = \sqrt{2\ln[(1-\varepsilon^2)^{1/2}\nu T]} + \gamma/\sqrt{2\ln[(1-\varepsilon^2)^{1/2}\nu T]}. \tag{10}$$

It is obvious that the above formula reduces to Davenport's formulation for $\varepsilon = 0$. When ε approaches to the zero, the process becomes a very narrow band one.

Figure 1 shows the ratio $\eta(x)/\nu(x)$ as a function of α_1. Figure 2 shows the improved CDFs, probability density functions (PDFs) from Varmarcke's formula for different α_1 and those from Poisson assumption under $\nu_0 T = 10^3$. The results show that there are considerable differences in the mean upcrossing rate and corresponding extreme value distribution, especially at low levels of x, as α_1 approaches to unity, i.e., the process becomes narrower. It is also seen that these differences are negligible as α_1 is <0.9. Figure 3 shows the mean, STD, and their ratio of the extreme value evaluated by Davenport's and Vanmarcke's formulas under $\nu_0 T = 10^2$, 10^3, and 10^4 for different α_1. It can be seen that the difference between two formulations is distinctive as the parameter α_1 increases to unity. For very narrow band processes, the improved Vanmarcke's formula offers a smaller mean and a larger STD of extreme value.

3 Alongwind building response

Consider an alongwind response of an isolated tall building with a regular cross section. The PSD of the alongwind generalized force of the fundamental mode is determined based on the Architectural Institute of Japan (AIJ) recommendations [7, 8]. This spectrum involves some assumptions including a liner fundamental mode shape, negligible

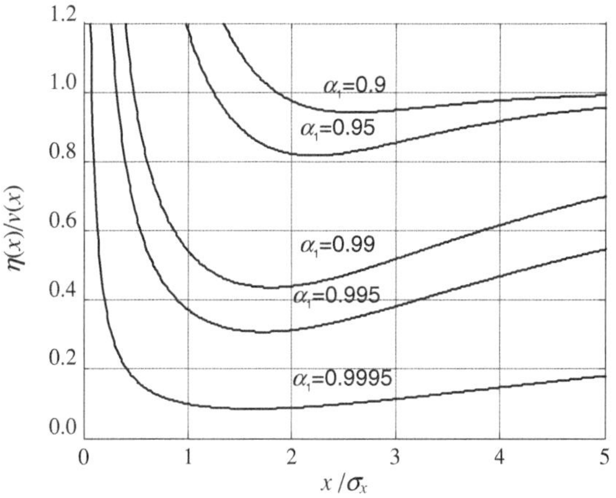

Fig. 1 $\eta(x)/v(x)$ as a function of bandwidth parameter α_1

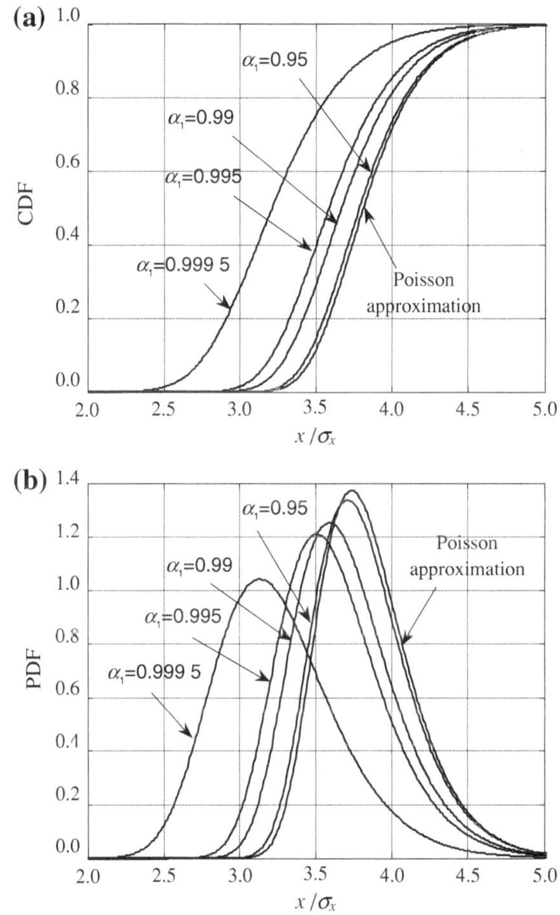

Fig. 2 Probability functions of extreme value. **a** CDFs ($v_0T=10^3$), **b** PDFs ($v_0T=10^3$)

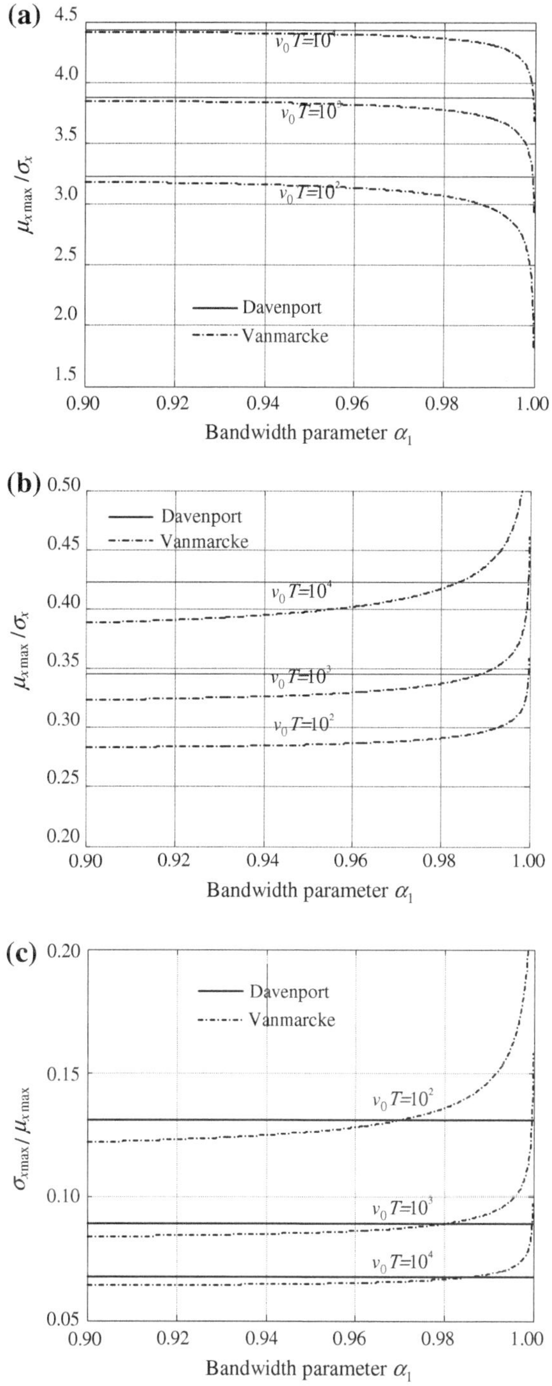

Fig. 3 Influence of bandwidth parameter on extreme value. **a** Mean of extreme value, **b** STD of extreme value, and **c** Coefficient of variation of extreme value

aerodynamic damping, and a power law profile of the mean wind speed and turbulence intensity. It is expressed as the product of aerodynamic admittance function and the wind fluctuation spectrum as

$$fS_Q(f) = \frac{0.84}{(1+2.1fH/U_H)(1+2.1fB/U_H)} \times \frac{4fL_H/U_H}{[1+71(fL_H/U_H)^2]^{5/6}}, \tag{11}$$

where *f* is the frequency; L_H is the turbulence length scale and given as $L_H = 100\sqrt{H/30}$; *H* is building height; *B* is building width; and U_H is the mean wind speed at building top.

Figure 4 shows the PSD of the generalized force $Q(t)$ with $H = 200$ m, $B = 0.2H$, and $U_H = 40$ m/s. Figure 5 portrays the calculated PSDs of the building top displacement and acceleration, i.e., $S_X(f)$ and $S_{\ddot{X}}(f) = (2\pi f)^4 S_X(f)$. The modal damping ratio and natural frequency are taken as $\xi = 1\ \%$ and f_0=0.23 Hz. As only the fundamental modal response is considered, the PSDs of base shear and based bending moments have the same features as the top displacement.

The building response can be decomposed into the broad band background and narrow band resonant components. It is seen that the PSD of the displacement contains significant background component which is negligible for the acceleration. The PSDs of the background and resonant components of the top displacement are given as

$$S_{X_b}(f) = S_Q(f)/[M^2(2\pi f_0)^4], \tag{12}$$

$$S_{X_r}(f) = \frac{S_Q(f)}{M^2(2\pi)^4[(f_0^2 - f^2)^2 + (2\xi f_0 f)^2]} - S_{X_b}(f), \tag{13}$$

where *M* is the generalized mass. Figure 6 shows the PSDs of the background and resonant components of the top displacement.

Based on the response spectrum, the spectral moments and bandwidth parameter can be calculated. These characteristics of the total response can also be expressed in terms of those of background and resonant components. Their upcrossing rate across zero mean level is defined as $\nu_{0s} = \sigma_{\dot{X}_s}/(2\pi\sigma_{X_s})$ $(s = b,\ r)$, where *b* and *r* denote background and resonant components; σ_{X_s} is RMS value of the background or resonant component; $\sigma_{\dot{X}_s}$ is RMS value of the derivative of the background or resonant component. Their corresponding bandwidth parameters can be expressed as $\alpha_{1s} = m_{1s}/(m_{0s}m_{2s})^{1/2}$, where $m_{is} = \int_0^\infty (2\pi f)^i S_{Xs}(f)df$ $(i = 0, 1, 2; s = b, r)$.

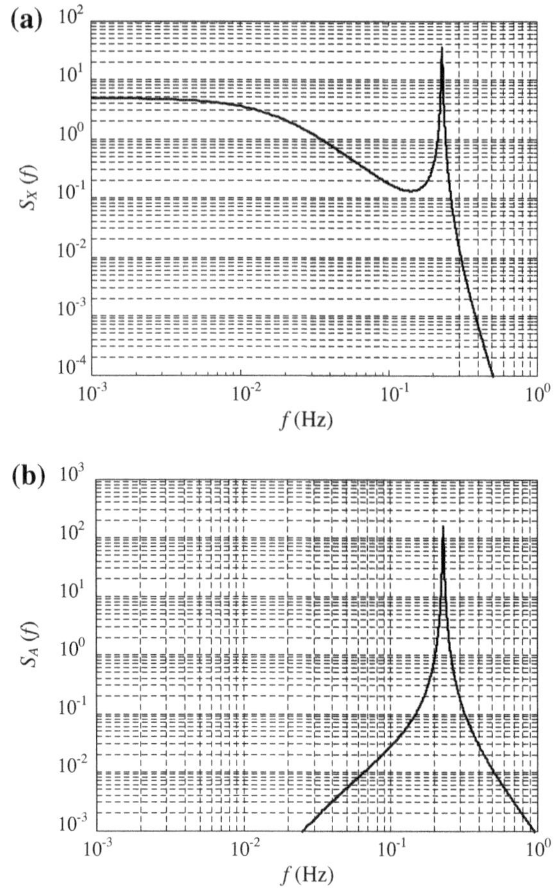

Fig. 5 PSDs of the alongwind top displacement. **a** Top displacement PSD, **b** Top acceleration PSD

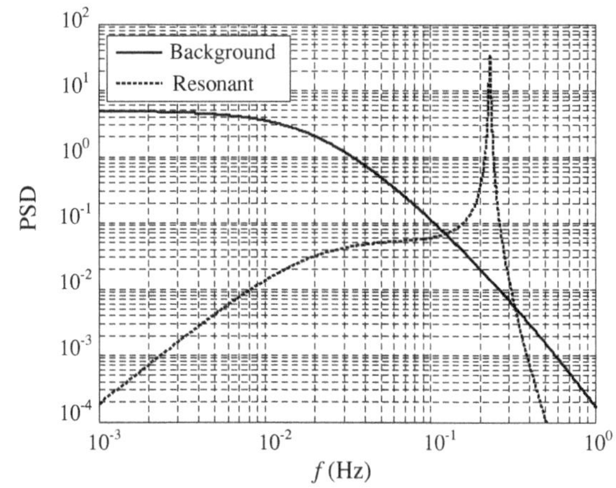

Fig. 6 PSDs of the background and resonant components of top displacement

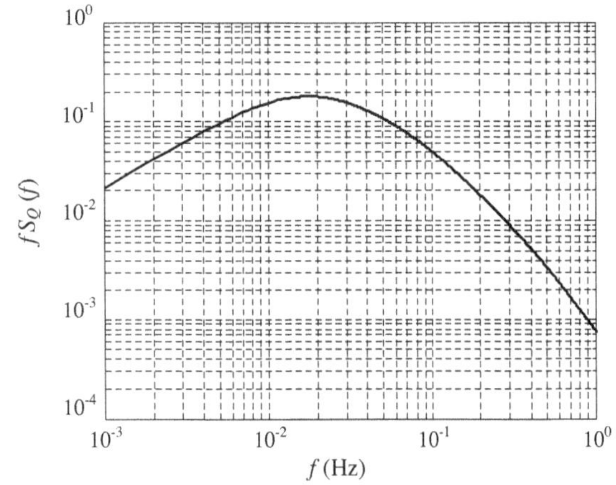

Fig. 4 PSD of the alongwind generalized force

Accordingly, the RMS, upcrossing rate across zero mean level and bandwidth parameter for the total response can be expressed as

$$\sigma_X = \sqrt{\sigma_{X_b}^2 + \sigma_{X_r}^2}, \tag{14}$$

$$\nu_0 = \sqrt{\nu_{0b}^2 + \nu_{0r}^2(\sigma_{X_r}/\sigma_{X_b})^2} \Big/ \sqrt{1 + (\sigma_{X_r}/\sigma_{X_b})^2}, \tag{15}$$

$$\alpha_1 = \left[\nu_{0b}\alpha_{1b} + \nu_{0r}\alpha_{1r}(\sigma_{X_r}/\sigma_{X_b})^2\right] \Big/ \left[\nu_0\left(1 + (\sigma_{X_r}/\sigma_{X_b})^2\right)\right], \tag{16}$$

where Eqs. (15) and (16) can be derived from the definitions.

Figure 7 shows the ratio of RMS resonant displacement to RMS background displacement as a function of damping ratio. It is seen that the portion of resonant component increases with the decrease in the damping ratio.

Figure 8 shows the bandwidth parameter α_1 as a function of damping ratio ξ for the displacement and acceleration. Figure 9 shows the relationship between the upcrossing rate across zero mean and damping ratio. The bandwidth parameter α_1 and uprcossing rate across zero mean for the background response are 0.627 and 0.043 respectively which are independent of the damping ratio. It can be seen that the displacement can be regarded as a broad band except very low-damping ratios where the resonant component becomes dominant. As expected, the acceleration and resonant displacement can be considered as narrow band for the damping ratio range of 1 %–3 %. It is noted that the upcrossing rate of the acceleration and resonant displacement across zero mean can be approximated as the structural natural frequency, i.e., $\nu_{0_r} \approx f_0$.

Figure 10 shows the comparison of the peak factors of alongwind response computed from Davenport's and Vanmarcke's formulations, i.e., without and with consideration of bandwidth parameter. The results illustrates that

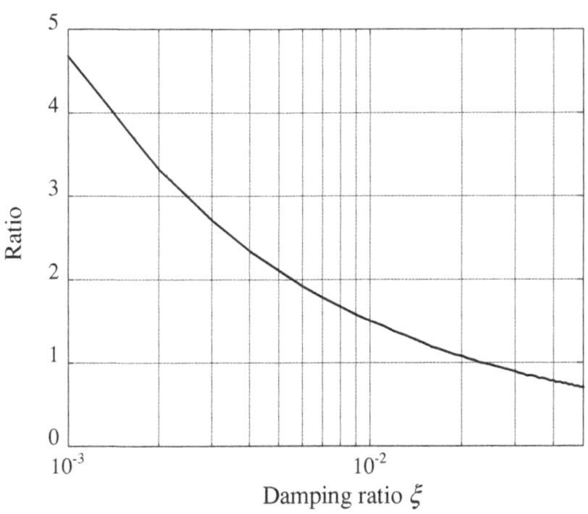

Fig. 7 Ratio of $\sigma_{X_r}/\sigma_{X_b}$

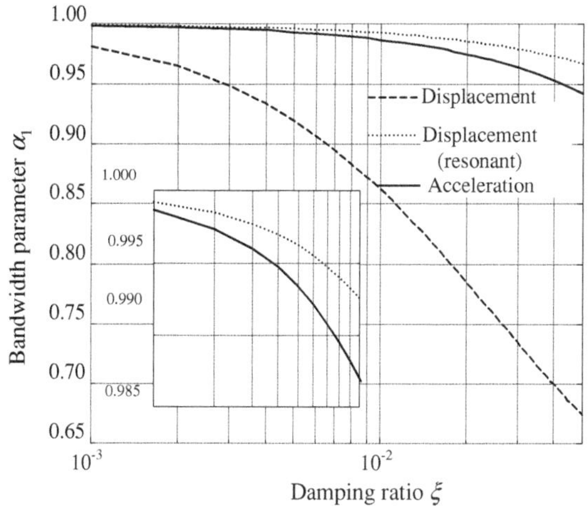

Fig. 8 Bandwidth parameter as a function of damping ratio

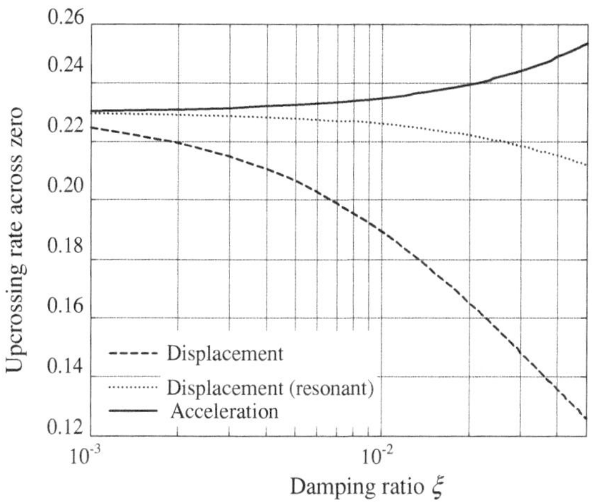

Fig. 9 Relationship between upcrossing rate across zero and damping ratio

the bandwidth parameter has a negligible influence on the displacement. For the acceleration, the consideration of bandwidth parameter results in 4 % difference at $\xi = 1$ %, and 12 % difference at $\xi = 0.1$ %.

In order to validate these formulations, Monte Carlo simulation of building response is also performed. A total of 100 samples of the time histories of the generalized force are generated using spectral representation method [9]. The building response is then calculated using step-by-step Newmark's method. The peak factors of top displacement and acceleration are qualified from the response time histories as shown in Fig. 10. It is seen that the peak factors obtained from the time domain simulation is slightly lower than that from Vanmarcke's formulation. Such a slight difference has also been reported in literature [3].

For the broad band background displacement with a bandwidth parameter of 0.627, both Davenport's and

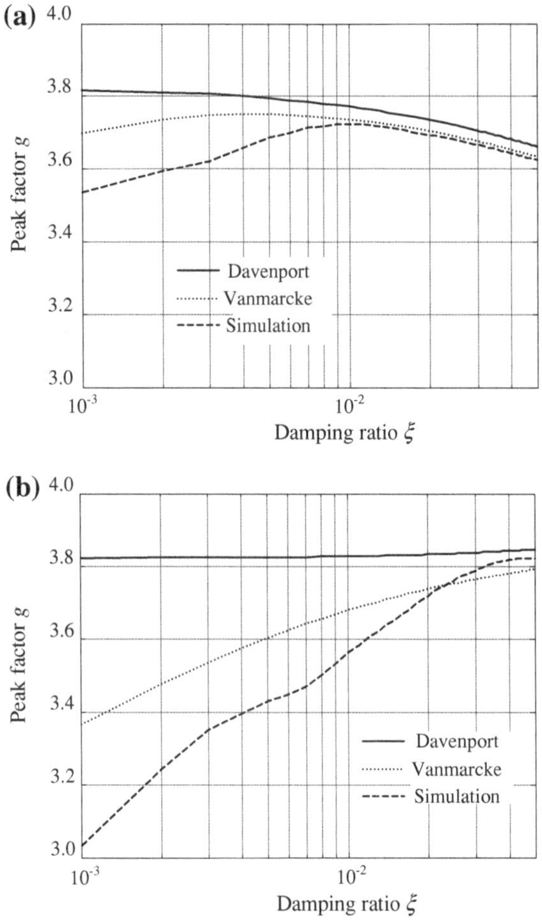

Fig. 10 Influence of bandwidth parameter on the peak factor of alongwind response. **a** Top displacement, **b** Top acceleration

Vanmarcke's formulas offer an identical peak factor of 3.22. However, they lead to a slightly different peak factor for the resonant displacement as shown in Fig. 11. As expected, the peak factor of the resonant displacement shows the similar characteristics as that of the acceleration.

In current practice, the mean extreme response is often estimated by directly combining the mean extreme values of background and resonant components as

$$\mu_{x\max} = \sqrt{g_b^2\sigma_{X_b}^2 + g_r^2\sigma_{X_r}^2}, \tag{17}$$

where g_b and g_r are the peak factors of background and resonant components.

As compared to the following equation

$$\mu_{x\max} = g_X\sqrt{\sigma_{X_b}^2 + \sigma_{X_r}^2}, \tag{18}$$

where g_X is the peak factor of total response, Eq. (17) is equivalent to introduce a peak factor for total response as

$$g_X' = \sqrt{g_b^2\sigma_{X_b}^2/\sigma_X^2 + g_r^2\sigma_{X_r}^2/\sigma_X^2}. \tag{19}$$

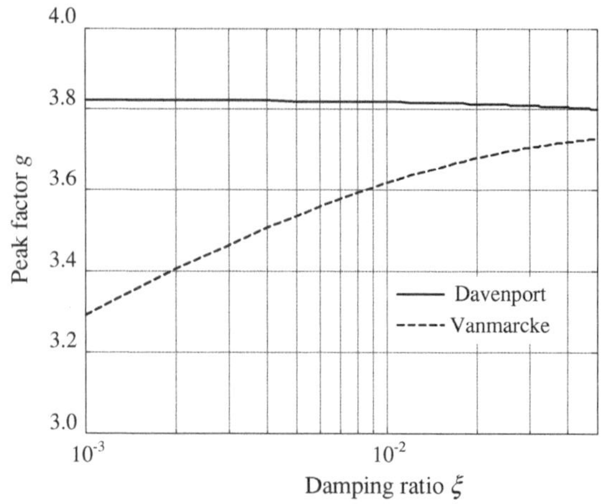

Fig. 11 Influence of bandwidth parameter on the peak factor of resonant top displacement

Figure 12 shows the ratio g_X'/g_X for the total displacement. It can be seen that when Eq. (17) is applied, Davenport's formula leads to a better estimation for the peak factor of total response. The reason is attributed to following coincidence: on one hand, approximate combination of Eq. (17) leads to a lower extreme value of total response when both background and resonant extreme vales are computed using Vanmarcke's formula; on the other hand, the extreme value of resonant component is overestimated by Davenport's formula.

As shown in Eq. (10), Cartwright and Longuet-Higgins [4] introduced a different bandwidth parameter ε in estimating the peak factor. Figure 13 shows the bandwidth parameter ε as a function of damping ratio for the top displacement and acceleration. Figure 14 compares the influence of this parameter on the responses. It can be seen that Cartwright and Longuet-Higgins' formulas gives an

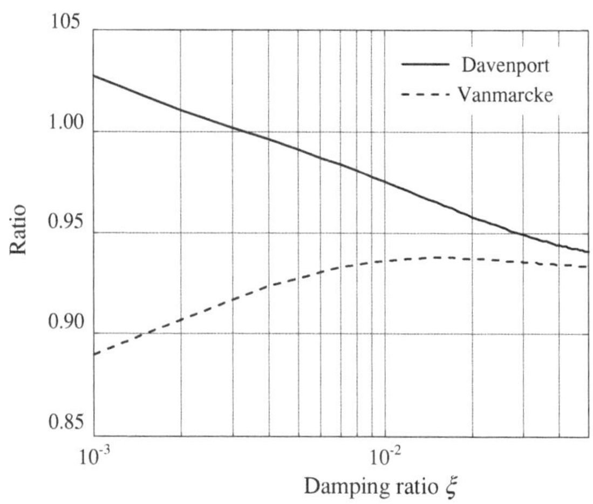

Fig. 12 Ratio of peak factors g_X'/g_X

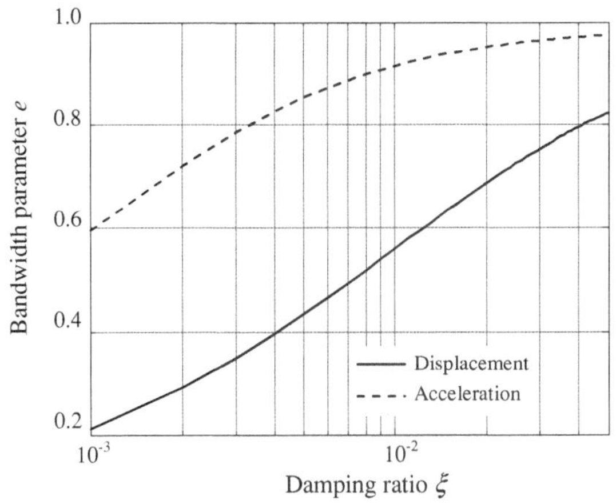

Fig. 13 Relationship between parameter ε and ratio ξ

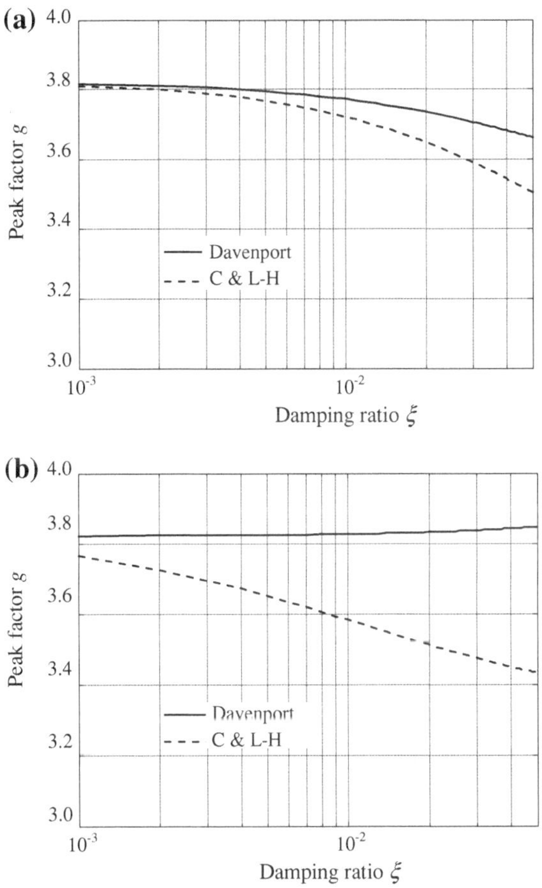

Fig. 14 Influence of bandwidth parameter ε on the peak factor of alongwind response. **a** Top displacement, **b** Top acceleration

opposite trend of the peak factor with the influence of the bandwidth parameter. Hence, the usage of the parameter ε is not effective on the estimation of the peak factor and extreme response.

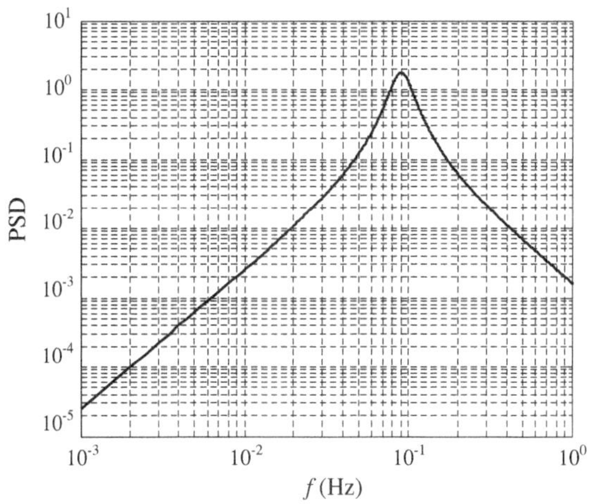

Fig. 15 PSD of the acrosswind generalized force

4 Acrosswind building response

The PSD of the acrosswind generalized force $Q(t)$ suggested by AIJ recommendations (AIJ 1996) [7] is also used for estimating the acrosswind response and discussing the bandwidth parameter on its peak factor and extreme value.

$$S_Q(f) = \sum_{j=1}^{N} \frac{4\kappa_j(1+0.6\beta_j)\beta_j \times (f/f_{sj})^2}{\pi \times \left\{[1-(f/f_{sj})^2]^2 + 4\beta_j^2(f/f_{sj})^2\right\}},$$
$$N = \begin{cases} 1, & D/B<3 \\ 2, & D/B>3 \end{cases}, \quad \begin{cases} \kappa \approx 0.85 \\ \kappa \approx 0.02 \end{cases}, \tag{20}$$

where β_j is related to the band width; f_{sj} represents the peak frequencies, i.e., vortex-shedding frequencies of the spectrum; and D is the building depth. They can be determined by the following empirical formulas as functions of the side ratio D/B (B is the building width)

$$f_{s1} = \frac{0.12}{[1+0.38(D/B)^2]^{0.89}} \frac{U_H}{B},$$
$$f_{s2} = \frac{0.56}{(D/B)^{0.85}} \frac{U_H}{B}, \tag{21}$$

$$\beta_1 = \frac{(D/B)^4}{1.2(D/B)^4 - 1.7(D/B)^2 + 21} + \frac{0.12}{D/B},$$
$$\beta_2 = 0.28(D/B)^{-0.34}. \tag{22}$$

Figure 15 shows the PSD of acrosswind generalized force with $H = 200$ m, $B = 0.2H$, $D/B = 1$, and $U_H = 40$ m/s. Figure 16 shows the PSDs of the building top displacement and acceleration with damping ratio $\xi = 1$ %.

Figure 17 shows the bandwidth parameter α_1 for the displacement and the acceleration as a function of damping ratio. Figure 18 shows the relationship between the upcrossing rate across zero and damping ratio. It can be seen

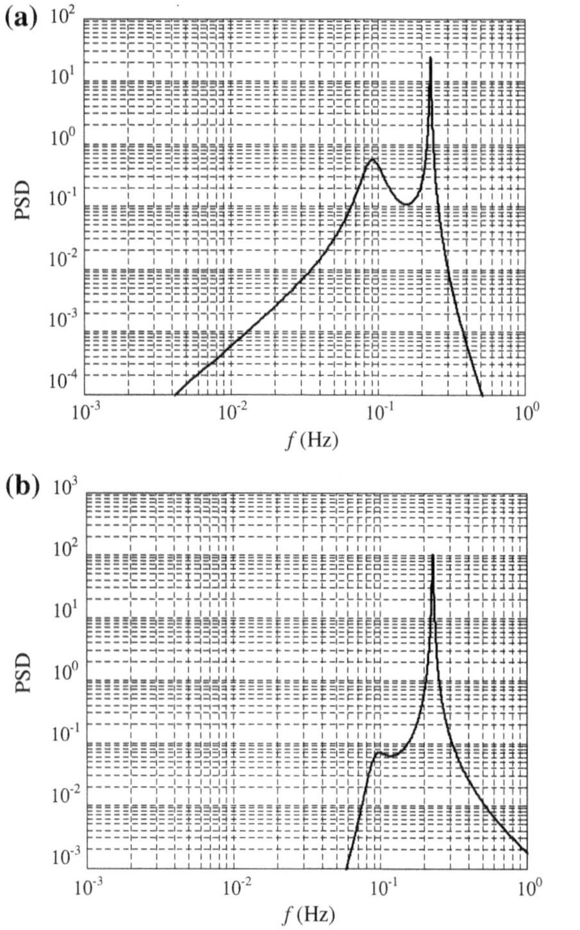

Fig. 16 PSDs of the acrosswind displacement and acceleration ($\xi = 1$ %). **a** Top displacement, **b** Top acceleration

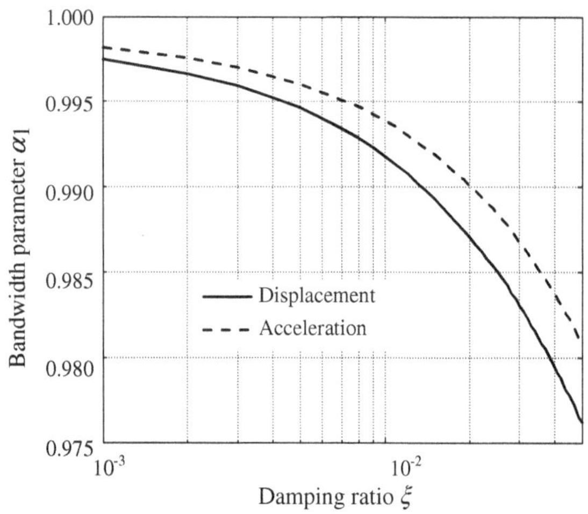

Fig. 17 Relationship between bandwidth parameter and damping ratio

that the displacement and acceleration can be considered as narrow band processes, and the upcrossing rate across zero level can be approximated by the natural frequency.

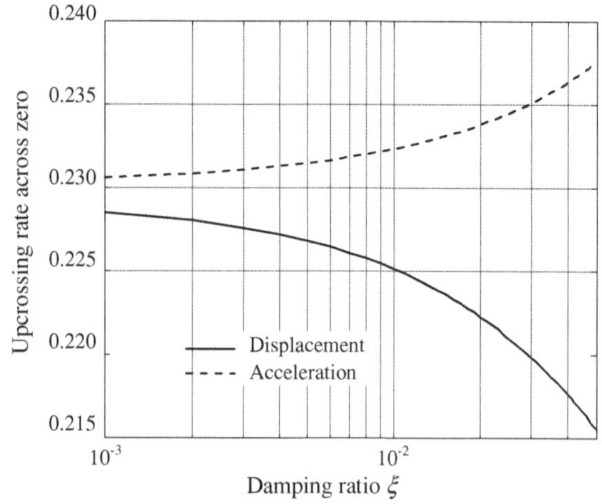

Fig. 18 Relationship between upcrossing rate and damping ratio

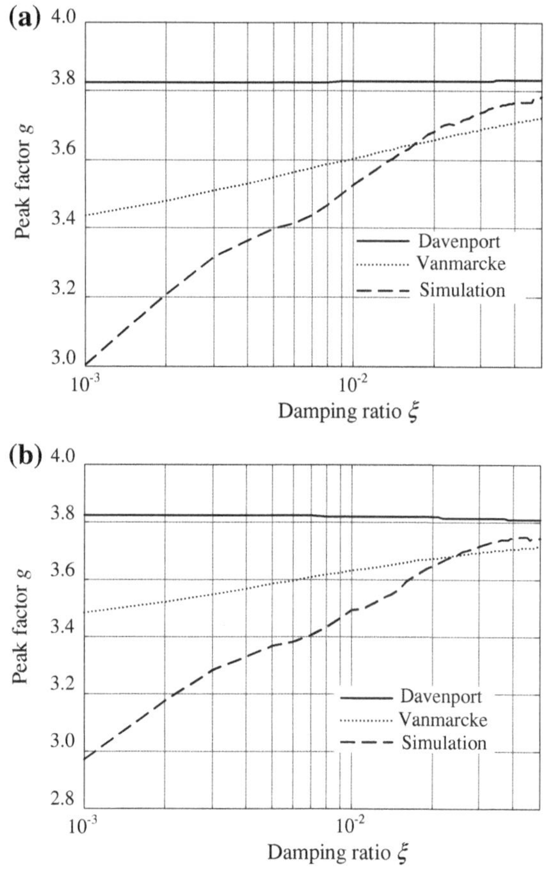

Fig. 19 Influence of bandwidth parameter on the peak factor of acrosswind response. **a** Top displacement, **b** Top acceleration

Figure 19 shows the influence of bandwidth parameter on the peak factor of acrosswind response. The results illustrate that the consideration of bandwidth parameter results in 6 % and 11 % differences at $\xi = 1$ % and 0.1 % for displacement, respectively. The differences are 5 % and 9 % at $\xi = 1$ % and 0.1 % for acceleration, respectively. It is seen that Vanmarcke's formula will provide a better prediction of the peak factor of wind-induced response of very lightly damped buildings.

5 Vortex-induced response

Consider a vortex-induced vibration of a spring-supported circular cylinder. A considerable variety of empirical analytical models have been developed to describe the vortex-induced vibration of cylinders [10]. In this study, the equation of vibration of a cylinder with a unit length is given as [11]

$$m(\ddot{y}+2\xi\omega_1\dot{y}+\omega_1^2y)=\frac{1}{2}\rho U^2D[Y_1(K)\times\left(1-\delta\frac{y^2}{D^2}\right)\frac{\dot{y}}{U} + Y_2(K)\frac{y}{D}+C_L(K)\sin(\omega t+\phi)], \quad (23)$$

where m is the mass per unit length of cylinder; ξ and ω_1 are damping ratio and natural circular frequency respectively; ρ is air density; D is diameter of cylinder; $K=\omega D/U$ is reduced frequency; Y_1, δ, Y_2, and C_L are parameters and functions of K and need to be determined through the observations.

This model allows the consideration of nonlinear wind-excited "negative damping". At lock-in, $\omega\approx\omega_1$, $Y_2\approx 0$, and $C_L\approx 0$, since the last two terms are found to be small compared to the first term reflecting the aerodynamic damping effects. Y_1 and δ are independent parameters; the former determines the initial negative damping, and the latter dominates the time to achieve the steady-state vibration. The steady-state amplitude of vortex-induced vibration, y_0, is given as [11]

$$\frac{y_0}{D}=2\left[\frac{Y_1-8\pi S_{cr}S_t}{\delta Y_1}\right]^{1/2}, \quad (24)$$

where S_t is Strouhal number; and $S_{cr}=\xi m/(\rho D^2)$ is Scruton number.

In this study, the Strouhal and Scruton numbers are taken as 0.2 and 1.5, respectively. Y_1 and δ are assumed to be 26.3 and 2,928.4, respectively. Accordingly, $Y_0/D=0.0312$.

Using the normalized displacement $y'=y/D$, Eq. (23) can be can be rewritten as

$$\ddot{y}'+2\omega_1\left[\xi-\frac{U}{4D\omega_1}\frac{\xi}{S_{cr}}Y_1\left(1-\delta y'^2\right)\right]\dot{y}'+\omega_1^2y'=0, \quad (25)$$

The time history of vortex-induced vibration can be simulated by solving Eq. (25) through Newmark's step-by-step method. Figure 20 shows an example with $\xi=1$ % and $\omega_1=\pi$ rad/s. Using the time history data when vortex-induced vibration reaches its steady state, the peak factor is

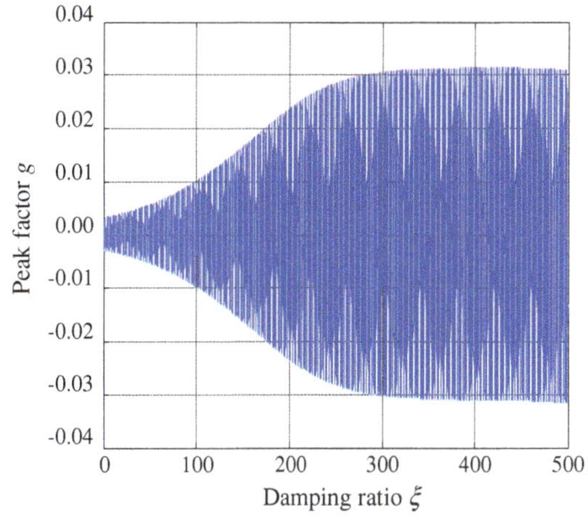

Fig. 20 A sample of vortex-induced vibration

determined to be around 1.42, which is close to that of a sinusoidal response, $\sqrt{2}$. In this situation, the bandwidth parameter α_1 should be very close to unity. It should be noted that at the wind speed around lock-in region, vortex-shedding frequency is close to the natural frequency of cylinder. The vibration will show the beating phenomena. The peak factor of vortex-induced vibration will be in the range of 1.42 to around 3.5. The advanced theory that takes into account the bandwidth parameter is able to interpret the trend of extreme value and peak factor in this response.

6 Conclusion

In wind engineering community, Davenport's closed-form formula was extensively applied in extreme value estimation of a Gaussian stationary response. Due to the "Poisson approximation" involving this formulation, it may fall short for very narrow band processes and lower threshold levels of practical interest. On the other hand, Vanmarcke's formula accounted for the effects of bandwidth and led to an improved estimation of the extreme value of very narrow band processes.

The wind-induced building response was generally separated into background and resonant components. The background response was regarded as a broad band process, while the resonant response was a narrow band process. The bandwidth parameter of the total response was affected by the ratio of resonant and background components. The alongwind displacement usually contained considerable background component. On the other hand, the alongwind acceleration, and acrosswind displacement and acceleration were dominant by the resonant component. The bandwidth parameter was strongly affected by the building damping ratio. Vortex-induced vibration at steady-state stage had a bandwidth parameter which was very close to unity because its vibration was almost a sinusoidal wave.

The estimation of the extreme value and peak factor using Davenport's formulation without considering the bandwidth parameter served well for both alongwind and acrosswind responses. It may lead to an around 6 % overestimation for the acrosswind response of buildings with very lightly damping ratios, say, 1 %. The time domain simulations validated those results.

However, Vanmarcke's formulation with the consideration of bandwidth parameter and the lower level of crossing was more accurate in describing the probability distribution of the extreme value of very narrow band processes. One such an example was vortex-induced vibration. At the steady-state stage, the peak factor was around 1.42. Around the lock-in range, the peak factor was between 1.42 and around 3.5. That was much lower than that computed by Davenport's formula. The advanced theory using Vanmarcke's formula could explain the behavior of very narrow band response processes.

In engineering applications, the peak value of wind-excited response was often estimated by directly combining the peak values of its background and resonant components using square-root-of-sum-of-squares (SRSS) rule. Such an approximate combination generally led to a smaller combined peak response when both extremes of the background and resonant components were accurately determined following Vanmarcke's formula. On the other hand, this approximate combination led to a better result when the peak factors were given by Davenport's formula. However, when the resonant response was dominant, the combination of SRSS rule and Davenport's formula may fall short in predicting the extreme response of very lightly damped structures.

Acknowledgments The partial support by the National Science Foundation of China (Grant 51278433) is greatly acknowledged.

References

1. Davenport AG (1964) Note on the distribution of the largest value of a random function with application to gust loading. Proceedings of the Institution of Civil Engineers. pp187–196
2. Vanmarcke EH (1975) On the distribution of the first-passage time for normal stationary random processes. J Appl Mech 14:217–243
3. Lutes LD, Sarkani S (2004) Random Vibration: Analysis of Structural and Mechanical Systems. Elsevier, NewYork
4. Cartwright DE, Longuet-Higgins MS (1956) The statistical distribution of the maxima of a random function. Proc Royal Soc Lond A 237:212–232
5. Huang MF, Chan CM, Lou WJ, Kwok KCS (2012) Statistical extremes and peak factors in wind-induced vibration of tall buildings. J Zhejiang University Sci A 13(1):18–32
6. Cramér H, Leadbetter MR (1967) Stationary and Related Stochastic Processes. Wiley, NewYork
7. Architectural Institute of Japan (1996) AIJ Recommendations for Loads on Buildings, Japan
8. Kijewski T, Kareem A (1998) Dynamic wind effects: a comparative study of provisions in codes and standards with wind tunnel data. Wind Struct 1:77–109
9. Shinozuka M, Jan CM (1972) Digiatal simulation of random processes and its applications. J Sound Vib 25(1):111–128
10. Gabbai RD, Benaroya H (2005) An overview of modeling and experiments of vortex-induced vibration of circular cylinders. J Sound Vib 282:575–616
11. Simiu E, Scanlan RH (1996) Wind Effects on Structures: Fundamentals and Applications to Design, 3rd edn. Wiley, NewYork

8

Seismic stability evaluation of embankment slope based on catastrophe theory

Hongwei Zhu · Lingkan Yao · Yuan Luo

Abstract An evaluation method for the seismic stability of embankment slope was presented based on catastrophe theory. Seven control factors, including internal frictional angle, cohesion force, slope height, slope angle, surface gradients, peak acceleration, and distance to fault were selected for analysis of multi-level objective decomposition. According to the normalization formula and the fuzzy subject function produced by combination of catastrophe theory and fuzzy math, a recursive calculation was carried out to obtain a catastrophic affiliated functional value, which can be used to evaluate the seismic stability of embankment slope. Fifteen samples were used to verify the effectiveness of this method. The results show that compared with the traditional quantitative method, the catastrophe progression owns higher accuracy and good application potential in predicting the seismic stability of embankment slope.

Keywords Embankment slopc · Scismic stability · Catastrophe progression method · Comprehensive evaluation

H. Zhu (✉)
School of Civil Engineering, Southwest Jiaotong University, Chengdu 610031, China
e-mail: zhw-1-zhw@163.com

H. Zhu
Sichuan College of Architectural Technology, Deyang 618000, China

L. Yao · Y. Luo
China Key Laboratory of Highway Engineering of Sichuan Province, Southwest Jiaotong University, Chengdu 610031, China

1 Introduction

At 14:28, May 12, 2008, a great earthquake measured $M_s = 8.0$ hit Wenchuan, Sichuan Province of China. According to field investigation, 53,295 km-long highways were badly destroyed in the earthquake [1]. In order to provide references for reconstruction and further research [2, 3], we have made a lot of field investigation to many highways in thc carthquake area. The results of field investigation show that the dominant failure modes of embankments are lateral spreading, surface subsidence, collapse, and dislocation, as shown in Fig. 1.

Failure modes, mechanism, and sliding surface of earthquake-induced landslides are studied by means of field investigation and shaking table model tests. The earthquake-induced failure surfaces usually consist of tension cracks and shear which is different from gravity-induced failure surface made by shear zone only. Depth of dynamic sliding surface will be deeper along with the increase of peak ground acceleration, as shown in Fig. 2.

Because the stability of a slope is affected by geological and engineering factors, and many of the factors can not be obtained directly, we have to use uncertain method to deal with this kind of issues, such as fuzzy math [4], artificial neural network method [5], grey theory [6], support vector machine model [7], and extension method [8]. In these methods, the weight of each factor index directly determines the accuracy to a certain extent.

Catastrophe theory, which originated from the study of the French mathematician René Thom in the 1960s, becomes very popular due to the efforts of Christopher Zeeman [3] in the 1970s. It considers the special case where the long-run stable equilibrium can be identified with the minimum of a smooth, well-defined potential function (Lyapunov function) [9, 10].

Fig. 1 Embankment slope landslide in Wenchuan earthquake

In mathematics, catastrophe theory is a branch of bifurcation theory for study of dynamical systems and a particular special case of more general singularity theory in geometry as well. Bifurcation theory studies and classifies phenomena characterized by sudden shifts in behavior arising from small changes in circumstances, analyzing how the qualitative nature of equation solutions depends on the parameters that appear in the equation. This may lead to sudden and dramatic changes, such as the unpredictable timing and magnitude of a landslide [11].

Small changes in certain parameters of a nonlinear system can cause equilibria to appear or disappear, or to change from attracting to repelling and vice versa, leading to large and sudden changes of the behavior of the system. However, examined in a larger parameter space, catastrophe theory reveals that such bifurcation points tend to occur as part of well-defined qualitative geometrical structures. The main feature of the catastrophe progression method is that the weight of indices is not considered, which can avoid the effect of human subjective factor in practice [12].

In this article, the catastrophe progression method is used to evaluate the seismic stability of the embankment slope, and a reasonable result was achieved, indicating that the catastrophe progression method is feasible in predicting the seismic stability of embankment slope.

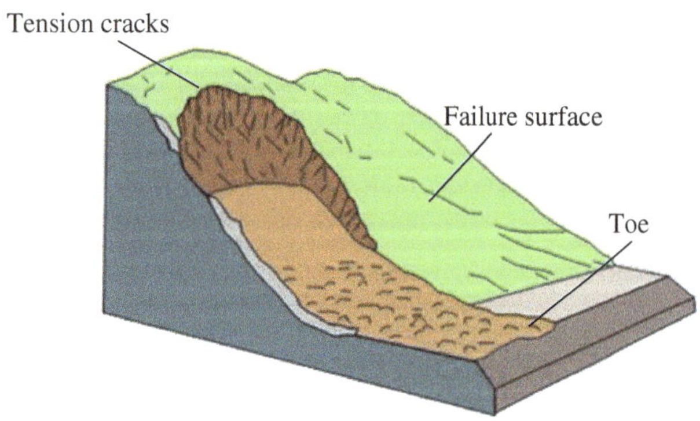

Fig. 2 Sliding of road embankment body

2 Key evaluation technique and steps of catastrophe progression method

In the catastrophe progression method, we first divide the evaluation system into sub-indexes, then normalize the control variables, and finally calculate the catastrophe affiliated functional value according to the complementary principle and non-complementary principle [13, 14]. The main evaluation steps are as follows.

2.1 Establishment of the hierarchy analysis model

First of all, an overall evaluation system is divided into sub-indices and all indices are grouped in accordance with the purpose of evaluation. In each hierarchy, the indices form a different catastrophe system. The weights of indices are not concerned, but the relative importance of each index is considered. Because the number of control variables could not exceed 4, the number of index in each hierarchy will not exceed 4 either.

2.2 Determination of catastrophe system classification for each hierarchy

There are seven types of catastrophe systems, wherein the most four common types are folded catastrophe, cusp catastrophe, swallowtail catastrophe, and butterfly catastrophe. The potential functions are shown as follows [15, 16]:

Folded catastrophe:

$$f_1(x) = x^3 + ax. \tag{1}$$

Cusp catastrophe:

$$f_2(x) = x^3 + ax^2 + bx. \tag{2}$$

Swallowtail catastrophe:

$$f_3(x) = x^5 + ax^3 + bx^2 + cx. \tag{3}$$

Butterfly catastrophe:

$$f_4(x) = x^6 + dx^4 + ax^3 + bx^2 + cx. \tag{4}$$

In above equations, $f_i(x)$ is the potential function of the state variable x, and a, b, c, and d are control variables of x.

2.3 Normalizing the control variables of the catastrophe model

Lack of proportionality in the indices can be eliminated using a standard transformation method so that the evaluation indices are dimensionless.

For *the-larger-the-better* indices can be calculated by the following functions:

$$r = \frac{x}{x_{\max}}, \tag{5}$$

and for *the-smaller-the-better* indices can be calculated by the following functions:

$$r = \frac{x_{\min}}{x}, \tag{6}$$

where $x_{\max}$ and $x_{\min}$ are the maximum and minimum value of the measured x, respectively.

If the values of x are within the range [0, 1], the measured values does not need to be normalized and can be directly used in the catastrophe progression computation.

2.4 Normalization formula

Let $f(x)$ be the potential function of the catastrophe system, and the critical points of the potential function form an equilibrium surface according to the catastrophe theory. We can obtain the equation by calculating the first derivative of $f(x)$, and obtain its singularity set by setting $f''(x) = 0$. The normalization formula is derived from the decomposition forms of the equation of the bifurcation point set. Different states of the control variables in the catastrophe system are then transformed into state variables using normalization formula. Based on catastrophe theory, the decomposition forms of the equation and normalization formulas of three conmen catastrophe systems are obtained as follows [17, 18].

For the cusp catastrophe system, the bifurcation point set equation is

$$a = -6x^2, \quad b = 8x^3, \tag{7}$$

and its normalization formulas arc

$$x_a = \sqrt{|a|}, \quad x_b = \sqrt[3]{|b|}. \tag{8}$$

For the swallowtail catastrophe system, the decomposition forms of the bifurcation point set equation are

$$a = -6x^2, \quad b = 8x^3, \quad c = -3x^4, \tag{9}$$

and the normalization formulas are

$$x_a = \sqrt{|a|}, \quad x_b = \sqrt[3]{|b|}, \quad x_c = \sqrt[4]{|c|}. \tag{10}$$

For the butterfly catastrophe system, the decomposition forms of the bifurcation point set equation are

$$a = -10x^2, \quad b = 20x^3, \quad c = -15x^4, d = 4x^5, \tag{11}$$

and the normalization formulas are

$$x_a = \sqrt{|a|}, \quad x_b = \sqrt[3]{|b|}, \quad x_c = \sqrt[4]{|c|}, \quad x_d = \sqrt[4]{|d|}. \tag{12}$$

In Eqs. (1)–(12), x_a, x_b, x_c, and x_d are values of x corresponding to a, b, c, and d.

2.5 Comprehensive evaluation through normalization formulas

Following the complementary principle and non-complementary principle, the catastrophe progression of each control variable can be computed from the initial fuzzy subordinate function. We find that each control variable of a system tends to reach the average value, so

$$x = (x_a + x_b + x_c + x_d)/4. \tag{13}$$

However, the non-complementary principle implies that the control variables cannot offset each other. Therefore, the smallest values of the state variables corresponding to the control variables are chosen to be state variable of the whole system. Based on hierarchical calculation, the value of the overall catastrophe subordinate function can be found in the same way [19, 20].

3 Application of catastrophe progression method to evaluate the seismic stability of embankment slope

3.1 Evaluation system of seismic stability of embankment slope

The control variables influencing the seismic stability of embankment slope are divided into three hierarchies, namely, the strength index, geometry characters of slope, and the earthquake effect. The indices of strength are internal frictional angle and cohesion force, the indices of geometry characters of slope include slope height, slope angle, and surface gradients; and peak acceleration and distance to fault are selected to be the influencing factors of earthquake. The overall evaluation system of seismic stability of the embankment slope is shown in Fig. 3.

3.2 Determination of evaluation factors of embankment seismic stability

At present, the seismic fortification principle in the world is that the frame structure can achieve the seismic protective objective of no damage in small earthquake, repairable damage in moderate earthquake, and no collapse in severe

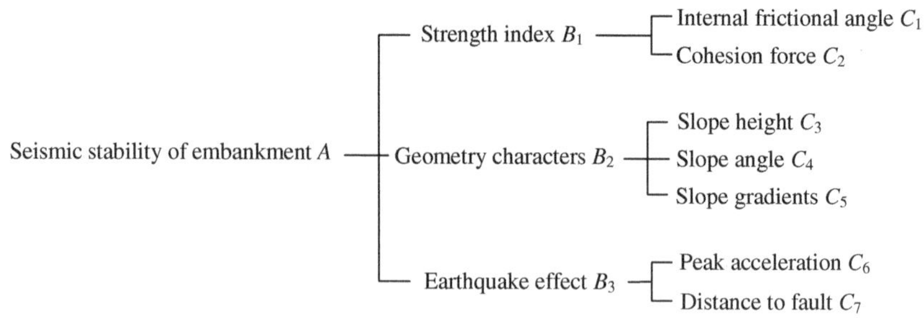

Fig. 3 Evaluation index system for seismic stability of embankment

Table 1 Evaluation standard of the seismic damage to embankment slope [6]

Grade	Damage degree	Evaluation standard	Reinforcing measures	Catastrophe progression value
I	Integrity	Slight cracks appear	Normal use	>0.95
II	Almost integrity	Small cracks and deformation appear	Can be used under regular maintenance	0.90–0.95
III	Slight damage	Obvious deformation and cracks appear, but the structure can maintain stability	Can be used in an emergency, but the structure should be reinforced after emergency	0.85–0.90
IV	Moderate damage	Great deformation and partial failure occurs, but without serious damage	Before used, reinforced measures should be done	0.8–0.85
V	Serious damage	Structures are seriously damaged and even collapsed	Reconstruction	<0.80

earthquake. According to the above principle, the evaluation standard of the seismic damage to embankment slope is presented in Table 1. In addition, based on lots of tentative calculation, the critical values of catastrophe progression are also obtained [21, 22].

The embankments investigated include normal embankment and high fill embankment. In the high fill embankment, the surface gradient varies and crushed stones are the main filling material. The actual seismic damage to embankment can be described as in Table 2.

3.3 Determination of utility function values of bottom factors

In order to illustrate the concrete application of the catastrophe method, we take sample 1 as an example. According to formulas (5) and (6), it is easy to get C_1, C_2, C_3, C_4, C_5, C_6, and C_7 as follows:

$$C_1 = 38/38 = 1, C_2 = 10/14 = 0.714, C_3 = 9/34 = 0.265, C_4 = 10/15 = 0.667, C_5 = 10/15 = 0.667, C_6 = 0.42/0.42 = 1, C_7 = 17/17 = 1.$$

The slope safety factor can be calculated by rigid limiting equilibrium method, and the catastrophe progression value is calculated by formulas (7)–(13).

Since the units of indexes are not the same, the data in Table 2 should be normalized. The utility function values of bottom factors and the catastrophe progression of the samples are calculated and shown in Table 3.

3.4 Result analysis

A comparison between the catastrophe progression and safety factor of embankment slope is shown in Fig. 4. We can see that the catastrophe progression and safety factor have an approximately similar change law: the safety factor increases (decreases) with the catastrophe progression value in a similar way. In addition, most of the safety factor values is over 1.2 and the catastrophe progression value over 0.8, which means that most part of the embankment is stable in Wenchuan earthquake except for the part collapsed. The result is in accord with field investigation.

The value of catastrophe progression reflects the combination relationship of the influencing factors, which can be used to analyze the synthetic effect of the influencing factors with different combinations on the seismic stability of embankment slope, and thus find the unfavorable combinations. Moreover, based on the field investigation and statistics, the catastrophe progression being employed as a quantitative criterion could avoid the complicated calculation and uncertainty assumption of safety factor.

The catastrophe progression method does not need to assign a weight for each index as required by the fuzzy mathematical method, and hence can avoid the human

Table 2 Measured data for seismic stability of embankment

Sample no.	Internal frictional angle C_1 (°)	Cohesion force C_2 (kPa)	Slope height C_3 (m)	Slope angle C_4 (°)	Surface gradient C_5 (°)	Peak acceleration C_6 (g)	Distance to fault C_7 (km)	Actual failure condition
1	38.0	10.0	34.0	15.0	15.0	0.42	17.00	Moderate damage
2	35.0	11.5	14.0	11.5	19.0	0.44	15.67	Almost integrity
3	35.0	12.0	25.0	17.5	10.0	0.45	14.13	Slight damage
4	31.0	8.5	9.0	32.0	21.0	0.47	13.95	Almost integrity
5	36.0	7.0	20.0	26.0	21.0	0.48	13.55	Slight damage
6	31.0	5.0	14.0	12.0	23.0	0.49	12.58	Almost integrity
7	35.0	10.0	25.0	11.0	10.0	0.51	12.27	Moderate damage
8	32.0	5.0	13.0	30.5	22.0	0.53	11.53	Almost integrity
9	35.0	10.0	10.0	25.0	24.0	0.56	10.01	Almost integrity
10	30.0	14.0	14.0	10.5	28.0	0.59	9.25	Almost integrity
11	35.0	6.0	28.0	11.0	20.0	0.58	9.39	Moderate damage
12	25.0	6.5	31.0	13.5	34.0	0.57	9.47	Serious damage
13	35.0	10.0	14.0	31.0	20.0	0.56	10.50	Slight damage
14	22.0	11.0	14.0	10.0	20.0	0.55	10.68	Almost integrity
15	34.0	12.0	16.0	14.0	20.0	0.54	11.10	Slight damage

Table 3 Utility function values of bottom factors

Sample no.	C_1	C_2	C_3	C_4	C_5	C_6	C_7	Safety factor	Catastrophe progression value
1	1.000	0.714	0.265	0.667	0.667	1.000	1.000	1.23	0.801
2	0.921	0.821	0.643	0.870	0.526	0.954	0.922	1.42	0.929
3	0.921	0.857	0.360	0.571	1.000	0.933	0.831	1.28	0.843
4	0.816	0.607	1.000	0.312	0.476	0.894	0.820	1.39	0.879
5	0.947	0.500	0.450	0.385	0.476	0.875	0.797	1.33	0.875
6	0.816	0.357	0.643	0.833	0.435	0.857	0.799	1.36	0.898
7	0.921	0.714	0.360	0.909	1.000	0.823	0.722	1.12	0.843
8	1.000	0.714	0.265	0.667	0.667	1.000	1.000	1.23	0.800
9	0.920	0.820	0.640	0.870	0.530	0.950	0.920	1.42	0.930
10	0.920	0.860	0.360	0.570	1.000	0.930	0.830	1.28	0.840
11	0.820	0.610	1.000	0.310	0.480	0.890	0.820	1.39	0.880
12	0.947	0.500	0.450	0.385	0.476	0.875	0.797	1.33	0.880
13	0.820	0.360	0.640	0.830	0.440	0.860	0.800	1.36	0.900
14	0.920	0.710	0.360	0.910	1.000	0.820	0.720	1.12	0.840
15	0.895	0.857	0.562	0.714	0.500	0.778	0.653	1.31	0.908

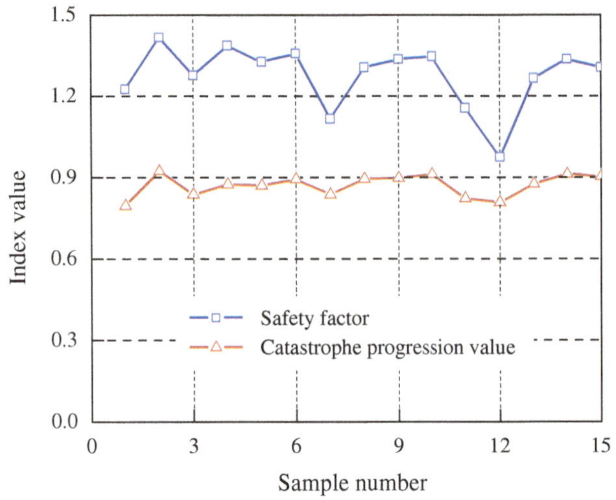

Fig. 4 The change law of catastrophe progression value and safety factor

subjectivity in determining the weights. Even so, subjectivity and fuzziness can not be eliminated completely from damage evaluation and description, thus resulting in some difference between the evaluation results and the actuality. From comparative analysis of the obtained results, however, the estimation model can meet for the demand of engineering practices on the whole.

4 Conclusion

In this article, we introduce the catastrophe progression method to evaluate the seismic stability of embankment, by selecting internal frictional angle, cohesion force, slope height, slope angle, surface gradient, peak acceleration, and distance to fault as evaluation indices. Each objective index is quantified and normalized without need to consider its weight. Compared with the traditional fuzzy mathematical method, the catastrophe progression method is simpler and can avoid the human subjectivity in determining the weights of indices. Comparative analysis of the on-site investigation results show that the method can evaluate the seismic stability of embankment with a high accuracy and therefore has an application potential in predicting the seismic stability of embankment.

Acknowledgments The research was financially supported by the open research fund of Key Laboratory of Highway Engineering of Sichuan Province, Southwest Jiaotong University (No. LHTE009201109).

References

1. Xie HP (2008) Wenchuan large earthquake and post-earthquake reconstruction-related geotechnical problems. Chin J Rock Mech Eng 27(9):1780–1791 (in Chinese)
2. Zhu HW, Yao LK, Liu ZS (2012) Research on seismic deformation characteristics of flexible wall. Chin J Rock Mech Eng 31(Suppl 1):2829–2838 in Chinese
3. Zhu HW, Yao LK, ZhangX H (2012) Comparison of dynamic characteristics between netted and packaged reinforced soil retaining walls and recommendations for seismic design. Chin J Geotech Eng 34(11):2072–2080 (in Chinese)
4. Zhang YB, Shi H (2000) Fuzzy mathematics method for analysis for slope stability. J Eng Geol 8(1):31–34 (in Chinese)
5. Li HZ, Deng GT (2006) Application of BP network to estimating stability of rock slope. J Geol Hazards Environ Preserv 17(3): 106–109 (in Chinese)
6. Wu ZY, Bao JS, Liu HS et al (2004) A method for evaluating dynamic safety factor rock slope seismic stability analysis. J Disaster Prev Mitig Eng 24(3):237–241 (in Chinese)
7. Li KG, Xu J, Li SC et al (2007) Research on evaluation of the slope stability based on the extension theory. J Chongqing Jianzhu Univ 29(4):75–78 (in Chinese)
8. Xi BX, Chen YJ, Shi XZ (2007) IRMR method for evaluation of surrounding rock quality in deep rock mass engineering. J Cent South Univ Sci Technol 38(5):987–992 (in Chinese)
9. Jones DD, Carl JW (1976) Catastrophe theory and fisheries regulation. J Fish Res Board Can 33(12):2829–2833
10. Kolata GB (1977) Catastrophe theory: the emperor has no clothes. Science 196(4286):287, 350–351
11. Rand D (1978) Exotic phenomena in games and duopoly models. J Math Econ 5(2):173–184
12. Puu T (1979) Regional modeling and structural stability. Environ Plan A 11:1431–1438
13. Tian YL, Zhou LH (2005) The study on the methods for predicting coal or gas outburst based on BP neural network. Syst Eng Theory Pract 12(12):104–106
14. Thom R (1975) Structural stability and morphogenesis. Benjamin, Reading
15. Yu DH, Peng JB (2005) Visual evaluation of neural network on classification of surrounding rocks for underground engineering. Chin J Geol Hazard Control 16(4):116–119 (in Chinese)
16. Gong FQ, Li XB, Gao K (2008) Catastrophe progression method for stability classification of underground engineering surrounding rock. J Disaster Prev Mitig Eng 39(5):1081–1087 (in Chinese)
17. Jiang JC (1999) Catastrophe theory and application in the safety engineering. J Nanjing Univ Chem Technol 21(1):2428 (in Chinese)
18. Sussman HJ, Raphael Z (1978) Catastrophe theory as applied to the social and biological sciences. Synthèse 37(2):117–216
19. Rosser JB Jr (2000) Aspects of dialectics and nonlinear dynamics. Camb J Econ 24(3):311–324
20. Li Y, Chen XH, Zhang PF (2007) Application of catastrophe progression method to evaluation of regional ecosystem health. Chin Popul Resour Environ 17(3):50–54 (in Chinese)
21. Li SP, Sun LY (2002) The assessment model of slope stability based on non-linear theory. Eng Geol Hydrol Geol 10(2):11–14 (in Chinese)
22. Huang YL (2001) Application of catastrophe progression method to evaluation of sustainable usage of water resource. Arid Environ Monit 15(3):167–170 (in Chinese)

9

System integration of China's first PEMFC locomotive

**WeiRong Chen · Fei Peng · Zhixiang Liu ·
Qi Li · Chaohua Dai**

Abstract In the face of growing environmental pollution, developing a fuel-cell-driven shunting locomotive is a great challenge in China for environmental protection and energy saving, which combines the environmental advantages of an electric locomotive with the lower infrastructure costs of a diesel-electric locomotive. In this paper, the investigation status and the development trend of the fuel-cell-driven shunting locomotive were introduced. Through innovation of the power system using fuel cells, an experiment prototype of a fuel-cell shunting locomotive was developed, which would reduce the effects on the environment of the existing locomotives. This was the first locomotive to use a proton exchange membrane fuel-cell (PEMFC) power plant in China. From October 2012, we started to test the fuel-cell power plant and further test runs on the test rail-line in Chengdu, Sichuan. The achieved encouraging results can provide fundamental data for the modification of the current individual fuel cell locomotives or further development of the fuel-cell hybrid ones in China.

Keywords Proton exchange membrane fuel cell · Locomotive transportation · Hydrogen storage · Permanent magnet synchronous motor

1 Introduction

Energy consumption plays an important role in our modern civilization and daily life, which is heavily dependent on fossil fuels. The increasing threat of the fast depletion of resources, such as petroleum, coal, and natural gas, forces people to seek renewable energy sources, such as solar, wind, geothermal, and hydroelectric. Among them, as a hydroelectric conversion assembly, fuel cells and fuel-cell power systems hold great promise as a clean technological approach; they meet all requirements for the future sustainable development for their high electrical efficiency, low emissions, and good part-load characteristics. Among the various kinds of fuel cells, proton exchange membrane fuel cells (PEMFCs) have the ordinary operation temperature (below 80 °C), which makes it greatly suitable for many kinds of applications from small portable electronic devices to automotive transport, with the power level ranging from several watts to hundreds of kilowatts [1–4].

With the rapid development of PEMFC technology in automotive industry and the electrification of railway systems as an alternative to diesel-electric locomotion undergoing serious consideration, large research efforts have been underway to develop the PEMFCs for applications in locomotive transportation in recent years. Besides, fuel-cell power for locomotives combines the environmental benefits of a catenary-electric locomotive with the higher overall energy efficiency and lower infrastructure costs of a diesel-electric; that is, fuel-cell locomotives are expected to be slightly more energy efficient than diesel locomotives, and the fuel infrastructure requirements of the former will be homologous to that of the latter. Therefore, they have a large emerging market, and their widespread adoption could lead to a reduced dependence on fossil fuels as well as encourage the development of a favorable hydrogen economy [5–7].

In comparison with the two types of traditional rail-traffic tools—catenary-electric and diesel-electric—the locomotives powered by PEMFC stacks have many great advantages as follows [6–8]:

W. Chen · F. Peng (✉) · Z. Liu · Q. Li · C. Dai
School of Electrical Engineering, Southwest Jiaotong University, Chengdu 610031, China
e-mail: kilmer_pf@126.com

- Power is derived from regenerative hydrogen instead of traction electric network which is eventually from fossil fuels, and the by-product of the electrochemical reaction in the PEMFC stacks is just pure water. Thus, the related issues in terms of urban air quality and national energy security affecting the rail industry and transportation sector can be resolved, and the generated heat in the meanwhile can also be used for water heating for the passengers.
- The relatively lower operation temperature of the PEMFC stacks reduces the costs of the heat transfer and precautionary measures in case of the high-temperature failure.
- The operational railway requirements of the PEMFC-powered locomotives are compatible with the existing electric railway and nonelectrified sectors, so the locomotives powered by PEMFC stacks could perform on the existing railway lines.
- Their effects on weather are minimal, which enables it to be capable of coping with emergency situations quickly and efficiently.
- No need for the use of the traditional tractive power supply system could avoid the deleterious consequences from the faults of pantograph-contact line and the traction system, thus improving the reliability in operational safety of the locomotive.

Several countries have made great efforts to the vigorous development of locomotives powered by PEMFC stacks, which shows big potential and extensive application foreground. Among them, several areas or countries, such as North America and Japan have successfully developed several prototype locomotives based on PEMFCs so far.

The world's first fuel-cell locomotive was born in North America for underground mining, in which PEMFC stacks with continuous rated power of 14 kW gross were the prime movers [9]. In 2007, a public–private project partnership composed of Vehicle Projects LLC, BNSF Railway Company, and the U.S. Army Corps of Engineers developed a prototype road-switcher locomotive for commercial and military railway applications in Canada [10]. It was a type of hybrid power locomotive, with 250 kW from its PEMFC power plant, and transient power well in excess of 1 MW; this hybrid locomotive is the heaviest and the most powerful fuel-cell land vehicle yet. Moreover, in 2009, Vehicle Projects and BNSF continued cooperation to develop a fuel-cell-powered shunting locomotive for testing in the USA [11].

Since 2000, East Japan Railway Company and Railway Technical Research Institute continued to make efforts to develop new energy train (NE Train), in order to reduce environmental load of railcar. The first-generation "NE Train" was first delivered from Tokyo Car Corporation in April 2003, configured as the world's first hybrid diesel/battery railcar, which was named KiYa E991-1 [12]. Then, the *NE Train* underwent modifications in 2006 in having the diesel generator replaced with a hydrogen fuel cell, becoming the world's first fuel-cell/battery hybrid railway vehicle, classified KuMoYa E995-1, which was used to power 95-kW traction motors and fitted with lithium-ion batteries with an increased storage capacity (19 kWh) [13, 14]. Table 1 summarizes the configuration of the locomotives mentioned above.

Table 1 Critical parameters of the existing fuel-cell locomotives

Item	LLC and BNSF		NE
Region	North America		Japan
Fuel-cell type	PEMFC		PEMFC
Usage	Mining	Shunting	Experimental railcar
Power level	14/17 kW	250 kW	2 × 95 kW
Hybrid power	No	Yes	Yes
Traction motor	Induced	Induced	Induced
Year	2002	2009	2006

Although there have been several types of fuel-cell locomotives put forward during the last decade, the domestic research and development is still on the threshold stage.

Based on the above, the main motives of this paper and its associated work are to design and develop a novel shunting locomotive powered by individual PEMFC stack, and evaluate the performance of the ultimate locomotive prototype as pilot study. Due to the benefit from the duty cycle and operational conditions of the shunting locomotive the need of the transient response performance is not as high as that of the automobile and other types of locomotives such as passenger or freight locomotives. Thus, under this circumstance, the power supply by individual PEMFC stack is feasible, although the PEMFC power system has inherent response time in the range of several hundred milliseconds to several seconds for large power applications [15–19].

Furthermore, some key factors and challenges that influence the operation and performance of the locomotive, such as temperature and the match between the power subsystem and the tractive subsystem, are also investigated. Some encouraging results have been obtained, which can provide fundamental data for the further research, modification, and optimization of the PEMFC locomotive. The photograph of the experimental shunting locomotive is shown in Fig. 1.

2 Locomotive layout and packaging

The integration of the complete PEMFC locomotive is shown in Fig. 2. The structure of the fuel-cell shunting locomotive consists of mechanical and electrical portions. The mechanical portion is made up of locomotive framework, bogies, traction apparatus, and air-braking subsystem; the electrical portion is composed of PEMFC power plant, high-voltage lithium-ion pack, traction inverter, permanent magnet synchronous motors (PMSMs), system controller, startup resistors array, and other auxiliary electrical system and actuators [20].

The middle machinery compartment houses the PEMFC power plant based on the framework of the traditional diesel locomotive along with the auxiliary cooling subsystem, ventilation subsystem, and traction driving subsystem. The hydrogen storage subsystem is composed of nine carbon-fiber composite tanks each with 128 L available volume, located near the power plant, which can store a total of 23 kg of hydrogen at 35 MPa [21]. The ballard 150 kW FCvelocity^{TM}–HD6 fuel-cell power module supplies power to the 600 V DC traction power bus and the existing locomotive auxiliary electrical system [22].

The locomotive prototype consists of four built-in subsystems: the PEMFC power plant, cooling subsystem, hydrogen storage subsystem, and tractive power supply subsystem. Each of the four subsystems was independently tested, and then tested as an integrated system, before being finally installed in the locomotive.

Fig. 1 Photograph of the experiment shunting locomotive

As seen from Fig. 2, the largest part of the fuel-cell power system is the hydrogen storage subsystem. It consists of nine 35 MPa carbon-fiber/aluminum cylinders that approximately store 23 kg of compressed hydrogen. The nominal maximum operation pressure of each cylinder is 50 MPa. When the impact pressure reaches or exceeds 80 MPa, hydrogen will be discharged to prevent from explosion hazards. Besides, in consideration of the approximate equiponderance between the hydrogen storage system and PEMFC power system, the above layout has minimal effect on the locomotive's center of gravity and symmetry with better power distribution between the two bogies located at the basis of the framework, which will be favorable for the efficient operation of the locomotive.

The PEMFC power plant, two traction inverters, electric cabinet, and cooling subsystem are housed in the remaining half of the machinery compartment. The air-braking subsystem already housed in the locomotive stands aside the hydrogen storage subsystem, which is used for the deceleration and braking of the locomotive. There is a low-power air compressor dedicated for the air pressurization, which is located at the lower left side of the locomotive framework. The air compressor will automatically startup when the operation pressure falls below 400 kPa and shutdown, when the operation pressure reaches 850 kPa in order to provide enough air for braking at low speed.

However, the fuel-cell power plant itself is equipped with air delivery system, which operates at a maximum air

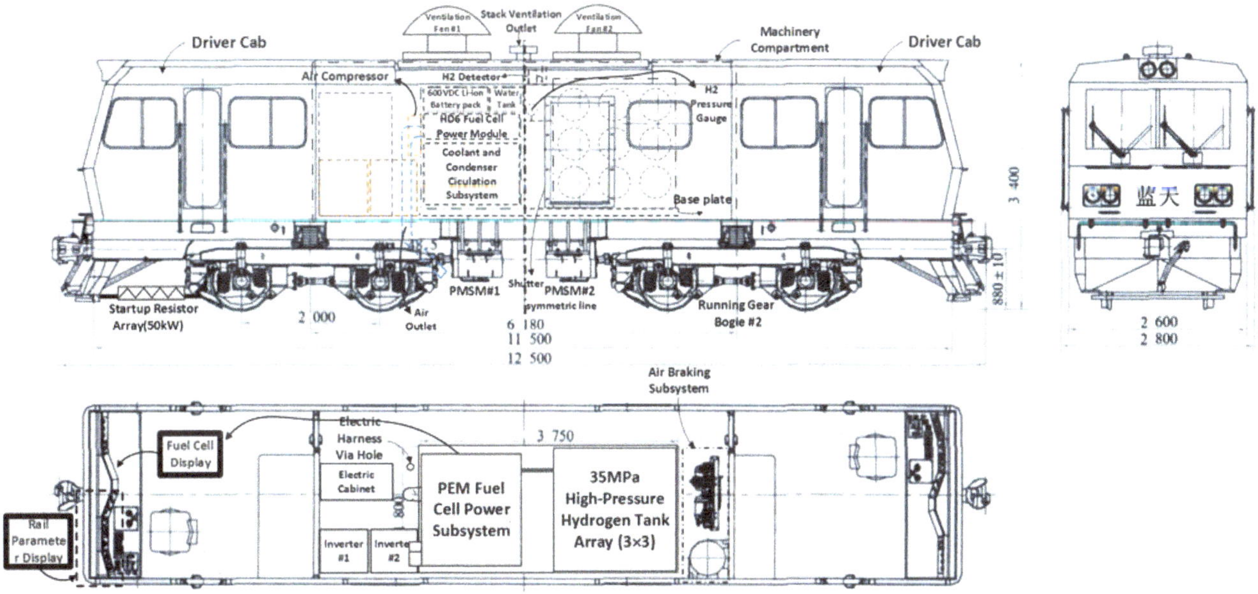

Fig. 2 System layout of the PEMFC locomotive including mechanical and electric portions

pressure of ~1.2 bar (relative pressure), with a maximum mass flow of ~150 g/s. Compared with the high pressure operating fuel cell, this "low-pressure" operation results in lower parasitic losses of gross power (~10 % for a 1.5–1.8 bar air system and near 20 % for higher-pressure air system) [7] and the need to employ just one-stage compression. Besides, in order to make the fuel-cell power system work more efficiently at the start-up stage, a 600 V lithium-ion pack is equipped to provide the startup power, which is located on the HD6 power module.

The generated pure water is exhausted with the residual air and a small fraction of formation heat through the "air outlet" pipe. The formation heat of the PEMFC power module is so large that the cooling subsystem is necessary to reject the redundant heat in order to maintain the optimal operating temperature of the PEMFC stack. The cooling subsystem consists of the primary and secondary radiators which reside in the lower left and lower middle sections of the power plant, respectively. The secondary radiator is applied for heat transfer of the stack condenser that is used to insure that enough process water be made available at all time for air humidification. The primary radiator will suck air and exhaust it through the shutter of the locomotive framework. Under normal conditions, the leaked hydrogen will be exhausted through the ventilation outlet of the PEMFC stack module, and the ventilation fans located on the upper side of the locomotive will assist in precluding confinement of any accidentally leaked hydrogen.

It is worth mentioning that the startup-resistors array (approximately 50 kW) located at the bottom of the locomotive (as shown in Fig. 2) is used to overcome the inherent surge phenomenon of the turbo charger in the air delivery system at low flow rate. The startup resistors array is divided into three groups to match with the power demand of the gear increment (about 15–20 kW/gear for each PMSM with velocity modulation).

3 Results and discussion

3.1 Overview of the test locomotive

The test shunting locomotive in this test module consists of only one locomotive, which is the smallest test unit that is able to individually operate. Also, in order to effectively use the existing locomotive technologies, we make various devices that are standardized, to be compatible with the latest electric locomotive. The locomotive body used is a stainless steel body which is the same as the traditional diesel locomotive that is run on local lines. As for the motors and locomotive controller, the latest PMSM and its driving technologies are used aiming at improving the efficiency and power factor in comparison with induction and wound-rotor synchronous motors [23–25]. Figure 2 shows an overview of the test locomotive, and Table 2 shows the technical specifications of the locomotive.

3.2 Performance test

At the time of this writing, the fuel-cell shunting locomotive has undergone several weeks of operational testing at the test rail-line in Chengdu, Sichuan, China. The locomotive work schedule involves the gear test and running test. The PEMFC locomotive performed all operational testings as a single unit, and thus, the entire work energy was provided solely from the PEMFC locomotive itself. The duty cycle of the running test as shown in Fig. 3 is an acceleration–deceleration duty cycle which simulates and evaluates the load–response performance.

From a functional perspective, the fuel-cell locomotive works well in all respects. The fuel-cell stack module and the associated cooling and fuel subsystems performed without any issue during the duty cycle test. During all work shifts, the power plant was able to provide power to the traction motors and/or provide current to all the auxiliary peripherals. Operation of the fuel-cell power plant was closely monitored, and data for key parameters were logged at a 0.5 s rate during operation. Of particular interest are the response performance of the mean operating power levels and the associated switch control of the startup resistors array. Figure 3 shows the snapshot of the typical acceleration–deceleration duty cycle for the integrated locomotive. The fuel-cell operating power level is dynamically predicted and determined by the system

Table 2 The test fuel-cell shunting locomotive specifications

Item	Value
Model	XQG45-600P
Unloaded mass (t)	45 ± 3 %
Driving mode	DC–AC transmission
Shaft type	B-B
Wheel diameter (mm)	840
Wheel track (mm)	1,435
Minimum bend radius (m)	80
Distance between shafts (mm)	2,000
Distance between bogies' center (mm)	6,180
Physical dimension ($l \times w \times h$, mm)	13,500 × 2,600 × 3,600
Continuous speed (km/h)	21
Maximum operating speed (km/h)	65
Design speed (km/h)	100
Continuous tractive force (kN)	36.5
Startup tractive force (kN)	50
Main motor	PMSMs
Tractive motor power (kW)	2 × 120
Brake type	Air brake + holding brake

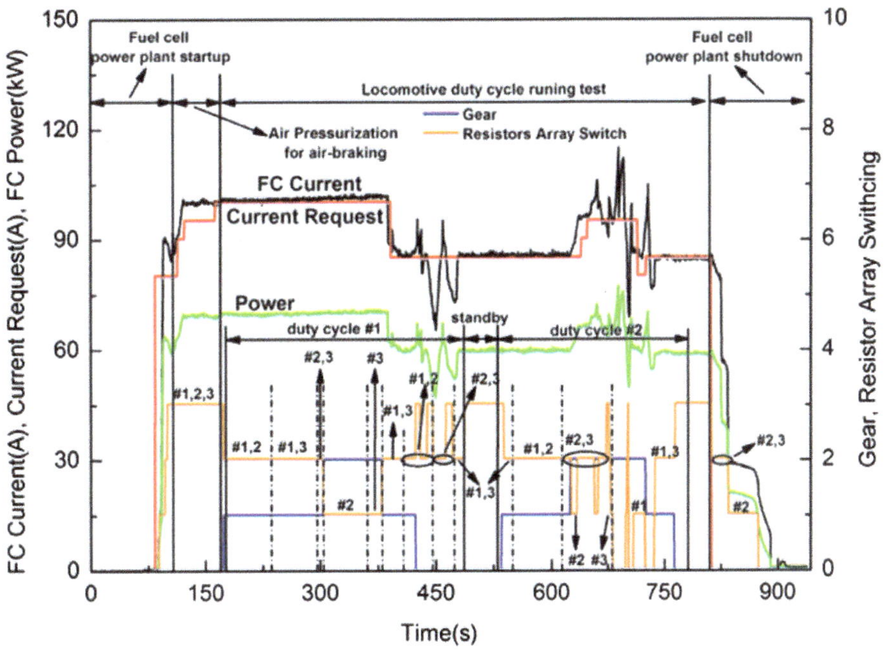

Fig. 3 Sample of the locomotive operation curve during a duty cycle of reciprocating running in the test rail-line in Chengdu, Sichuan, China, the maximum power requirement of which was 80 kW corresponding to gear 2

controller which determines the power set-point based on the current demand, tractive characteristics of the traction motors, and states of the startup resistors array.

As seen from Fig. 3, the sample locomotive integration test consisted of two duty cycles, each with independent acceleration–deceleration procedure. Each time when the locomotive starts up, the air pressurization for air-braking is necessarily carried out at first, so that there is sufficient compressed air for braking at low speed (≤20 km/h). After the preparation as mentioned above, the integration test could be carried out subsequently. As the test rail-line is just about 1 km, it could not test all the gears switching. Besides, as the speed modulation parameters of the two inverters are not optimal, when the locomotive decelerates, in particular in duty cycle #2, the real-time current drawn is fluctuant with the effect of energy braking, thus resulting in the fluctuating power consumption. The situation that there is some slope in the test rail-line makes the power fluctuation in the return acceleration–deceleration procedure–duty cycle #2 more serious.

Furthermore, the startup-resistors-array-switching subroutine is customized in order to keep the temperature rise of each resistor within reasonable limit, and minimize the heat-transfer power consumption. The progress of the startup resistors switching is also shown in Fig. 3.

4 Conclusion

In this paper, the first PEMFC shunting locomotive developed in China that combines the environmental advantages of an electric locomotive with the lower infrastructure costs of a diesel-electric locomotive was introduced. Moreover, the performance of the fuel-cell power plant and the integrated locomotive were experimentally investigated in a test rail-line in Chengdu, Sichuan, China. Depending on the primary PEMFC power source and relatively environmental high-voltage lithium-ion batteries pack for the startup power source, it can a totally zero-emissions vehicle, that is, with zero carbon in the energy duty cycle. Through the proper system design and development of the PEMFC shunting locomotive, utilization of hydrogen fuel cell in the rail environment with the characteristics of relatively simple duty cycle condition has proven technically feasible. After several weeks of operational testing, the achieved encouraging results can provide fundamental data for further modification or development of the fuel cell or even fuel-cell hybrid locomotives in China.

The body of the locomotive in particular for the fuel-cell power plant underwent a large number of detailed adjustments using important technological know-how learned during the course of running test, and thus we arrived at designing the current system and locomotive prototype. However, from a technical perspective, there are still many important issues that need to be modified or improved, such as follows:

- Substitution of the turbotype compressor with double flight screw-type compressor because of its lower noise level and little surge, although the cost of the former is lower.
- Although the turbo charger with the substitutive double flight screw-type compressor having low boost pressure

ratio is suitable for these locomotive applications and quieter material has been utilized in the framework construction of the locomotive, the air outlet is still necessary for optimal noise abatement.
- With the double flight screw-type compressor in place, the startup resistors array can be canceled, the control logic will be simplified, and the available power will be maximized.
- The parameters of the traction inverters for PMSMs driving need to be further debugged, in order to achieve more efficient PMSMs and the better matching between the fuel-cell power plant and the tractive motors.
- The condenser circulation system could be further improved to dynamically control the temperature of the condenser, as it is directly related to the humidification of the air entering the stack as mentioned above.

In light of the above, we will make further efforts and work on resolving the above issues to achieve the viability for the practical use of an ecofriendly PEMFC shunting locomotive.

Acknowledgments This work was supported by the National Natural Science Foundation of China (51177138); the Specialized Research Fund for the Doctoral Program of Higher Education (20100184110015); the International Science and Technology Cooperation and Exchange Research Plan of Sichuan Province (2012HH0007); the Science and Technology Development Plan of Ministry of Railways (2012J012-D); the Fundamental Research Funds for the Central Universities (SWJTU11CX030); and the Specialized Research Fund for the Doctoral Program of Higher Education (20120184120011).

References

1. Karl VK, Guenter RS (1995) Environmental impact of fuel cell technology. Chem Rev 95(1):191–207
2. Kishinevsky Y, Zelingher S (2003) Coming clean with fuel cells. IEEE Power Energy Mag 1:20–25
3. Phatiphat T, Bernard D, Stephane R et al (2009) Fuel cell high-power application. IEEE Ind Electron Mag 1(3):32–46
4. Varigonda S, Kamat M (2006) Control of stationary and transportation fuel cell systems: progress and opportunities. Comput Chem Eng 30:1735–1748
5. Jones LE, Hayward GW, Kalyanam KM et al (1985) Fuel cell alternative for locomotive propulsion. J Power Sources 10(7):505–516
6. Miller AR, Barnes DL (2002) Fuel cell locomotives. In: proceedings of fuel cell world, Lucerne, Switzerland
7. Miller AR, Hess KS, Baesrnes DL et al (2007) System design of a large fuel cell hybrid locomotive. J Power Sources 107:935–942
8. Chen WR, Qian QQ, Li Q (2009) Investigation status and development trend of hybrid power train based on fuel cell. J. Southwest Jiaotong Univ 44(1):1–6
9. Miller AR. (2000) Tunneling and mining applications of Fuel cell vehicles. Fuel Cells bull : 5-9
10. Hess K S, Miller A R, Erickson T L, et al. (2008) Demonstration of a hydrogen fuel-cell locomotive In: proceedings of locomotive maintenance officers association conference, Chicago
11. BNSF (2009) Vehicle projects build experimental shunting engine. Fuel Cells Bull 5:4
12. Taketo F, Nobutsugu T, Mitsuyuki O (2006) Development of an NE train. JR EAST Tech Rev 156(4):62–70
13. World-first hybrid rail vehicle "NE Train". (2003) Japan Railfan Mag 45(506): 86
14. World-first fuel-cell hybrid rail vehicle KuMoYa E995 (2008) Japan Railfan Mag 48(561): 53-55
15. Yu DC, Yuvarajan S (2005) Electronic circuit model for proton ex-change membrane fuel cells. J Power Sources 142(1):238–242
16. Engeti PN, Howze JW (2004) Development of an equivalent circuit model of a fuel cell to evaluate the effects of inverter ripple current. In: APEC'04. 1(1): pp 355-361
17. Wai RJ, Lin CY (2010) Active low-frequency ripple control for clean -energy power conditioning mechanism. IEEE Trans Ind Electron 57(11):3780–3792
18. Page SC, Anbuky AH, Krumdieck SP et al (2007) Test method and equivalent circuit modeling of a PEM fuel cell in a passive state. IEEE Trans Energy Convers 22(3):764–773
19. Choe SY, Ahn JW, Lee JG et al (2008) Dynamic simulator for a PEM fuel cell system with a PWM DC/DC converter. IEEE Trans Energy Convers 23(2):669–680
20. Luo HJ, Li WH, Li XQ (2011) XQG45-600P light rail vehicles using new energy fuel cell. Railw Locomot Car 31(3):53–55
21. SUNWISE Energy System (2013) http://www.sunwise.sh.cn. Accessed 05 June 2013
22. FCvelocity^{TM}-HD6 integration manual ballard power system (2011)
23. Xu JF, Li YL, Xu JP (2005) Present situation and perspect of applying permanent magnet synchronous motors to railway locomotive. J China Railw Soc 27(2):130–132
24. Comparison of permanent magnet synchronous motors applied to railway vehicle traction system. (2007) J China Railw Soc 29(5): 111-116
25. Krishnan R (2009) Permanent magnet synchronous and brushless DC motor drives. pp 34–38

10

MD simulation on the interactions between CH_2 groups and the (001) surface of tungsten

Xianfu Feng · Shuping Tian · Xinkui He · Fujun Gou · Shuiquan Deng · Yong Zhao

Abstract This work studies the angle dependence of the interactions between impinging CH_2 particles of 150 eV with the tungsten surface. The simulations show that the carbon atoms are much more easily bonded to the tungsten atoms than hydrogen atoms, though a few of the latter can also penetrate into the tungsten material. When the incidence angle is greater than 75°, the incident CH_2 particles are reflected without break-ups. Below this angle, a W–C layer of about 0.5 nm is formed with another C, H-rich layer depositing on top of it. The molecular dynamics (MD) approach has proved to be a powerful tool to solve the structural problems at atomic length scale of various materials. Some of its possible applications to the railway track materials have also been discussed.

Keywords Molecular dynamics · First-wall material · Tungsten · Plasma-material interaction

1 Introduction

Thc plasma-facing components (PFCs) are designed to work in an extreme environment exposing to high energy particles and heat flux from the plasma [1]. However, as the designed extreme conditions in the reaction chamber of ITER (International Thermonuclear Experimental Reactor) are still difficult to realize, theoretical simulations become crucial at this stage [2]. Tungsten steel, a composite material containing C and other transition metals, is now widely accepted as one candidate for the first-wall material because of its high melting point, low sputtering rate, good mechanical properties, resistance to radiation, etc. [2, 3]. The performance of plasma-facing surface under intense transient thermal loads is another critical issue, since the formation of a melt layer may favor the generation of highly activated dust particles [1, 4]. Graphite has been selected as one material in the middle and the lower part of the vertical target of the divertor and as the collection panel in the thermonuclear fusion reactor, for their good outgassing property, low Z character and capability of handling high heat fluxes [2, 5]. However, graphite is prone to physical and chemical sputter due to the low sputtering energy threshold [6], which results in many dust particulates. The formation mechanisms, existing forms and their consequences of various kinds of dust particulates resulting from the plasma first-wall interactions have been systematically studied by many authors [7–11]. Among these dust particulates, hydrocarbon radicals have also attracted intensive studies [7–11]. These works shows that carbon atoms can easily combine with hydrogen or its isotope atoms in the plasma chamber leading to the formation of hydrocarbon groups H_xC_y, which reduces dramatically the performance of the first-wall materials [7–9, 11, 12]. Ohya et al. [13, 14] have investigated the interactions of hydrocarbons with the surfaces of the possible first-wall materials (W, C, W–C), as well as their reflections, dissociations and the redepositions by using Monte Carlo (MC) method, and molecular dynamics (MD) method. Because of the torus-like shape of the plasma chamber, the confined high-energy particles are expected to fly in directions shear to the surface of the first wall. This work, thus focuses on the incident angle dependence of the interactions between CH_2

X. Feng · X. He · S. Deng (✉) · Y. Zhao
Superconductivity and New Energy R & D Center, Southwest Jiaotong University, Chengdu 610031, China
e-mail: s.deng@swjtu.edu.cn

S. Tian · F. Gou
Institute of Nuclear Science and Technology, Sichuan University, Chengdu 610064, China

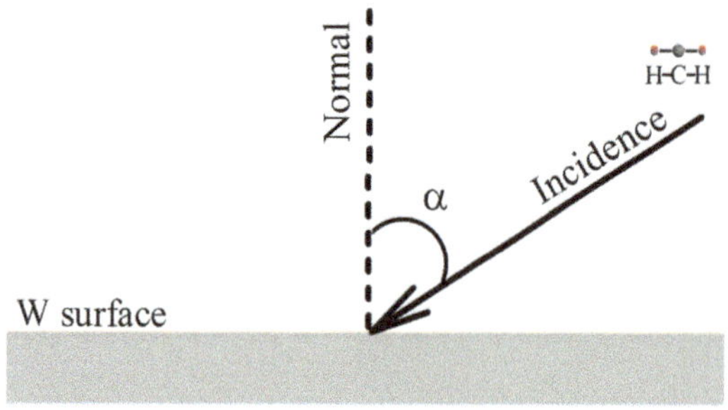

Fig. 1 A scheme of incident angle of CH_2 particles

groups and tungsten surface using MD simulation method. In addition, we have used higher energy flux than the earlier work [13] to investigate the effects of incident energy of such radicals.

2 Computational details

In our MD calculations, all particles in the system are assumed to abide by Newtonian mechanics. The velocity-Verlet method [15] was used for the integration of Newton's equations of motion.

The force on each atom was calculated from the analytical derivative of the modified Tersoff–Brenner potential [16]. This potential consists of the attractive and repulsive terms as shown in Eqs. (1) and (2), respectively.

$$V^{A}(r) = \frac{SD_0}{S-1}\exp\left[-\beta\sqrt{2/S}(r-r_0)\right], \quad (1)$$

$$V^{R}(r) = \frac{D_0}{S-1}\exp\left[-\beta\sqrt{2S}(r-r_0)\right], \quad (2)$$

where D_0, r_0, S indicates the dimer bond energy, the dimer bond distance and the adjustable parameter, respectively. The parameter β in Eqs. (1), (2) can be derived from the ground-state oscillation frequency of the dimer. This potential function has been designed specifically for treating the W–C-H system in non-equilibrium state and has been tested in many cases [4, 13, 14, 17].

An $8 \times 8 \times 10$ supercell of the $\alpha - W$ (body-centered cubic) has been chosen as the initial configuration. This model contains 1,280 W atoms, which has a depth of about 3.165 nm and a top surface area of about 6.412 nm^2. The

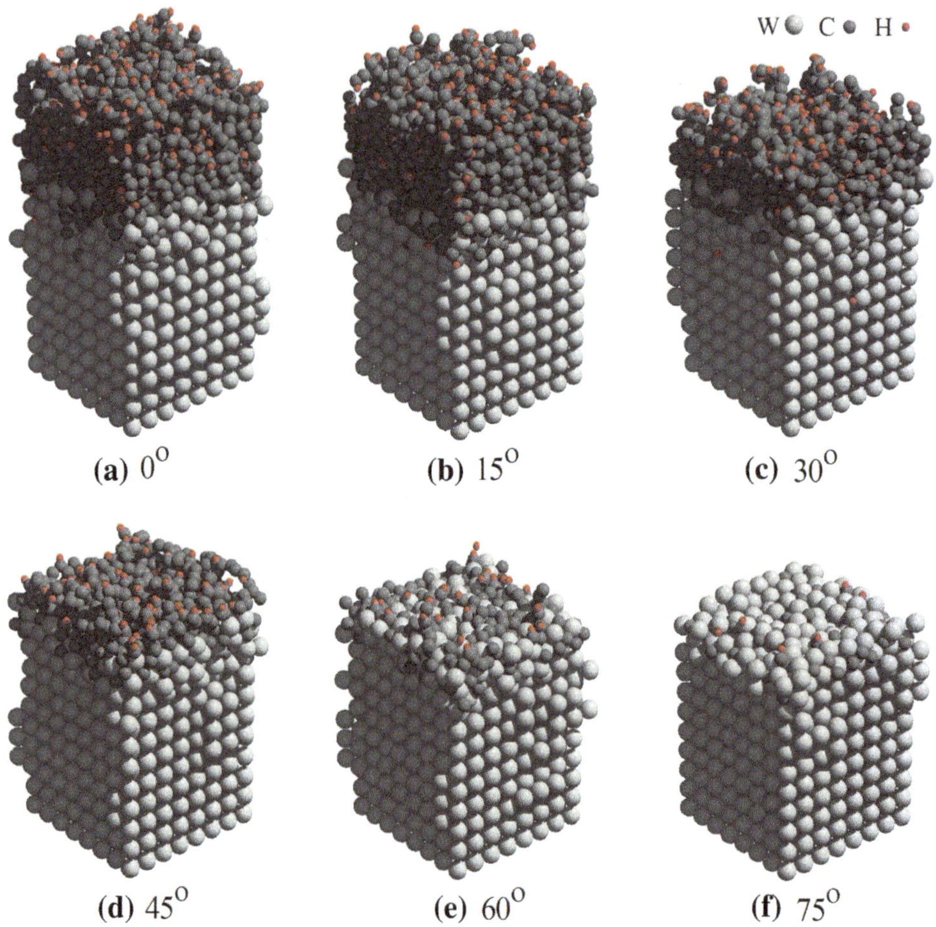

Fig. 2 *Snapshots* of the three simulated surfaces at an incident fluence of 4.09×10^{16} cm^{-2}

(001) plane of $\alpha - W$ is the top surface, i.e., the plasma-facing surface. Periodic boundary conditions are assumed in the *a*, *b* directions of the supercell in solving the equations of motion. The bottom layers with a thickness of two-unit cell length are fixed in the simulations. In the simulation, the coupling time between the incident particles and the sample is set to be 0.5 ps with a time step of 0.0005 ps. The initial sample temperature is set to be 300 K. In order to maintain a thermal equilibrium of the system, the BERENDSEN's heat bath technique was used with a BenerdsenTau of 0.01 ps [18]. In this work, the linear H–C–H configuration of CH_2 groups is assumed with the bond axes randomly distributed with respect to the incident direction. In all the cases, the incident energy of CH_2 particles is set to be 150 eV. The incident angle as defined in Fig. 1 is as 0°, 5°, 15°, 30°, 45°, 60°, 75°, and 85°, respectively. The incident fluence is chosen to be 4.09×10^{16} cm^{-2}, a value comparable to those used in other similar simulation works [13, 14].

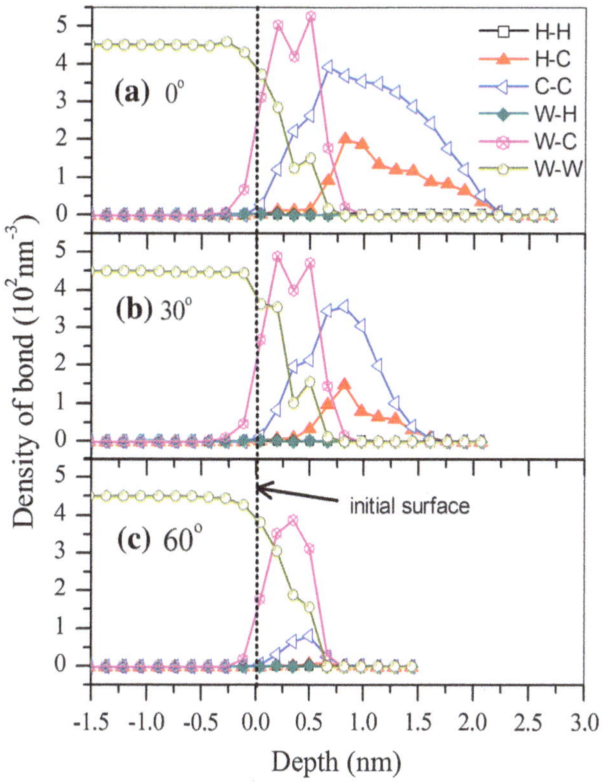

Fig. 3 The bond density distribution along the vertical direction with respect to the initial W (001) surface

3 Results and discussion

In Fig. 2, we show snapshots of the bombarded surface of the sample at the chosen incident fluence with three different incident angles. As shown in Fig. 2, both C atoms and H atoms can penetrate into the tungsten surface resulting in a new layer composed of W, C, and H atoms. Our simulations indicate that when the incident angle is less than 45°, a thin "hydrocarbon" layer consisting of mainly C and H atoms is formed on top of the W–C–H layer. When the incident angle is greater than 45°, only W–C–H layer is formed without the upper covering "hydrocarbon" layer.

Figure 3 shows the distributions of the various densities of bonds with respect to the depth in reference to the initial surface. As shown in Fig. 3, the original W surface is smeared with the occurrence of W–W bond distribution above the initial surface. The density of the W–C bonds shows peaks in all of the three cases above the initial surface, which drops rapidly below the initial surface. When the incident angle is less than 30°, the formation of C–C and C–H bonds becomes significant, which indicates the breaking of the incident CH_2 groups (see also Fig. 5). In all the cases, the H–H and W–H bonds are negligible in comparison with the other bonds. It is obvious that the bombardment process results in the displacement of W atoms (see Figs. 2, 3) and the formation of W–C bonding (see Fig. 3) etc.

Figure 4 shows the relations between the numbers of deposited H (a), C (b) atoms and the number of respective incident particulates at different incident angles. As shown in Fig. 4, when the incident angle is smaller than or equal to 30° for H and 45° for C, respectively, the deposition is more effective. For both C and H atoms, when the incident angle is larger than 45° for H, 75° for C the deposition rate becomes negligibly small regardless of the density of the incident particles. These results indicate that when the incident direction leans toward the tangent of the surface of the W matrix, the kinetic energy of the incident particles exceeds their bound energy, which invalidates any trapping of the incident particles, and thus the reflection dominates. Figures 3 and 4 clearly show that C atoms are much more easily trapped by the W matrix indicating a stronger bonding between W and C atoms than that between W and H atoms.

Figure 5 shows the incident angle dependence of the relative scattering ratio (RSR) of some important particles produced in the interaction process and the remained CH_2 groups. For clearance, the logarithmic ordinate is used for the RSRs of the indicated particles. As can be seen from Fig. 5, the RSR of C atoms is the lowest in a wide range of incident angles among the scattered particles. Furthermore it has very weak incident angle dependence. In contrast, the RSR of H atoms is the largest among the investigated particles, except for CH_2 radicals whose RSR surpasses all those of the other particles when the incident angle exceeds ~35°, which indicates the dominance of reflection in this range. Considering the following relevant process in the

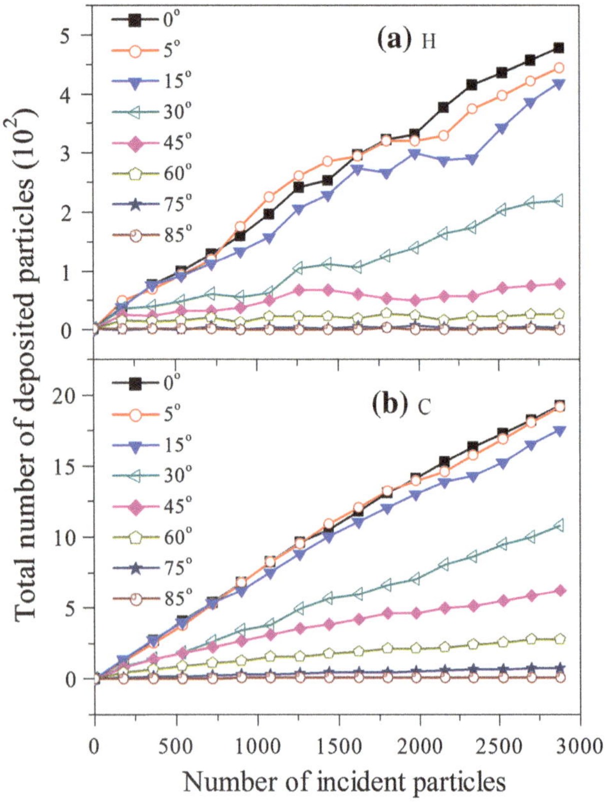

Fig. 4 The relations between the numbers of deposited H (**a**), C (**b**) atoms and the *number* of the respective incident particulates at different incident angles

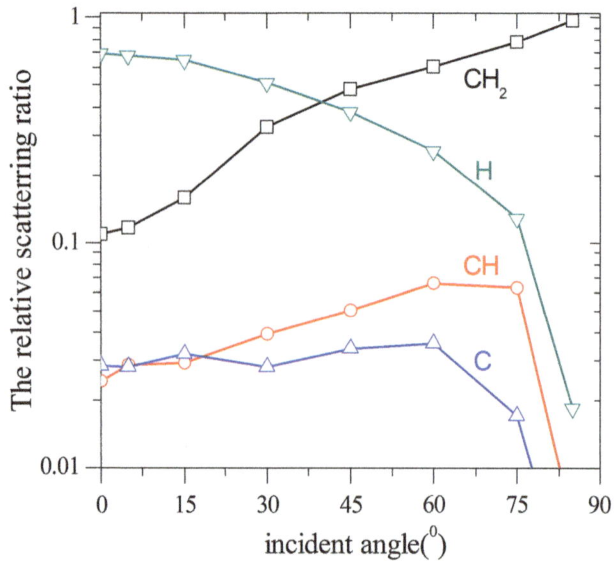

Fig. 5 The scattering ratio of the number of scattered particles to the total number of scattered particles

interactions between the incident CH_2 groups and the W surface,

$$CH_2 \rightarrow H + CH,$$
$$CH_2 \rightarrow 2H + C,$$
$$CH \rightarrow C + H.$$

One can expect that the RSR of C atoms should be approximately two times smaller than that of H atoms, if the above three processes have the similar probability and the relevant particulates are scattered out in the same efficiency. As can be seen from Fig. 5, the RSR of H atoms far exceeds that of C atoms by approximately one order. This may also be possible if the first process dominates, however, considering the results as shown in Figs. 3 and 4, it can be neglected. Taking into account the results shown in Fig. 3, we attribute the origin to be the stronger W–C bonding than the W–H boning. It should be pointed out that the dramatic drop of the scattering ratios for H, C, and CH particles at high incident angles (>75°) results from the reflectance of CH_2 particles, which reduces the source of all C, H, and CH particles.

4 Conclusions

Our simulations show that the deposition ratio of C and H atoms has a strong dependence on the incident angle. When the incident angle exceeds a threshold value the composite particle CH_2 is mainly reflected. The carbon atoms are much more easily bonded to the tungsten atoms than hydrogen atoms, though a few of the latter can also penetrate into the tungsten sample. When the incidence angle is greater than 75°, the kinetic energy of the particles exceeds significantly the bound energy. Below this angle, an effective W–C bonding occurs, which results in a W–C layer of about 0.5 nm with another C, H-rich layer depositing on top of it. It is clear that our studies on the interactions between the plasma and the first-wall materials are still not complete. Some interesting issues such as the consequences of choosing different hydrocarbon groups, the different crystal plane of $\alpha - W$, e.g., (011), (111), etc., the different incident energy of the particulates, etc., are also very important in understanding the interactions between the plasma and the first-wall materials. It is thus worthy of further studies to clarify these issues.

It should be pointed that our preliminary study has mainly involved the interactions between the possible particles produced in the fusion chamber and the first-wall materials, i.e., tungsten steel. The role of carbon and the other elements in enhancing and improving the properties of a specific material has not been studied, though it is very important, otherwise we even do not have the concept of steel. The point raised here is that chemical composition does not uniquely decide the properties of a steel, in almost all cases, the treatment processes such as heat treatment processes are also very crucial. This implies that the microstructure of a material is the dominating factor.

For example, the microstructure of rail steel is composed of pearlite, while that of low carbon steel used for railway sleepers consists of ferrite and cementite [19]. The properties, such as the hardness, wear resistance, tensile strength, etc., depend actually on the multi-scale structures, say, the composition, the proportion of different structural domains and the details of each domain. The MD simulation, a powerful tool to study the structure at the atomic length scale, has proved to be an efficient approach, as we have shown in this work, to reveal the relations between the microstructures of various complicated steels and their properties. Our future work will be done along this line for both fusion reactor materials and railway track materials.

Acknowledgments This work is financially supported by the Science Foundation for International Cooperation of Sichuan Province (2014HH0016), the Fundamental Research Funds for the Central Universities (SWJTU2014: A0920502051113-10000), National Magnetic Confinement Fusion Science Program (2011GB112001).

References

1. Bolt H, Barabash V, Krauss W et al (2004) Materials for the plasma-facing components of fusion reactors. J Nucl Mater 329–333:66–73
2. Loarte A, Lipschultz B, Kukushkin AS et al (2007) Chapter 4: power and particle control. Nucl Fus 47(6):203–263
3. YueLin Liu, HongBo Zhou, Ying Zhang et al (2011) Interaction of C with vacancy in W: a first-principles study. Comput Mater Sci 50(11):3213–3217
4. Zhongshi Yang, Xu Q, Junqi Liao et al (2009) Study on C-W interactions by molecular dynamics simulations. Nucl Instrum Methods Phys Res Sect B 267(18):3144–3147
5. Miyahara A, Tanabe T (1988) Graphite as plasma facing material. J Nucl Mater 155–157:49–57
6. Grote H, Bohmeyer W, Kornejew P et al (1999) Chemical sputtering yields of carbon based materials at high ion flux densities. J Nucl Mater 266–269:1059–1064
7. Janev Ratko K, Reiter Detlev (2002) Collision processes of hydrocarbon species in hydrogen plasmas: I. the methane family. Berichte des Forschungszentrums Jülich, Zentralbibliothek. ISBN 0944-2952
8. Winter J (1998) Dust in fusion devices—experimental evidence, possible sources and consequences. Plasma Phys Controll Fus 40(6):1201
9. Winter J, Gebauer G (1999) Dust in magnetic confinement fusion devices and its impact on plasma operation. J Nucl Mater 266–269:228–233
10. Winter J (2000) Dust: a new challenge in nuclear fusion research? Phys Plasmas 7(10):3862
11. Sugai H, Mitsuoka Y, Toyoda H (1998) Observation of surface dissociation of low-energy polyatomic ions relevant to plasma processing. J Vac Sci Technol A 16(1):290–293
12. Fukumoto M, Kashiwagi H, Ohtsuka Y et al (2009) Hydrogen behavior in damaged tungsten by high-energy ion irradiation. J Nucl Mater 386–388:768–771
13. Ohya K, Inai K, Kikuhara Y et al (2011) Molecular dynamics study on hydrocarbon interaction with plasma facing walls. J Nucl Mater 417(1–3):637–642
14. Ohya K, Kikuhara Y, Inai K et al (2009) Simulation of hydrocarbon reflection from carbon and tungsten surfaces and its impact on co-deposition patterns on plasma facing components. J Nucl Mater 390–391:72–75
15. Swope WC, Andersen HC, Berens PH, Wilson KR (1982) A computer simulation method for the calculation of equilibrium constants for the formation of physical clusters of molecules: application to small water clusters. J Chem Phys 76(1):637
16. Juslin N, Erhart P, Träskelin P et al (2005) Analytical interatomic potential for modeling nonequilibrium processes in the W–C–H system. J Appl Phys 98(12):123520
17. Kensuke Inai, Kikuhara Y, Kaoru Ohya (2008) Comparison of carbon deposition on tungsten between molecular dynamics and dynamic Monte Carlo simulation. Surf Coat Technol 202(22–23):5374–5378
18. Berendsen HJC, Postma JPM, van Gunsteren WF et al (1984) Molecular dynamics with coupling to an external bath. J Chem Phys 81(8):3684
19. Resources and news for science education: http://resources.schoolscience.co.uk/Corus/16plus/steelch3pg4.html. Accessed 9 Aug 2003

11

A software-defined networking approach for handover management with real-time video in WLANs

Peter Dely · Andreas Kassler · Lawrence Chow · Nicholas Bambos · Nico Bayer · Hans Einsiedler · Christoph Peylo · Daniel Mellado · Miguel Sanchez

Abstract To achieve high performance and reliability in video streaming over wireless local area networks (WLANs), one must jointly consider both optimized association to access points (APs) and handover management based on dynamic scanning of alternate APs. In this article, we propose a new architecture within the software-defined networking (SDN) framework, which allows stations to be connected to several APs simultaneously and to switch fast between them. We evaluate our system in a real-time testbed and demonstrate that our SDN-based handover mechanism significantly reduces the number and duration of video freeze events and allows for smaller playout buffers.

Keywords Openflow · Handover · Video streaming

1 Introduction

The upcoming wireless local area network (WLAN) standards IEEE 802.11ac and 802.11ad will offer PHY layer rates of several Gbit/s using high-order coding and modulation schemes, wide channels, and high carrier frequencies. To achieve those high rates, short- and high-quality radio links and a dense network deployment are necessary. As a consequence, even with mobility at moderate speeds, handovers between access points (APs) need to be executed frequently to insure that a mobile station is always using a close-by AP with a high signal quality. Thus, mobility management will play a more pronounced role in future WLANs.

Several IEEE standards for mobility management have been published. For example, IEEE 802.21 [1] describes signaling messages exchanged to trigger handovers between WLANs and other networks. IEEE 802.11r [2] specifies how the authentication process and the encryption key negotiation during a handover can be accelerated. While those standards offer some messaging primitives to control handovers, they are not sufficient to enable seamless mobility in WLANs. For example, those standards do not specify when and how to detect new APs, when to schedule a handover, and how to control the wired distribution system to route the traffic to the right AP.

Those open issues, however, are a key to enable seamless mobility in WLANs. In particular, video streaming and conferencing applications require fast handovers, since such applications use small playout buffers and a too long a handover duration would result in a frozen video. To this end, we propose to allow a station to be associated to several APs simultaneously and to use the concept of software-defined networks (SDNs) [3] and the Openflow protocol [4] to steer the handover. SDNs allow us to integrate application characteristics into the handover decision in an elegant way. As the main contribution of this article, we describe an SDN-based system architecture for fast handovers in WLANs and evaluate this architecture in a WLAN testbed using video streaming applications.

P. Dely · A. Kassler (✉)
Computer Science Department, Karlstad University,
65188 Karlstad, Sweden
e-mail: andreas.kassler@kau.se

P. Dely
e-mail: peter.dely@kau.se

L. Chow · N. Bambos
Department of Management Science and Engineering,
Stanford University, Stanford, CA 94305, USA

N. Bayer · H. Einsiedler · C. Peylo
Telekom Innovation Laboratories, Deutsche Telekom,
Berlin, Germany

D. Mellado · M. Sanchez
The Charles III University of Madrid, Madrid, Spain

Optimization of WLAN handovers has been examined in several studies before. For example, Refs. [5] and [6] investigated how to optimize the scanning procedure for new APs. Those methods are orthogonal to our proposal which targets the actual handover and can be used to improve the scanning duration. Nah et al. [7] proposed a method for scheduling the scanning procedure in WLANs, while considering the buffer level of the video player. In comparison to this study, their evaluation is based on network simulation, and the re-configuration of the wired backhaul network is not considered.

The use of SDN/OpenFlow for improving streaming video has been explored [8]. However, in contrast to our study, Ref. [8] stipulates that multiple interfaces are required at the station and explores the concurrent use of different radio technologies (WLAN and WiMAX). Scanning for new APs is not explicitly addressed in this study. CloudMAC [9] is a recent proposal to distribute the MAC processing in WLANs using SDNs. CloudMAC also allows implementing fast handovers between APs, but is mainly targeting network initiated handovers, while this article uses client initiated handovers.

References [10] and [11] formulate and solve optimization problems to compute the best moment for performing a handover. While Ref. [10] uses an application independent formulation, which aims to maximize the long-term throughput, Ref. [11] explicitly considers the characteristics of streaming video and the playout buffer.

The remainder of the article is organized as follows: In Sect. 2, we describe in detail the proposed system architecture. Section 3 outlines the implementation of the architecture. In Sect. 4, we present evaluation results. We wrap-up the article in Sect. 5.

2 System architecture

2.1 Overview

Figure 1 provides an overview of the system architecture. The system consists of a streaming server acting as the video source. The streaming server is connected to WLAN APs via an OpenFlow switch. A controller dynamically configures the forwarding table of the OpenFlow switch to direct traffic from the streaming server to the station via the right AP.

2.2 Mobile station

As Fig. 2 shows, the mobile station is equipped with one IEEE 802.11 WLAN card and optionally with one dedicated card for scanning for new APs. On top of the physical card, several virtual WLAN cards are created. Each virtual

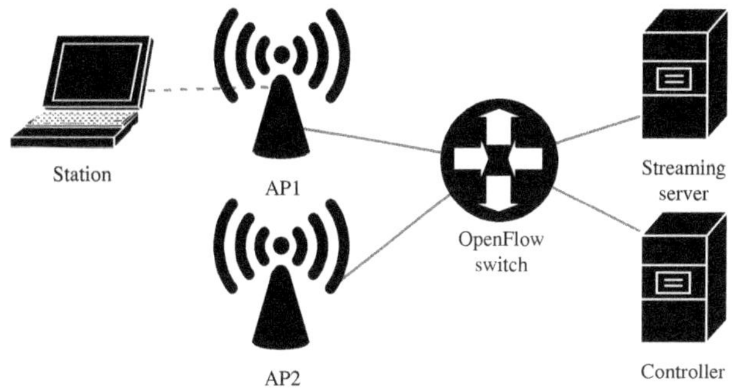

Fig. 1 Overall system architecture

WLAN card has a unique MAC address and can be associated to one AP. As one station has multiple virtual WLAN cards, it can be connected to several APs simultaneously. It is not required that all APs are operated on the same channel.

The virtual WLAN cards are connected to a software switch, the OpenFlow switching element. In addition, the software switch contains a super virtual WLAN device and a transmission buffer. The super virtual WLAN is the interface to the OS networking stack and has a device-wide IP and MAC address. The transmission buffer can be set into a blocked state, in which all data packets (not control packets) from higher layers are stored in the buffer, but not forwarded to the OpenFlow switching element. If the transmission buffer is in the unblocked state, then it simply passes packets down to the switching element. Blocking traffic at the transmission buffer insures that no packets are lost when the station is performing a handover. Packets that are lost during the wireless transmission (e.g., due to bit errors) are retransmitted using normal operations of the IEEE 802.11 MAC protocol. Retransmissions are always targeted to the same AP.

The switching element contains rules that specify (1) which virtual WLAN card outgoing packets should be forwarded to and (2) what source MAC address should be used for outgoing packets. Those rules are configured from the switching controller application using OpenFlow. As each virtual WLAN card is associated to one AP, choosing a virtual WLAN card for transmission results in a transmission to a specific AP. Rewriting of the source MAC address is required, since MAC frames generated by the operating system carry the MAC address of the super virtual WLAN device as source. To be transparent to the underlying network, the MAC address of the respective virtual WLAN card is used instead.

The station contains the handover manager which consists of a scheduler and a scanning module. The scheduler decides when to transmit or receive frames via which AP and when to scan for new APs. In other words, the scheduler decides when to perform a handover. In addition,

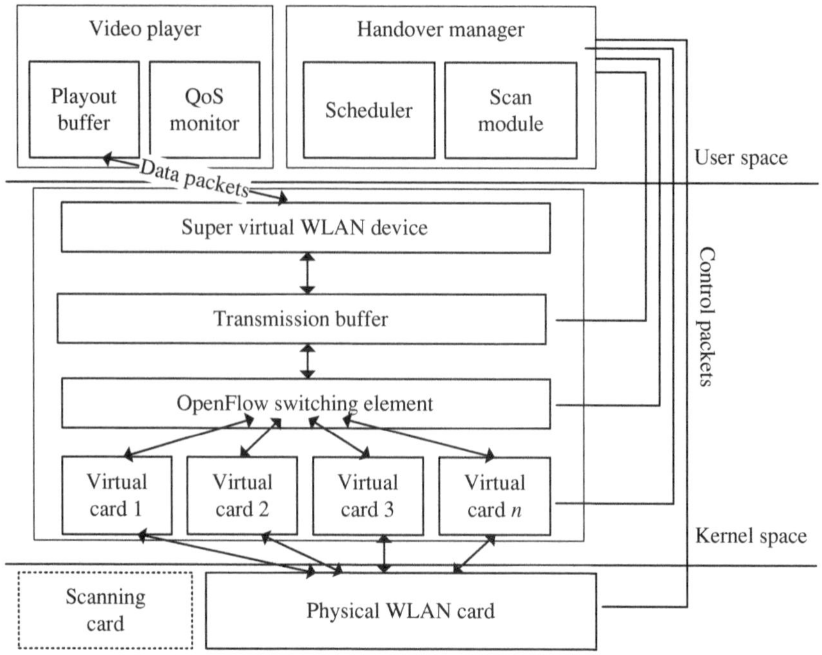

Fig. 2 Station architecture

the handover manager generates the signaling messages to reconfigure the routing in the backhaul network for downstream packets (see Sect. 2.5). The decision when to schedule a handover or a scan for new APs can use information from the Quality of Service (QoS) monitor and the playout buffer of the video player application. For example, a handover could be scheduled when the video buffer level is getting low or the QoS level is degrading.

2.3 Pre-authentication and pre-association

A station can associate to an AP outside its coverage area using out-of-band signaling. Using an existing wireless connection and the wired backhaul, it can exchange association and authentication messages with the control server, which relays them to the AP. This mechanism allows a station to associate to all APs of a limited geographic area, before the station arrives at this area. As we will discuss in Sect. 2.5, this allows handovers to be accelerated. Note, that our approach of pre-authentication and pre-association is fundamentally different from IEEE 802.11r, as we allow exchanging messages via an AP which is not in reach using the connection of a close-by AP. This can be performed before the handover. Therefore, the number of authentication message exchanged is not a performance limiting factor in our scheme and the authentication does not increase the duration of the actual handover.

2.4 Scanning for new APs

Before a station can initiate a handover to a new AP, it needs to detect it. To detect new APs, the scan module in the handover manager performs a standard IEEE 802.11 active scanning procedure, i.e., it broadcasts probe request frames. APs on the respective channel answer with probe response messages. The scan procedure can either be performed on the regular WLAN card, which is also used for data communication, or on a dedicated scanning card. If the regular WLAN card is used for scanning, then a trade-off between the scan frequency and the achievable throughput and QoS emerges. A scan for new APs needs to be scheduled often enough to detect new APs fast enough. However, too frequent scanning would result in a reduction of throughput and QoS, as during scanning no data can be transferred.

A dedicated scanning card would avoid this trade-off, but increase hardware costs. Future mobile devices, such as smart phones and tablets, are presumably equipped with several flexible radios. With such devices, a dedicated scanning card would be a viable option. In Sect. 4, we will evaluate the trade-off between scanning frequency and AP detection speed as well as the benefit of a dedicated scanning card.

2.5 Handover process

Based on the scanning results, the scheduler decides when to initiate a handover to which AP. In the current version, the scheduler triggers a handover, if the Received Signal Strength Indicator (RSSI) of any AP is larger than the RSSI of the current AP plus a hysteresis margin. Figure 3 shows a sequence diagram of the messages exchanged during a handover. The station puts its transmission buffer into the blocked state. Hereafter, the station transmits a "Handover Initiate" message to its currently used AP (AP1), which

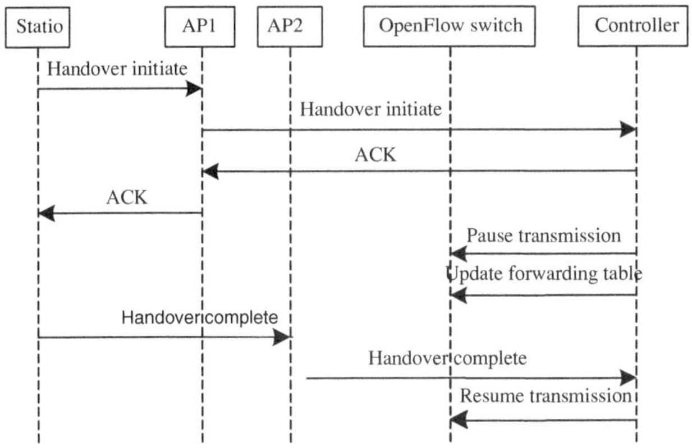

Fig. 3 Sequence of messages exchanged during a handover from AP1 to AP2

forwards the message to the controller. The controller (1) answers to the station with an ACK message, (2) sends a pause message to the switch and (3) using the OpenFlow protocol updates the forwarding table of the switch to send future downstream data packets via AP2 and to forward upstream traffic coming from the station via AP2 toward the streaming server.

As soon as the station has received the ACK message, it tunes its WLAN card to the channel of the new AP. In case no ACK is received within a timeout period, the station retries up to five times. If no retry is successful, for example, because AP1 is already out of reach, then the station nevertheless attempts to change the channel and sends a "Handover Initiate" message via AP2. The OpenFlow switch, just like the station, contains a per station transmission buffer, which is set into the blocking state upon the reception of the pause transmission command. Once the station has tuned its card to the new channel, it sends a "Handover Complete" message to the controller via AP2 and unblocks its transmission buffer. The controller then resumes the downlink traffic by unblocking its transmission buffer.

The handover process does not require exchanging any association or authentication messages, as pre-authentication and pre-association insures that the station is already associated to the new AP before the actual handover is started.

3 Implementation

We have implemented the proposed system in Linux. The station uses the virtual WLAN card function provided by ath5k and ath9k drivers. By default, Linux does not allow virtual interfaces to be used on different channels, which we changed accordingly. Furthermore, we add an interface to the WLAN driver, which allows initiating a fast channel switch. The pre-authentication/association scheme is implemented by extending the mac80211-subsystem and hostapd. For example, a new control command was added to hostapd, which allows external applications to create an association for a given station MAC address and capability list (PHY rates, 40 MHz channels etc.). Similarly, the mac80211-subsystem was modified to create an association without exchanging IEEE 802.11 authentication/association frames over the air.

The OpenFlow switching element is based on OpenVSwitch [12]. The "Super virtual WLAN card" is a bridge device connected to OpenVSwitch. All user-space applications (handover manager, video player, control server) were written in Python. The scanning module uses a raw-socket to inject probe messages into the WLAN card driver. The video player application is based on the Gstreamer framework and has a socket interface, which allows the handover manager to query its current buffer level and QoS statistics such as frame drop rate. We use VLC as streaming server and HTTP streaming.

4 Evaluation

We have evaluated the proposed system in a WLAN testbed deployed in an office building in downtown Berlin (see Fig. 4).

The testbed consists of 8 embedded systems with a Dual Core Atom D525 1.80 GHz CPU, which act as APs. Some of the APs are deployed in offices and meeting rooms (e.g., AP1), while others are located on corridors (e.g., AP5). The APs are positioned so that a mobile node in the corridor at any time is within range of at least one AP.

Each AP is equipped with two wireless cards, of which only one is used in our experiments. The cards are based on the Atheros AR5418 IEEE 802.11abgn chipset. To reduce interference from other WLANs deployed in the area, the network is operated in the less occupied 5 GHz band. The APs are connected via Ethernet and a Generic Routing Encapsulation (GRE) tunnel to a virtual machine host, in which the OpenFlow switch, the streaming server and the control server are hosted. The OpenFlow switch uses OpenVSwitch.

4.1 Micro-benchmarks

4.1.1 AP scanning duration

The mobile station needs to scan for new APs occasionally to detect new target APs for a handover. While scanning, the station needs to process a list of channels and send out probe request messages on each channel. After sending a probe request message, the station needs to remain on the channel for a while to wait for probe response messages. We remain on the channel for 5 ms, which turned out to be a good compromise between the probability of missing a probe response and the duration of the scan. In our

Fig. 4 Map of the testbed

experiments, switching from one channel to another channel requires on average 1.2 ms if we empty the transmission WLAN queue before a switch and recalibrate the card after the switch ("Normal channel switching"). If we just tune the synthesizer to the new frequency ("Fast channel switching"), then the channel switch takes on average 1 ms. Figure 5 plots the duration for scanning 2–8 channels. The scan duration increases linearly with the number of channels to scan. Each extra channel adds about 8 ms (1.2 ms for switching, 5 ms for waiting and listening and 1.8 ms for processing the frames) to the total scan duration. To reduce the scan duration in a real deployment, the station could only scan channels which are used by the network operator (often channels 1, 6 and 11). The total scan durations for the normal and fast channel switch mode are almost identical as the majority of the time is spent on listening for probes and other processing activities, and the actual channel switch does not play a major role.

4.1.2 Handover duration

As discussed in Sect. 2.5, a handover consists of a series of messages to be exchanged. We call the time between the transmission of the handover initiate message and the reception of the ACK message the ACK duration. The total

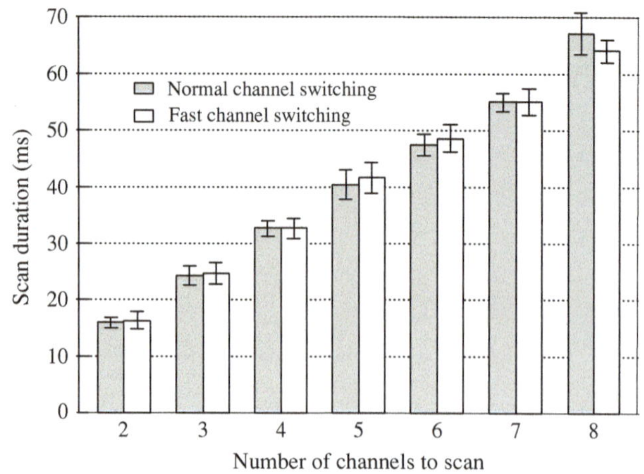

Fig. 5 Duration of a scan for new APs

handover duration spans the period from the transmission of the handover initiate to the transmission of the handover complete message. We measured the duration taken by our optimized handover for 100 handovers and plot the Empirical Cumulative Distribution Function in Fig. 6. The average ACK and handover durations are 9.4 and 34.7 ms, respectively. For a small fraction, the handover duration exceeds 50 ms. In those cases, if one of the control messages (e.g., the handover initiate) is lost, then it needs to be retransmitted after a timeout. In contrast, Ref. [13] reports handover times of 6–9 s for IEEE 802.11i secured WLANs.

4.1.3 Frequency of handovers

We walked along the corridors marked with the red line in Fig. 4 ten times and recorded the signal strength of the APs every 0.5 s. The decision to trigger a handover is based on the RSSI of the APs. If any AP has a higher signal strength than the current AP plus a hysteresis margin, then a handover is

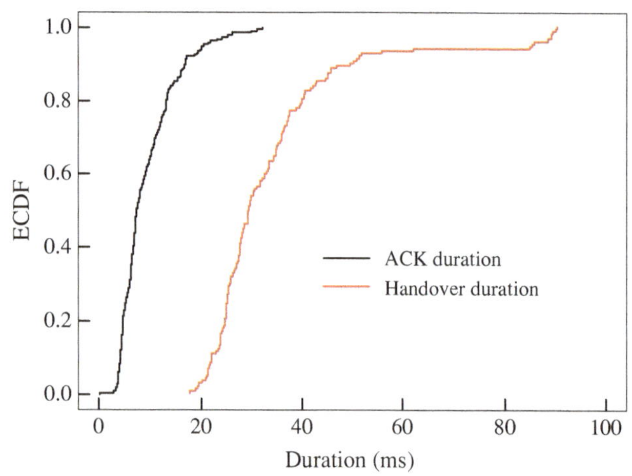

Fig. 6 Empirical Cumulative Distribution Function of handover and ACK duration

performed. As Fig. 7 shows, by increasing the hysteresis margin, the average time between two handovers increases. If the hysteresis margin is 0, then a handover is triggered on average every 3.6 s.

Increasing the hysteresis margin reduces the number of handovers, but comes at price: the station is using optimal APs less frequently. With a hysteresis margin of 20 dB, the station uses the best AP only 60 % of the time, while it stays connected to suboptimal APs 40 % of the time.

It depends on the application characteristics, whether a handover is beneficial or not. For example, low bandwidth application might not benefit from the higher throughput after handover to a better AP, but might suffer from the disruption the handover can cause. This shows that an intelligent handover scheduler is necessary and simple hysteresis schemes might not be sufficient.

4.2 Video streaming

Next, we compare how well our proposed system supports streaming video. Streaming video is an application that is gaining popularity rapidly and poses high requirements on the performance of the handovers. For our experiments, we streamed a pre-recorded MPEG-2 video using HTTP-streaming to the client laptop. A person is walking with the laptop five times from point A to B and back (see Fig. 4), while streaming the video. One walk from point A to B takes approximately 90 s. The video player has a playout buffer of 400 ms and logs the buffer fill level every 20 ms to a file. If within 400 ms, then no new video frames are received, and the video freezes until the buffer is filled up again. Such video freezes impair the perceived quality and thus should be avoided.

In the following, we compare (1) Linux with wpa_supplicant, (2) Linux with wpa_supplicant and optimized scanning, (3) our system with a dedicated scanning card, and (4) our system without a dedicated scanning card. The optimized scanning procedure in (2) only scans on the four channels which are actually used by our network, while the normal scanning procedure of (1) scans all channels. In our system, without a dedicated scanning card, we scan for a new AP every 2 s, while with a dedicated scanning card, we initiate a new scan every 500 ms. For all experiments, the hysteresis threshold was 0 dB.

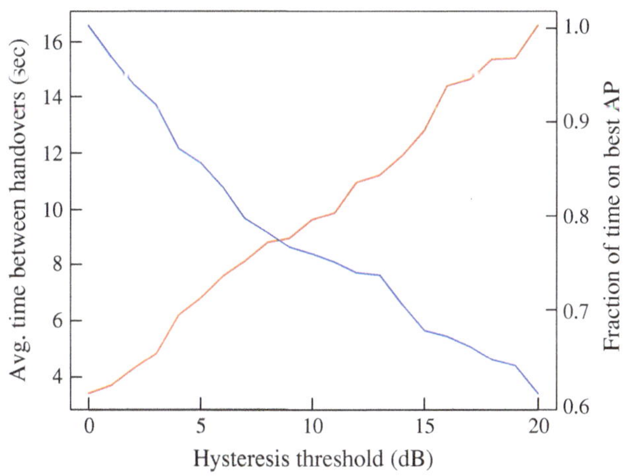

Fig. 7 Average time between two handovers and the loss of signal quality due to a higher hysteresis

4.2.1 Smoothness of video playout

Table 1 shows that with the standard Linux system 28.3 % and 12.48 % of the time the video is in freeze mode. With our proposed system, the video is frozen only 0.73 % and 3.25 %, respectively. Interestingly, with the dedicated scan card, the freeze time is slightly higher than without. This can be mainly attributed to two factors: first, with a dedicated scanning card, updates on the APs' quality are obtained more frequently, and hence handovers happen more often. Such behavior could be avoided with a more intelligent handover scheduler. Second, in our experiments with the dedicated scanning card, two relatively long freeze events occurred as result of failed handovers, which have large impact on the average video freeze time. This phenomenon can also be observed from Fig. 8, which shows the distribution of the freeze event durations. Figure 8 further shows that the standard Linux implementation leads to many long freeze events, which have severe impact on the user perception. In particular, if the channel scan is not optimized, freeze events last 10 s or more. With our system, however, the number of freeze events is small and the duration is usually short and thus not very disturbing to the viewer.

4.2.2 Necessity of a dedicated scanning card

The total number of freeze events is lower if a dedicated scanning card is in operation, while the total length of freeze events is higher. With our optimized scanning procedure, the benefits in terms of the total number of freeze events are marginal and also the subjective quality differences experienced during the measurements were low. On the one hand, a dedicated scanning card might decrease the system complexity, as the question when to schedule a new scan is easier to be answered then. On the other hand, a dedicated scanning card increases hardware cost and energy consumption. Considering all those factors, a dedicated scanning card is only useful, if it allows being integrated into the system with low effort, for example, through software defined radios.

Table 1 Video freeze events

Mode	Freeze time fraction (%)	Total number of freeze events	Maximum freeze time (s)	Mean freeze time (s)
Proposal/no dedicated scan card	0.73	15	2.24	0.46
Proposal/dedicated scan card	3.25	11	6.24	2.24
Linux/optimized scanning	12.48	55	8.43	2.01
Linux/standard scanning	28.30	79	32.06	2.75

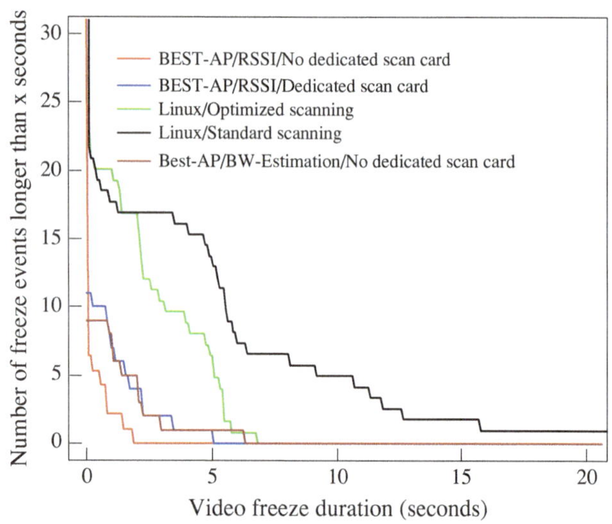

Fig. 8 Distribution of freeze event durations

4.2.3 Optimal playout buffer size

The playout buffer adds delay, which should be kept small, in particular for real-time applications such as video conferencing. Decreasing the playout buffer size would increase the number of freeze events, while increasing the buffer size would offer more protection against freeze events. Figure 8 shows, that to offer protection against all freeze events, the buffer size should be approximately 3–5 s large. Such large buffers are suitable for on-demand video, but the large delay would render real-time applications unusable. However, even with the 400 ms buffer used in our experiments, the number of noticeable freeze events is relatively low and the perceived quality is good.

4.2.4 Maximum possible station velocity

The proposed architecture is targeted to WLAN environments. As the WLAN PHY is not designed for fast moving users (e.g., in cars), high velocities are not possible. For example, Ref. [14] showed that even at moderate speeds of 50–60 km/h the packet error rate of an IEEE 802.11a channel exceeds 10 %. Besides the PHY, also communication distance of typical WLANs is a limiting factor. Considering a straight movement with 60 km/h and a communication range of 30 m, a handover would have to be performed at least every 3.5 s. Thus, the proposed system is only suitable for pedestrian-type speeds.

5 Conclusions

In this article, we have presented and evaluated an SDN-based architecture for optimizing handovers in WLANs. We have investigated as regards when and how to detect new APs using an optimized scanning procedure, when to schedule a handover, and how to use OpenFlow to control the wired distribution system to route the traffic to the right AP. The evaluation has shown that our architecture significantly improves the quality of streaming video over WLANs. As future study, we plan to investigate how to couple the handover decision more tightly with the video player application. This would allow for more intelligent scheduling of scanning or handovers.

References

1. IEEE Std 802.21-2008 (2009) IEEE standard for local and metropolitan area networks—part 21: media independent handover
2. IEEE Std 802.11r-2008 (Amendment to IEEE Std 802.11-2007 as amended by IEEE Std 802.11k-2008) (2008)
3. Open Networking Foundation (2012) Software-defined networking: the new norm for networks. http://www.opennetworking.org
4. The OpenFlow Consortium (2012) Openflow switch specification 1.1
5. Mishra A, Shin M, Arbaush WA (2004) Context caching using neighbor graphs for fast handoffs in a wireless network. In: Proceedings of INFOCOM 2004, Hongkong
6. Wu HT, Tan K, Zhang YG et al (2007) Proactive scan: fast handoff with smart triggers for 802.11 wireless WLAN. In: Proceedings of INFOCOM 2007, Alaska
7. Nah JW, Chun SM, Wang S et al (2009) Adaptive handover method with application-awareness for multimedia streaming service in wireless Lan. In: Proceedings of ICOIN 2009, Chiang Mai
8. Yap KK, Huang TY, Kobayashi M et al (2009) Lossless handover with n-casting between wifi-wimax on openroads. In: Proceedings of ACM MOBICOM 2009, Beijing
9. Vestin J, Dely P, Kassler A et al (2012) Cloudmac: towards software defined WLANs. Mobile Comput Commun Rev 16(4):42–45
10. Dely P, Kassler A, Bayer N et al (2012) Optimization of WLAN associations considering handover costs. EURASIP J Wirel Commun Netw 2012(1):1–14

11. Chow L, Collins B, Bambos N et al (2012) Playout-buffer aware hand-off control for wireless video streaming. In: Proceedings of GLOBECOM 2012, Anaheim
12. Pfaff B, Pettit J, Koponen T et al (2009) Extending networking into the virtualization layer. In: Proceedings of ACM HotNets, New York
13. Martinovic I, Zdarsky FA, Bachorek A et al (2007) Measurement and analysis of handover latencies in IEEE 802.11i secured networks. In: Proceedings of EW2007, Paris
14. Sibecas S, Corral C, Emami S et al (2002) On the suitability of 802.11a/RA for high-mobility DSRC. In: Proceedings of IEEE vehicular technology conference, Birmingham

12

Study on calculation of rock pressure for ultra-shallow tunnel in poor surrounding rock and its tunneling procedure

Xiaojun Zhou · Jinghe Wang · Bentao Lin

Abstract A computational method of rock pressure applied to an ultra-shallow tunnel is presented by key block theory, and its mathematical formula is proposed according to a mechanical tunnel model with super-shallow depth. Theoretical analysis shows that the tunnel is subject to asymmetric rock pressure due to oblique topography. The rock pressure applied to the tunnel crown and sidewall is closely related to the surrounding rock bulk density, tunnel size, depth and angle of oblique ground slope. The rock pressure applied to the tunnel crown is much greater than that to the sidewalls, and the load applied to the left sidewall is also greater than that to the right sidewall. Meanwhile, the safety of the lining for an ultra-shallow tunnel in strata with inclined surface is affected by rock pressure and tunnel support parameters. Steel pipe grouting from ground surface is used to consolidate the unfavorable surrounding rock before tunnel excavation, and the reinforcing scope is proposed according to the analysis of the asymmetric load induced by tunnel excavation in weak rock with inclined ground surface. The tunneling procedure of bench cut method with pipe roof protection is still discussed and carried out in this paper according to the special geological condition. The method and tunneling procedure have been successfully utilized to design and drive a real expressway tunnel. The practice in building the super-shallow tunnel has proved the feasibility of the calculation method and tunneling procedure presented in this paper.

Keywords Ultra-shallow tunnel · Asymmetric rock pressure · Surrounding rock · Rock treatment · Tunneling method

1 Introduction

The fast development of economy requires much more rapid and convenient transportation systems. Therefore, expressways and high speed railways have been built in China. Since the most land of western and northern China belongs to mountainous area, it is extremely difficult to build railways and expressways in these regions. If a transportation line such as highway or railway will be built in mountainous areas, tunnels are frequently adopted to overcome height barriers in the line. When the conditions of the area where transportation lines pass through is complicated in geology and topography, special geological problems might be encountered during the design and construction of transportation tunnels. If the ground has inclined topography, namely the ground surface appears in oblique form, then tunnel is easily subject to asymmetric rock pressure and its support structure must be excogitated according to asymmetric rock pressure.

In order to realize rational design and safe construction of mountainous tunnels, many studies and experiments have been carried out by scholars across the world. Goodman et al. [1] investigated the modeling techniques of tunnels in jointed rock, and presented an experimental and a numerical method to analyze the tunnel excavation in jointed rockmass. Shen and Barton [2] analyzed the disturbed zone around tunnels in jointed rock mass; they

X. Zhou (✉) · J. Wang
Key Laboratory of Transportation Tunnel Engineering of Ministry of Education, School of Civil Engineering, Southwest Jiaotong University, Chengdu 610031, China
e-mail: Zhouxjyu69@163.com; Zhouxjyu69@sina.com

B. Lin
The 2nd Institute of Civil and Architecture Engineering, China Railway Eryuan Engineering Corporation Ltd., Chengdu 610031, China

classified the disturbed zone into failure, open, and shear ones around a tunnel in jointed rock mass. Zhou et al. [3] made an insight into the rock pressure on tunnel with shallow depth in geologically inclined bedding strata, and set up formulas to calculate the asymmetric rock pressure applied to a tunnel with shallow depth in stratified rock. Later, Zhou and Yang [4] discussed the asymmetric rock pressure applied to the shallow tunnel in strata with inclined ground surface, and proposed a method to calculate the asymmetric rock pressure applied to the tunnel in strata with inclined ground surface by key block theory. Yang et al. [5] analyzed the calculation of rock pressure applied to three tunnels with large transection and small neighborhood in shallow surrounding rock; they suggested that the conventional method to calculate the rock pressure applied to shallow tunnel with small neighborhood should be improved. He et al. [6] studied the asymmetrical load effect on tunnels in geologically inclined bedding strata. They found that the rock pressure at the left wall of a tunnel is greater than that at the right wall. As the dip angle of bedding strata increases, the asymmetrical load gradually tends to be symmetric. In Zhou's [7] recent research, he studied the method to calculate the rock pressure for shallow asymmetric tunnel, and presented a method to determine a proper depth for shallow asymmetric tunnel in strata with inclined ground surface.

Most of studies available focused on the stability and deformation of tunnels and surrounding rock in jointed rockmass. Only a few concerned the calculation method of rock pressure applied to shallow tunnel [8, 9]. When designing tunnel structures, the rock pressure applied to tunnel lining is often obtained by numerical simulation or field monitoring. However, this is often hard for designers to use. To date, the calculation method of asymmetric rock pressure applied to super-shallow tunnel in strata with inclined ground surface has not been addressed [10]. As we know, the most difficult problem during design and construction of a tunnel in strata with inclined topography is to deal with the extremely thin overburden depth. This means that an ultra-shallow-tunnel will be excogitated and built in strata with inclined geomorphology; consequently, the influence of rock excavation will extend to ground surface and possibly cause casualties during tunneling [11, 12].

This paper aims to find out a simple way to calculate the asymmetric rock pressure for design of tunnel lining in super-shallow surrounding rock. The driving procedure for the bored tunnel is determined, taking into account the poor geological condition. Finally, a typical ultra-shallow tunnel for an expressway in Sichuan, China was taken as an example to make its structural design and safe construction procedure, which verifies the effectiveness of the proposed method.

2 Analysis of asymmetric rock pressure applied to super-shallow tunnel

In order to analyze the asymmetric load on the tunnel support with an ultra-shallow depth in strata with inclined ground surface, a mechanical model is set up according to the geological and topographical condition of a vehicular tunnel named Zagunao with two lanes in an expressway in Sichuan, the southwest province of China, as shown in Fig. 1.

According to the normal structural design of expressway tunnel with two lanes, the upper part of its reinforced

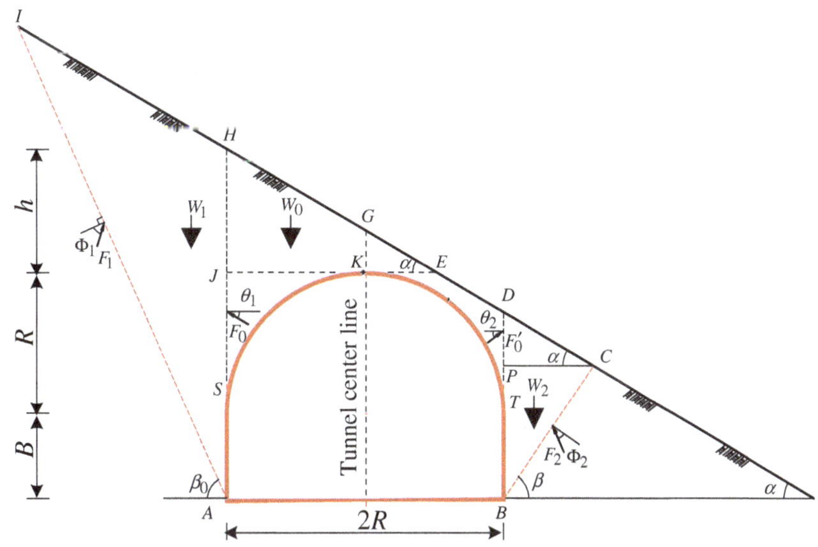

Fig. 1 Mechanical model of ultra-shallow tunnel

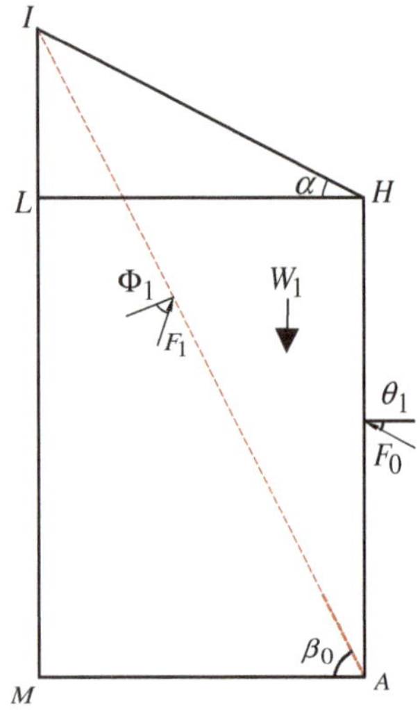

Fig. 2 Geometric relation of angle α and β_0

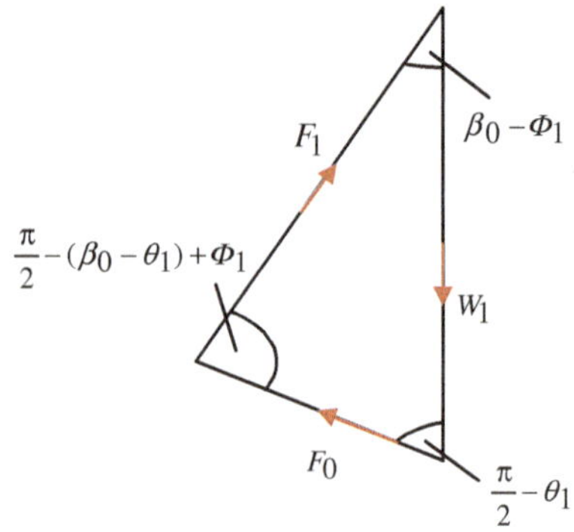

Fig. 3 Vector triangle of forces F_1, W_1, and F_0

concrete lining from its spring line appears in semi-circular arch with a radius of R and a width of $2R$. Since the tunnel has an ultra-shallow overburden depth, excavation of surrounding rock must cause ground subsidence and its influence will extend to the surface. If it is supposed that there exist two breaking planes in the surrounding rock on each side of the tunnel, namely plane BC and plane AI, then the surrounding rock in ΔBCD and ΔAHI may tend to descend downward due to its rock gravity; synchronously, the surrounding rock above the tunnel crown namely in ΔJEH may also be caused to descend along breaking planes BD and AH by gravity as the tunnel will be driven. However, friction resistances must exist on each fracture plane; frictions will impede the slide of the surrounding rock in each triangular block. Let the surrounding rock in blocks exist in a critical equilibrium state; then their mechanical relations can be derived from Fig. 1.

In ΔAHI, the gravity W_1 of the surrounding rock circumscribed by block AHI is derived from the geometric relation as shown in Fig. 2, i.e.,

$$W_1 = \frac{1}{2}\gamma_1 (h+R+B)^2, \tag{1}$$

where α denotes the slope angle and β_0 the breaking angle between planes AI and MA, in degrees; γ_1 stands for the bulk density of the surrounding rock in block AHI, in kN/m^3; and other symbols are delineated in Figs. 1 and 2.

In addition, three forces F_1, W_1, and F_0 constitute a vector triangle as shown in Fig. 3, where F_1 denotes the friction resistance applied to plane AI, F_0 the friction resistance to plane AH, and W_1 the gravity of the surrounding rock enclosed by block AHI. We can deduce from Fig. 3 by sine theorem that

$$\frac{F_0}{\sin(\beta_0 - \phi_1)} = \frac{W_1}{\sin\left[\frac{\pi}{2} - (\beta_0 - \theta_1) + \phi_1\right]}. \tag{2}$$

Substituting Eq. (1) into Eq. (2), we obtain

$$F_0 = \frac{\gamma_1 (h+R+B)^2}{2\cos\theta_1} \cdot \frac{1}{\tan\beta_0 - \tan\alpha} \times \frac{\tan\beta_0 - \tan\phi_1}{1 + \tan\beta_0 \tan\phi_1 + \tan\theta_1(\tan\beta_0 - \tan\phi_1)}. \tag{3}$$

Equation (3) shows that the friction resistance applied to breaking plane AH varies with the breaking angle β_0, so F_0 may get its maximum or minimum value as β_0 varies within 90°. Thus, if let

$$\lambda = \frac{1}{\tan\beta_0 - \tan\alpha} \times \frac{\tan\beta_0 - \tan\phi_1}{1 + \tan\beta_0 \tan\phi_1 + \tan\theta_1(\tan\beta_0 - \tan\phi_1)}, \tag{4}$$

then Eq. (3) is converted into

$$F_0 = \frac{\gamma_1 (h+R+B)^2}{2} \cdot \frac{\lambda}{\cos\theta_1}. \tag{5}$$

In order to get the extreme value of F_0, let

$$\frac{\partial F_0}{\partial \beta_0} = 0. \tag{6}$$

Then the following expression is derived:

$$\tan^2\beta_0 - 2\tan\phi_1 \tan\beta_0 - \frac{\tan\phi_1 - \tan\alpha - (\tan\theta_1 + \tan\alpha)\tan^2\phi_1}{\tan\phi_1 + \tan\theta_1} = 0. \tag{7}$$

Equation (7) represents a quadratic equation with one variable β_0, and its maximum root is derived as follows:

$$\tan\beta_0 = \tan\phi_1 + \sqrt{\frac{(\tan\phi_1 - \tan\alpha)(1 + \tan^2\phi_1)}{\tan\phi_1 + \tan\theta_1}}. \tag{8}$$

It is known from Eq. (8) that the maximum β_0 for breaking plane AI is controlled by angle ϕ_1, α, and θ_1.

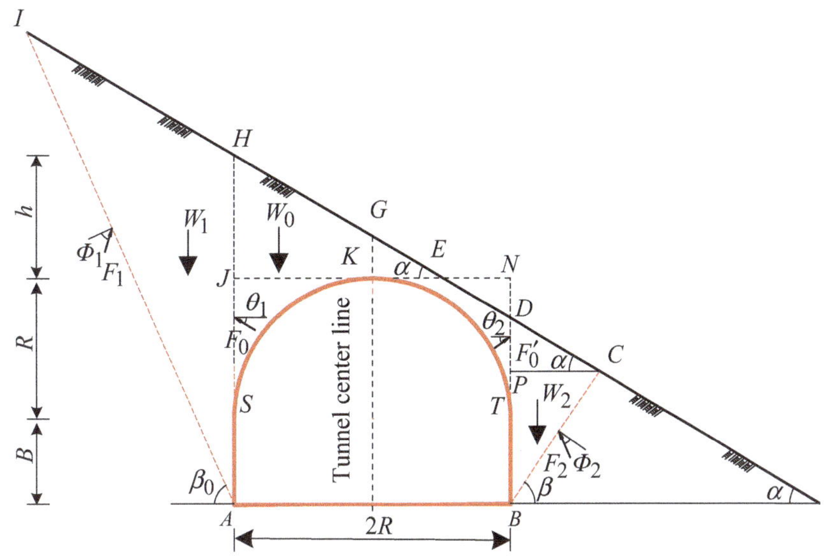

Fig. 4 Geometric relation of ΔBDC and ΔDPC

As for the triangular block BCD shown in Fig. 1, the surrounding rock may also slide along the fracture planes BC and BD due to its gravity W_2. In order to anatomize the friction resistance applied to breaking plane BD, the gravity W_2 must be obtained. Therefore, if extend straight line KE to point N, and line TD to N, then point N is the intersection between line EN and DN, as shown in Fig. 4.

In ΔHJE, since $HJ = h$ and $JE = h{\cdot}c \tan\alpha$, we have

$$EN = 2R - h\cdot c\tan\alpha. \tag{9}$$

Furthermore, in ΔEDN, we have $ND = 2R\tan\alpha - h$, and $BD = B{+}R - (2R\tan\alpha - h)$. Thus, there is

$$BD = B + h + R(1 - 2\tan\alpha). \tag{10}$$

In addition, both ΔDPC and ΔBCP are in right triangles; therefore, following relations may exist, i.e.,

$$\tan\alpha = \frac{DP}{PC}, \tag{11}$$

$$\tan\beta = \frac{PB}{PC}, \tag{12}$$

$$PB = BD - DP. \tag{13}$$

Substituting Eqs. (11) and (12) into Eq. (13), we get

$$PC = \frac{B + h + R(1 - 2\tan\alpha)}{\tan\alpha + \tan\beta}. \tag{14}$$

Therefore, the gravity W_2 of the triangular block BCD is derived as

$$W_2 = \frac{\gamma_2}{2}\frac{[B + h + R(1 - 2\tan\alpha)]^2}{\tan\alpha + \tan\beta}, \tag{15}$$

where γ_2 denotes the bulk density of the surrounding rock within the block BCD.

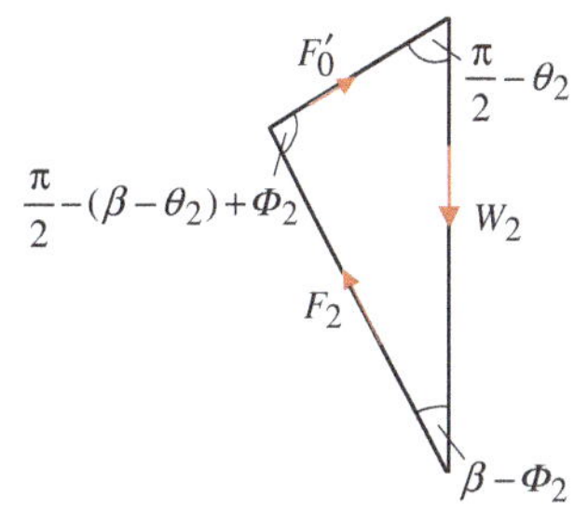

Fig. 5 Vector triangle of forces F_2, W_2, and F_0'

Similarly, the three forces F_2, W_2, and F_0' also constitute a vector triangle as shown in Fig. 5 and their relation is as follows:

$$\frac{F_0'}{\sin(\beta - \phi_2)} = \frac{W_2}{\sin\left[\frac{\pi}{2} - (\beta - \theta_2) + \phi_2\right]}. \tag{16}$$

Substituting Eq. (15) into Eq. (16), we get

$$F_0' = \frac{\gamma_2[B + h + R(1 - 2\tan\alpha)]^2}{2\cos\theta_2}\cdot\frac{1}{\tan\beta + \tan\alpha} \times \frac{\tan\beta - \tan\phi_2}{1 + \tan\beta\tan\phi_2 + \tan\theta_2(\tan\beta - \tan\phi_2)}. \tag{17}$$

It is known from Eq. (17) that the friction resistance F_0' applied to the breaking plane BD varies with the breaking angle β, so it may get an extreme value as β varies within 90°. Let

$$\lambda' = \frac{1}{\tan\beta + \tan\alpha} \times \frac{\tan\beta - \tan\phi_2}{1 + \tan\beta\tan\phi_2 + \tan\theta_2(\tan\beta - \tan\phi_2)}. \tag{18}$$

Then Eq. (17) is converted into

$$F_0' = \frac{\gamma_2[B+h+R(1-2\tan\alpha)]^2}{2} \cdot \frac{\lambda'}{\cos\theta_2}. \tag{19}$$

To derive the extreme value of F_0', let

$$\frac{\partial F_0'}{\partial \beta_0} = 0. \tag{20}$$

Then we obtain

$$\tan^2\beta - 2\tan\phi_2 \tan\beta - \frac{\tan\phi_2 + \tan\alpha - (\tan\theta_2 + \tan\alpha)\tan^2\phi_2}{\tan\phi_2 + \tan\theta_2} = 0. \tag{21}$$

Equation (21) is also a quadratic equation with one variable β, so its maximum root is derived as follows:

$$\tan\beta = \tan\phi_2 + \sqrt{\frac{(\tan\alpha + \tan\phi_2)(1+\tan^2\phi_2)}{\tan\phi_2 + \tan\theta_2}}. \tag{22}$$

Equation (22) shows that the failure angle β is closely related to angles ϕ_2, θ_2, and α. As for the vertical load applied on the crown of the tunnel, W_0 contributes the great proportion. If the surrounding rock above its crown tends to slide along the assumed planes *AH* and *BD*, it will be resisted by the vertical components of friction resistance F_0 and F_0'. Their vertical components are derived as follows:

$$F_{0v} = \frac{\gamma_1}{2}(B+h+R)^2 \lambda \tan\theta_1, \tag{23}$$

$$F_{0v}' = \frac{\gamma_2}{2}[B+h+R(1-2\tan\alpha)]^2 \lambda' \tan\theta_2. \tag{24}$$

Then the downward load applied to the crown of the tunnel is

$$Q = W_0 - (F_{0v} + F_{0v}'), \tag{25}$$

where W_0 stands for the gravity of the surrounding rock above the tunnel crown, in kN; its mathematical expression can be derived from the geometrical relation as shown in Fig. 1. The gravity of the surrounding rock above the tunnel crown is obtained by deducting the area of tunnel transection from the area of trapezoid *ABDH*, namely the area above the tunnel crown is as follows

$$A = \frac{(AH+BD)}{2} \cdot AB - AB \cdot AS - \frac{\pi R^2}{2}. \tag{26}$$

Then gravity W_0 is obtained as follows:

$$W_0 = 2\gamma_0 Rh + 2\gamma_0 R^2(1-\tan\alpha) - \frac{\gamma_0}{2} \cdot \pi R^2, \tag{27}$$

where γ_0 stands for the bulk density of the surrounding rock above the tunnel crown, in kN/m^3.

Substitution of Eqs. (23), (24), and (27) into (25) yields

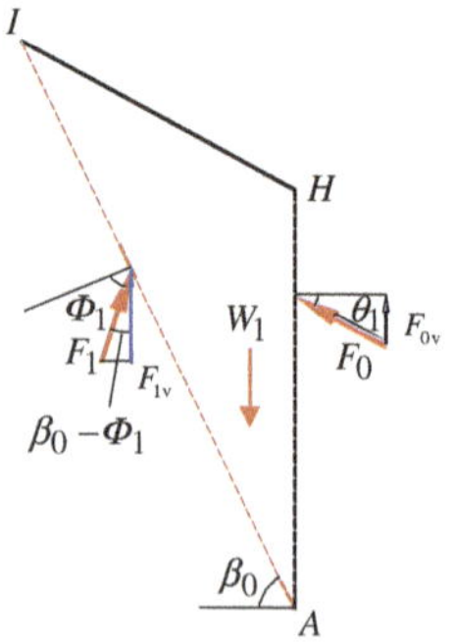

Fig. 6 Vertical components of F_1 and F_0

$$Q = 2\gamma_0 Rh + 2\gamma_0 R^2(1-\tan\alpha) - \frac{\gamma_0}{2}\cdot \pi R^2 - \frac{\gamma_1}{2}(B+h+R)^2\lambda\tan\theta_1 - \frac{\gamma_2}{2}[B+h+R(1-2\tan\alpha)]^2\lambda'\tan\theta_2. \tag{28}$$

Then the vertical downward pressure applied to the crown is

$$q = \frac{Q}{2R}. \tag{29}$$

Substituting Eq. (28) into Eq. (29), we obtain the rock pressure applied on the tunnel crown as follows:

$$q = \gamma_0 h + \gamma_0 R(1-\tan\alpha) - \frac{\gamma_0}{4}\pi R - \frac{\gamma_1}{4R}(B+h+R)^2\lambda\tan\theta_1 - \frac{\gamma_2}{4R}[B+h+R(1-2\tan\alpha)]^2\lambda'\tan\theta_2. \tag{30}$$

Equation (30) shows that the vertical pressure applied to the tunnel crown caused by gravity W_0 is related to parameters such as γ_0, h, α, β, R, θ_1, and θ_2; and its direction is always downward.

According to Fig. 1, the surrounding rock enclosed by ΔAHI and ΔBCD exert a lateral rock thrust on two side walls of the tunnel. As for ΔAHI, the lateral load applied to the side wall is derived as follows.

In ΔAHI as shown in Fig. 6, according to the division of forces, the vertical component of friction resistance can be obtained. According to Fig. 3, there exists sine theorem, namely

$$\frac{F_1}{\sin\left(\frac{\pi}{2}-\theta_1\right)} = \frac{F_0}{\sin(\beta_0 - \phi_1)}. \tag{31}$$

Then friction resistance F_1 is

$$F_1 = \frac{\gamma_1}{2}(B+h+R)^2 \frac{\lambda}{\sin(\beta_0-\phi_1)}. \tag{32}$$

In addition, the sine theorem also holds in Fig. 5, and there is a relation as follows:

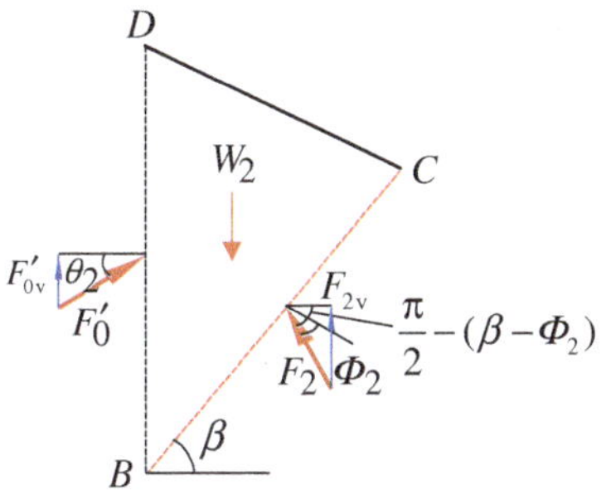

Fig. 7 Vertical components of F_0' and F_2

$$\frac{F_2}{\sin(\frac{\pi}{2}-\theta_2)}=\frac{F_0'}{\sin(\beta-\phi_2)}. \tag{33}$$

And the friction resistance F_2 is

$$F_2=\frac{\gamma_2}{2}[B+h+R(1-2\tan\alpha)]^2\frac{\lambda'}{\sin(\beta-\phi_2)}. \tag{34}$$

The vertical components of friction resistance F_1 and F_2 tends to prevent the surrounding rock both in block *AHI* and *BCD* from sliding along the assumed fracture plane. The vertical components of F_1 and F_2 are as follows:

$$F_1\text{v}=F_1\cos(\beta_0-\phi_1), \tag{35}$$

$$F_2\text{v}=F_2\sin\left[\frac{\pi}{2}-(\beta-\phi_2)\right]. \tag{36}$$

Substitution of Eqs. (32) and (34) into Eqs. (35) and (36) yields the following expressions:

$$F_{1\text{v}}=\frac{\gamma_1}{2}(B+h+R)^2\frac{\lambda}{\tan(\beta_0-\phi_1)}, \tag{37}$$

$$F_{2\text{v}}=\frac{\gamma_2}{2}[B+h+R(1-2\tan\alpha)]^2\frac{\lambda'}{\tan(\beta-\phi_2)}. \tag{38}$$

As for block *AHI*, the downward load Q_1 is derived as

$$Q_1=W_1-(F_{1\text{v}}+F_{0\text{v}}). \tag{39}$$

Thus, the downward load Q_1 caused by gravity W_1 is derived as

$$Q_1=\frac{\gamma_1}{2}(B+h+R)^2\times\left[\frac{1}{\tan\beta_0-\tan\alpha}-\frac{\lambda}{\tan(\beta_0-\phi_1)}-\lambda\tan\theta_1\right]. \tag{40}$$

If Q_1 is totally applied to the left side wall, then the lateral pressure (e_l) applied to the left side wall *AS* is derived as

$$e_l=\frac{\gamma_1}{2B}(B+h+R)^2\eta, \tag{41}$$

where η is the lateral pressure coefficient, and its specific expression is as follows:

$$\eta=\frac{1}{\tan\beta_0-\tan\alpha}-\frac{\lambda}{\tan(\beta_0-\phi_1)}-\lambda\tan\theta_1. \tag{42}$$

For the lateral load applied to the right side wall of the tunnel, the forces and their relation are shown in Fig. 7.

According to the relation between gravity and friction resistances in Fig. 7, the downward load Q_2 is derived as

$$Q_2=W_2-(F_{2\text{v}}+F_{0\text{v}}'). \tag{43}$$

If Eqs. (24) and (36) are substituted into Eq. (43), then the downward load Q_2 caused by gravity W_2 is also obtained as

$$Q_2=\frac{\gamma_2}{2}[B+h+R(1-2\tan\alpha)]^2\times\left[\frac{1}{\tan\beta+\tan\alpha}-\frac{\lambda'}{\tan(\beta-\phi_2)}-\lambda'\tan\theta_2\right]. \tag{44}$$

Then, the lateral pressure (e_r) applied to the right side wall *BD* is

$$e_r=\frac{\gamma_2}{2B}[B+h+R(1-2\tan\alpha)]^2\eta', \tag{45}$$

where η' denotes lateral pressure coefficient, and its mathematical expression is as follows:

$$\eta'=\frac{1}{\tan\beta+\tan\alpha}-\frac{\lambda'}{\tan(\beta-\phi_2)}-\lambda'\tan\theta_2. \tag{46}$$

To sum up, the method to calculate the rock pressure applied to an ultra-shallow tunnel in strata with inclined ground has been derived from above analysis. It is apparent that the lateral pressure applied to two side walls of the tunnel is not identical. Therefore, a tunnel with ultra-shallow depth undergoes asymmetric rock pressure.

3 Calculation of asymmetric rock pressure for an expressway tunnel

In order to analyze the rock pressure applied to an typical ultra-shallow tunnel, Zagunao tunnel in an expressway of Sichuan province in China is analyzed; its typical geological transection is shown in Fig. 8. There are two tunnels in the expressway; both are separated from each other. The net distance from outside wall of the left tunnel to that of the right tunnel is only 20 m. The main feature of this expressway tunnel is that its entry portal section lies in strata with a shallow and ultra-shallow overburden due to the requirement of line development.

Each tunnel has two vehicular lanes. The total length of the entry portal section with shallow depth is 25 m long in its longitudinal direction. In this shallow section, the left belongs to deep tunnel with an overburden from 13 to 36 m, but the right one has a very thin overburden that

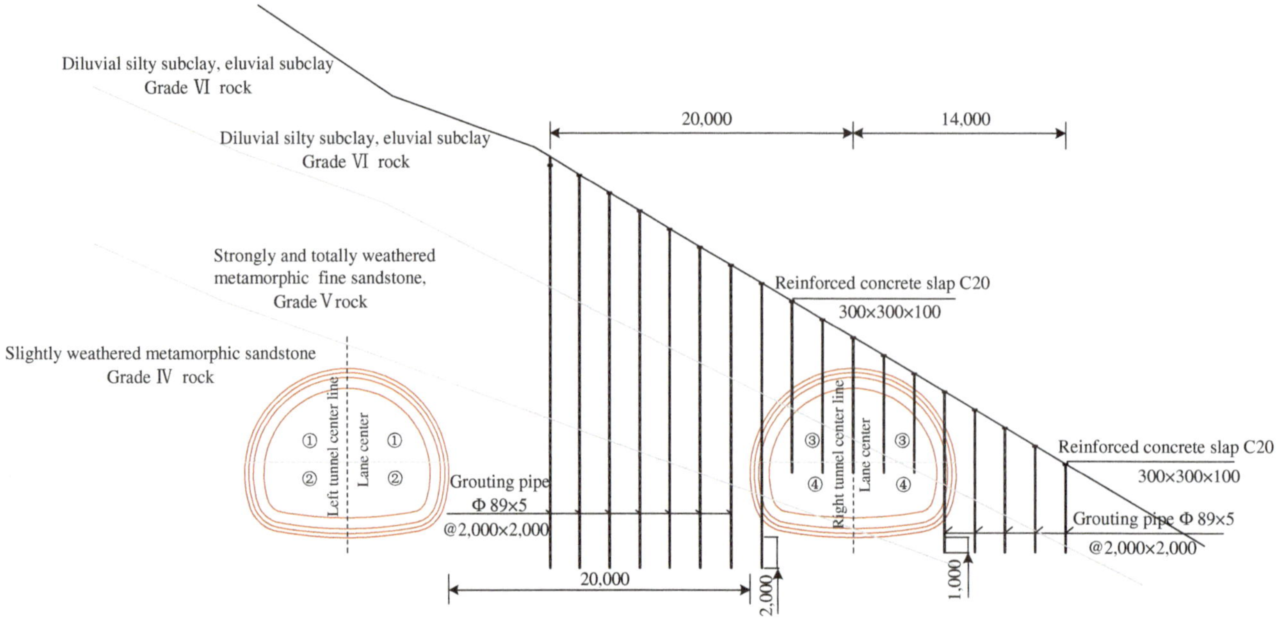

Fig. 8 Geological transection of an ultra-shallow tunnel in entry portal section (unit: mm)

Table 1 Basic parameters of the surrounding rock and tunnel

γ_0 (kN/m^3)	γ_1 (kN/m^3)	γ_2 (kN/m^3)	θ_1 (°)	θ_2 (°)	ϕ_1(°)	ϕ_1(°)	α(°)	B (m)	R (m)	h (m)
15	16	14	20	18	40	35	28	4.11	6.77	6.0

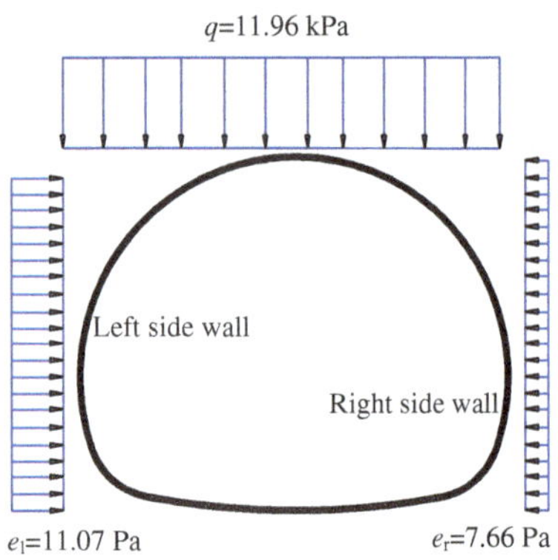

Fig. 9 Rock pressure applied to tunnel support

varies from 2 to 18 m. The surrounding rock mainly falls into diluvial silty subclay, eluvial subclay, and strongly weathered metamorphic fine sandstone. According to the Rock Quality Designation (RQD) and Basic Quality (BQ) system for rock mass classification [13, 14], and the *Code for Design of Highway Tunnel*, the surrounding rock belongs to grades IV, V, and VI [15, 16].

In order to make a concise analysis and simplify the computation process of the surrounding rock pressure, let all surrounding rock belong to grade VI. The physical and mechanical parameters of the surrounding rock and the basic size of the tunnel are shown in Table 1. Although this hypothesis seems to be simple, the result is still representative.

Substituting all the parameters in Table 1 into the above-stated equations, we derived the rock pressure applied to the ultra-shallow tunnel, as shown in Fig. 9.

It is clear from Fig. 9 that the right tunnel is subject to asymmetric pressure; and its crown is subject to the maximum vertical pressure with downward direction. The rock pressure applied to its left sidewall is greater than that to the right one in magnitude, but smaller than that on its crown. In addition, if these parameters are substituted into Eqs. (8) and (22), then the breaking plane angle are obtained as $\beta_0 = 56.28°$ and $\beta = 63.86°$. This indicates that excavation of the surrounding rock inside the tunnel contour may possibly cause overburden rock to slide or subside along the assumed breaking plane as shown in Fig. 10.

In order to illustrate the assumed plane of the surrounding rock, the arc invert of the tunnel is simplified to a straight floor but this does not affect the analysis of rock pressure.

According to the calculated results of the rock pressure applied to tunnel and the assumed fracture plane, the structural design of the ultra-shallow tunnel and its ground

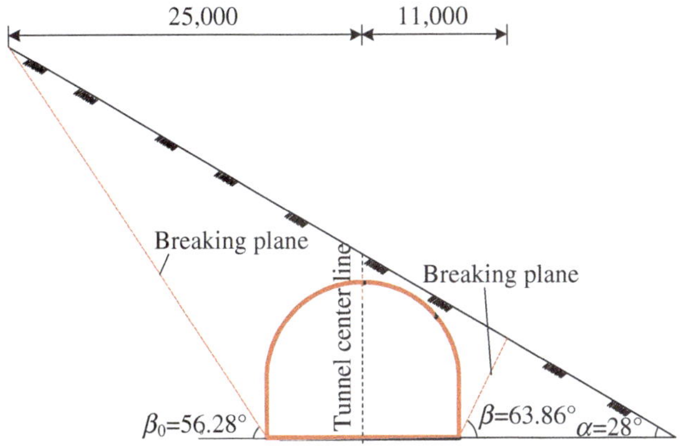

Fig. 10 Slide plane of the surrounding rock (unit: mm)

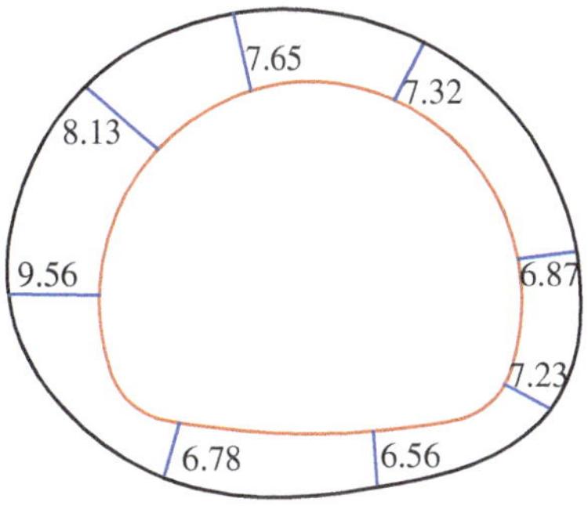

Fig. 12 Safety factor of primary support

treatment for the right tunnel have been carried out. The tunnel support and its designed parameters are shown in Fig. 11 in detail.

The tunnel is supported with composite lining which consists of primary support and inner lining. According to the load and structure model in structure mechanics [15, 16], the internal forces in the primary and secondary support are calculated. By assuming the rock pressure calculated by the method presented in this paper is totally applied to the primary support, the safety factors in the typical cross section of the primary support is obtained by allowable stress design and ultimate strength design method [15, 16]. The obtained results for primary support are shown in Fig. 12. Since the minimum safety factor reaches 6.56, the designed primary support is safe when it is subject to asymmetric rock pressure. In addition, if all the pressure is applied to the inner lining, then safety factors of the secondary lining are also calculated by using the same model as shown in Fig. 13. It is apparent that the secondary lining is also secure when it is subject to asymmetric rock pressure.

According to the rock pressure calculation of the ultra-shallow tunnel, the surrounding rock above the tunnel vault tends to slide along the two assumed fracture planes, and it will result in ground subsidence; therefore, ground treatment should be conducted in light of the above calculated

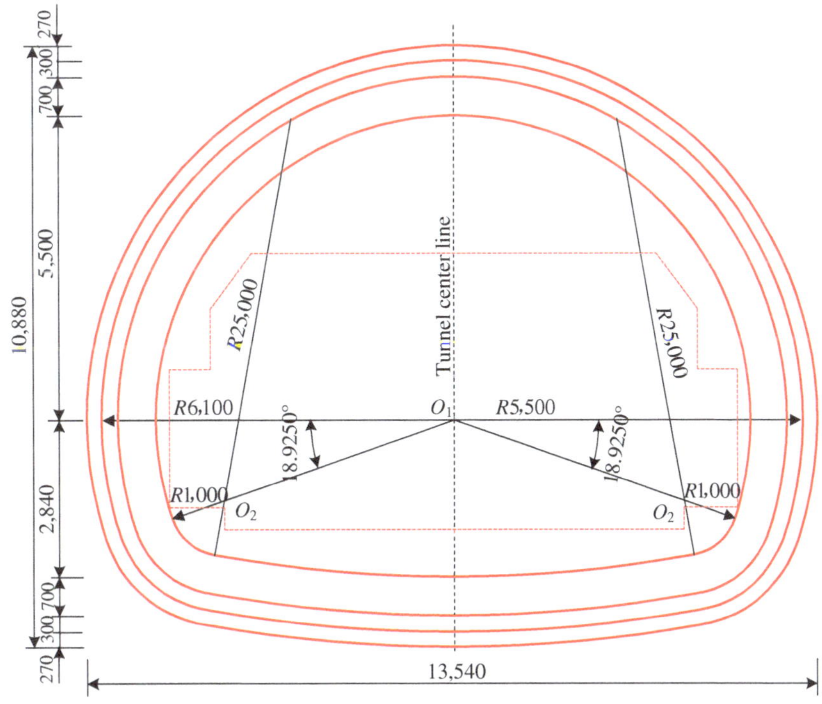

Fig. 11 Tunnel support and its transection (unit: mm). Grouting pipe ϕ42 mm × 5 mm, L = 4,500 mm @400 mm; Hollow grouting bolt ϕ25, L = 6,000 mm @300; I-shaped steel arch, I20, longitudinal @600 mm. Shotcrete, C25, δ = 270, ϕ8 steel mesh @150 × 150. Allowable camber 300 mm. Geotextile 350 g/m^2, δ = 1.2 mm PVC-P. Reinforced concrete, C25, δ = 700 mm

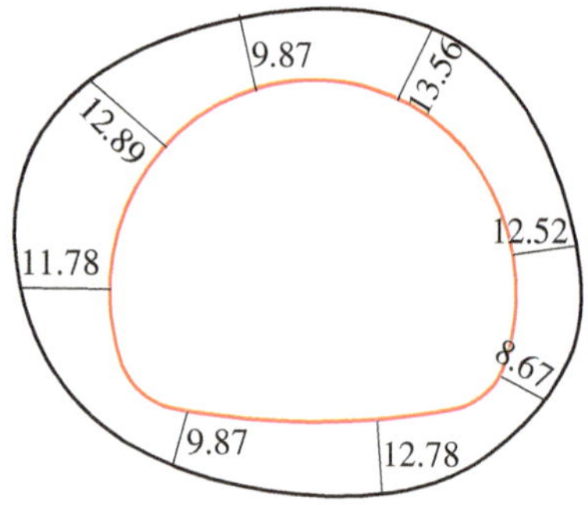

Fig. 13 Safety factors of inner lining

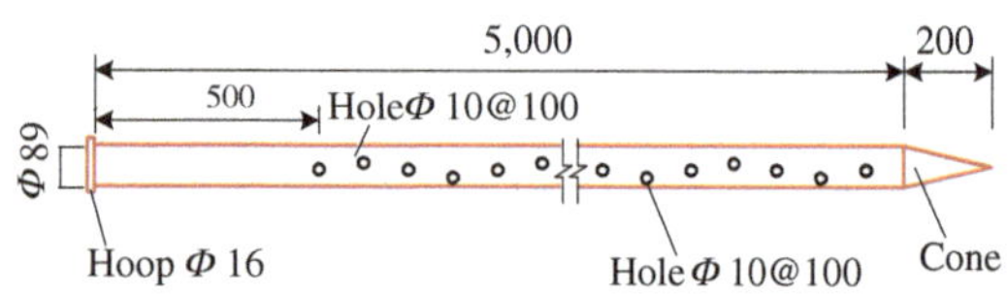

Fig. 14 Detail of grouting pipe (unit: mm)

results so as to lessen the rock settlement and construction risk during tunneling [17, 18]. According to the calculated maximum fracture angle of plane shown in Fig. 10, the ground treatment range must be greater than the subsidence scope; in addition, the supposed slide planes must also be consolidated. In consideration of the geology and geomorphology of the tunnel, cement grouting method is adopted to reinforce the surrounding rock with steel pipes from the ground surface. Since sandstone possesses large porosity, grouting can effectively enhance its shear strength; the detail of the grouting pipe is shown in Fig. 14. Since the portal section of the right tunnel is 25 m long in its longitudinal direction, cement grouting was carried out only in this section.

In order to analyze the effect of cement grouting in strata, the friction angle of the slide plane will be raised after cement grouting in the surrounding rock with steel pipes from ground surface. This means that their value will be enhanced. If the computational friction angle of the surrounding rock rises to $\phi_1 = 55°$ and $\phi_2 = 50°$, then $\theta_1 = 38.5°$ and $\theta_2 = 35°$. If these parameters are substituted into Eqs. (8) and (22), then the maximum fracture angles are obtained as $\beta_0{}' = 68.47°$ and $\beta' = 69.51°$. By comparing this result with the angle of slide plane before grouting, we obtain their relations as follows

$$\beta_0' = 1.22\beta_0, \tag{47}$$

$$\beta' = 1.09\beta. \tag{48}$$

This apparently shows that the shear strength of the surrounding rock gets enhanced after the ground has been reinforced with cement grouting. A comparison of ground treatment effect is shown in Fig. 15. From Fig. 15 we can infer that the vertical and lateral rock pressure applied to the tunnel support decrease largely as well after the surrounding rock is treated with steel pipe grouting from ground surface.

During the operation of in situ grouting, the grouting pressure must be kept within 1.0–1.5 MPa. In view of the in situ surface condition and ground geology at the tunnel site, the actual ground treatment and the details of designed parameters and grouting scope are shown in Fig. 8. Since the right tunnel more approaches to the ultra-shallow depth in the strata in comparison with the right one, ground

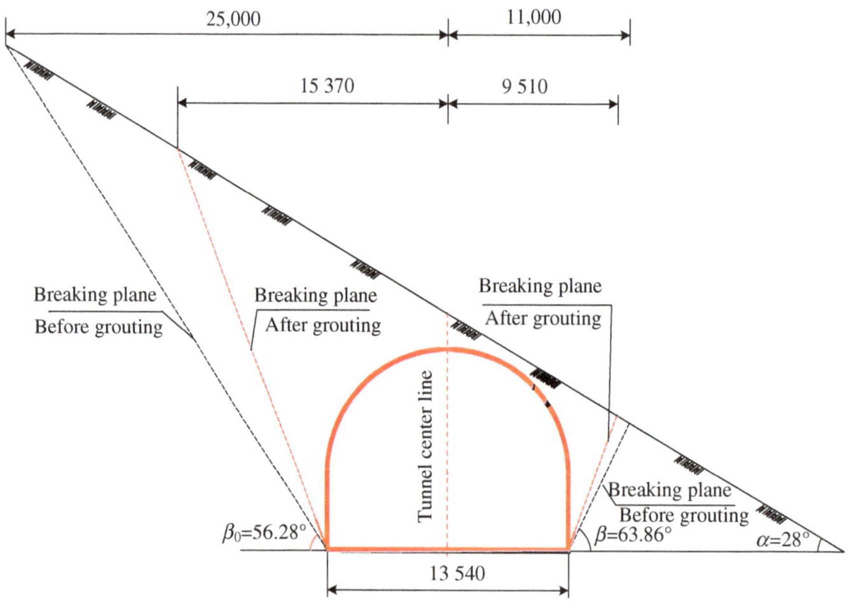

Fig. 15 Effect of ground treatment by cement grouting (unit: mm)

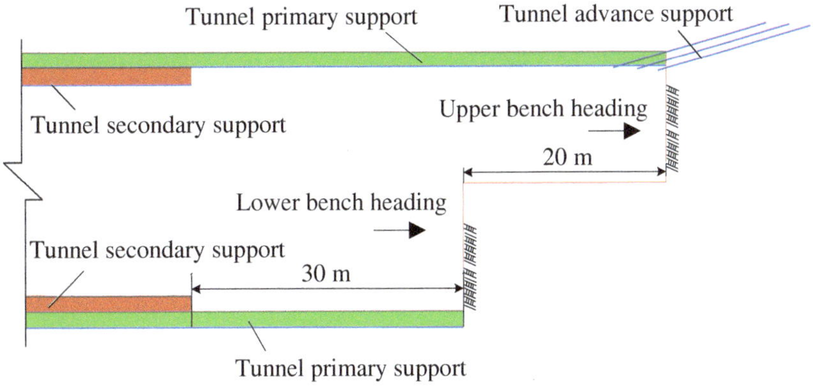

Fig. 16 Conventional tunneling process

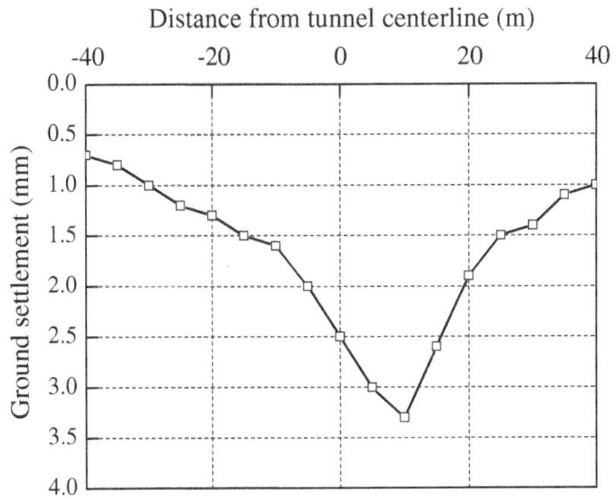

Fig. 17 Ground settlement due to tunnel excavation

treatment was only performed in its surrounding rock. As the left tunnel is deeply buried and the net distance between the two tunnels reaches 20 m, there is no need to treat the surrounding rock around the right tunnel.

4 Construction procedure of the ultra-shallow tunnel

The surrounding rock around the two expressway tunnels falls into three grades in terms of its uniaxial compressive strength and lithological integrity, namely, grades IV, V, and VI according to its RQD [12, 13]. The main lithology comprises sub-clay and metamorphic sandstone. The upper strata belongs to diluvial silty subclay and eluvial subclay, the middle strata belongs to totally and strongly weathered metamorphic fine sandstone, and the lower strata exists in slightly weathered metamorphic sandstone. Furthermore, the maximum area of tunnel transection amounts to 123 m^2, and the maximum tunnel span reaches 13.54 m. Since the left tunnel has higher overburden depth than the right one, in order to reduce tunneling risk such as rock collapse and cave-in, the two tunnels are driven with the conventional drill and blast method. The excavation sequence of the surrounding rock inside tunnel contour is divided into four main steps. First, the left tunnel is excavated, and then the right one. The detailed construction process and its support system are both shown in Fig. 8. The whole working face of each tunnel is divided into two parts, first the upper part is headed, and then the lower part: The circled digits in Fig. 8 represent the tunneling sequence of the surrounding rock within the tunnel contour by conventional tunneling method [18–20]. The longitudinal profile of the tunnel construction process is shown in Fig. 16.

During the construction period, in situ ground settlement was measured in order to analyze the ground subsidence. The monitored final settlement of ground is shown in Fig. 17.

It is known from Fig. 17 that the maximum value of ground settlement reaches 3.7 mm; it does not occur at the center point above the tunnel crown, and just takes place at the point 5 m away from centerline. This phenomenon is principally caused by uneven pressure in the surrounding rock. It must be mentioned that the ground settlement was measured after the ground treatment had been completed. The measurement reflects that performing ground treatment and tunnel primary support has achieved a desired effect. Therefore, the ground treatment, tunnel support, and tunneling procedure are practicable.

Through the above technical scenario, the construction of Zagunao tunnel, a typical ultra-shallow expressway tunnel in Sichuan Province, has been successfully completed. During its whole construction process, no disaster or safety accident occurred. This, in turn, verified that the design and construction program for the ultra-shallow expressway tunnel was feasible; furthermore, the goal of safe and economical construction for the ultra-shallow tunnel was fulfilled.

5 Conclusions

This paper mainly deals with the calculation of rock pressure for ultra-shallow tunnel and its ground treatment, including its construction procedure. For this kind of tunnel, the rock pressure applied to the tunnel support is related to parameters such as ground slope angel α, rock fracture angle β, tunnel size B, R, H, and its depth h. After theoretical analysis on the calculation method of asymmetric rock pressure, the ground treatment of ultra-shallow tunnel and its tunneling procedure, some useful conclusions can be drawn as follows.

(1) The vertically downward load caused by gravity of the surrounding rock above the tunnel crown is much greater than the lateral load applied to tunnel side walls. When calculating the lateral pressure applied to super-shallow tunnel sidewalls, its value is mainly derived from the lateral surrounding rock, not from the surrounding rock above the tunnel crown. This is an innovative outcome in comparison with conventional methods.

(2) The rock pressure applied to the left sidewall is greater than that applied to the right one, and the lateral forces are not identical; therefore, the tunnel is subject to asymmetric pressure under ultra-shallow depth in topographically inclined strata.

(3) As for expressway tunnels, if there are two tunnels in a trunk line, and the external tunnel lies in an ultra-shallow condition, then its support should be designed according to the ultra-shallow condition, and the internal tunnel can be designed in terms of in situ geological condition. The construction method is determined in accordance with surrounding rock conditions. In order to control ground settlement, and keep safe tunneling, the priority of excavation should be given first to internal tunnel, and then to the external one.

(4) If surrounding rock is poor and unfavorable, it is necessary to consolidate the strata using techniques such as ground rockbolt, cement and mortar grouting, steel pipe grouting, jet grouting with high pressure, pile, ground freezing. Meanwhile, ground surface and water should also be treated with a proper method as well so as to keep tunneling safe.

(5) The structural design and construction technique for the ultra-shallow tunnel in an expressway in Sichuan province turns out to be a great success, and can be used to guide the design and construction of other super-shallow tunnels under similar situations.

Acknowledgments The work is financially supported by the National Natural Science Foundation of China (No. 51378436) and the Fundamental Research Funds for the Central Universities (SWJTU11ZT33).

References

1. Goodman RE, Heuze HE, Bureau GJ (1997) On modeling techniques for the study of tunnels in jointed rock. 14th symposium on rock mechanics, pp 441–479
2. Shen B, Barton N (1997) The disturbed zone around tunnels in jointed rock masses. Int J Rock Mech Min Sci 34(1):117–125
3. Zhou XJ, Li ZL, Yang CY, Gao Y (2006) Rock pressure on tunnel with shallow depth in geologically inclined bedding strata. J Southwest Jiaotong Univ (English edition) 14(1):52–62
4. Zhou XJ, Yang CY (2007) Asymmetric rock pressure on shallow tunnel in strata with inclined ground surface. J Southwest Jiaotong Univ (English edition) 15(3):203–207
5. Yang XL, Jin QY, Ma J (2012) Pressure from surrounding rock of three tunnels with large section and small spacing. J Cent South Univ (English Edition) 19:2380–2385
6. He BG, Zhang ZQ, Chen Y (2012) Unsymmetrical load effect of geologically inclined bedding strata on tunnels of passenger dedicated lines. J Mod Transp 21(1):24–30
7. Zhou XJ (2011) Study on calculation of rock pressure and determination of depth for shallow asymmetric tunnel. Adv Mater Res 261–263:1034–1038
8. Zhou XJ, Gao B, Gao Y (2005) Safety study on tunnel with shallow depth in geologically oblique bedding strata. Prog Saf Sci Technol 5:869–875
9. Zhou XJ, Gao Y, Li ZL, Yang CY (2006) Experimental study on the uneven rock pressure and its distribution applied on a tunnel embedded in geologically bedding strata. Mod Tunn Technol 43(1):12–21 (in Chinese)
10. Bhawani Singh, Goel RK (2011) Engineering rockmass classification. Butterworth-Heinemann
11. Wu AQ, Liu FZ (2012) Advancement and application of the standard of engineering classification of rock masses. Chin J Rock Mech Eng 31(8):1513–1523 (in Chinese)
12. Hack R (1997) Rock mass strength by rock mass classification. South African rock engineering congress. Johannesburg, pp 346–356
13. Jaeger JC, Cook NGW, Zimmerman RW (2007) Fundamentals of rock mechanics, 4th edn. Blackwell Publishing Ltd, Oxford
14. Sheng MR (2000) Rock mass mechanics. Tongji University Press, Shanghai (in Chinese)
15. Vocational standard of the P.R.C. Code for design of highway tunnel (JTGD70-2004). China communication press, Beijing, 2004 (in Chinese)
16. Vocational standard of the P.R.C. Code for design of railway tunnel (TB10003-2005). China railway press, Beijing 2005 (in Chinese)
17. Goel RK, Bhawani Singh, Zhao J (2011) Underground infrastructures: planning, design and construction. Butterworth-Heinemann
18. Gioda G, Locatelli L (1999) Back Analysis of the measurements performed during the excavation of a shallow tunnel in sand. Int J Numer Anal Meth Geomech 23:1407–1425
19. Chehade FH, Shahrour I (2008) Numerical analysis of the interaction between twin tunnels: influence of the relative position and construction procedure. Tunn Undergr Space Technol 23:210–214
20. Miura K, Yagi H, Shiroma H, Takekuni K (2003) Study on design and construction method for the New Tomei-Meishin expressway tunnels. Tunn Undergr Space Technol 18:271–281

13

The electronic structure of Nb_3Al/Nb_3Sn, a new test case for flat/steep band model of superconductivity

Yanlong Ding · Shuiquan Deng · Yong Zhao

Abstract In this work, we choose Nb_3Al/Nb_3Sn as a new test case for flat/steep band model of superconductivity. Based on the density functional theory in the generalized gradient approximation, the electronic structure of Nb_3Al/Nb_3Sn has been studied. The obtained results agree well with those of the earlier studies and show clearly flat bands around the Fermi level. The steep bands as characterized in this work locate around the M point in the first Brillouin zone. The obtained results reveal that Nb_3Al/Nb_3Sn fits more to the "Flat/steep" band model than to the van-Hove singularity scenario. The flat/steep band condition for superconductivity implies a different thermodynamic behavior of superconductors other than that predicted from the conventional BCS theory. This observation sets up an indicator for selecting a suitable superconductor when its large-scale industrial use is needed, for example, in superconducting maglev system or ITER project.

Keywords Nb_3Al/Nb_3Sn · Electronic structure · Flat/steep band model · First-principle

1 Introduction

Superconductors have found widespread applications ranging from civil to military uses. The superconductor-based Magnetic Resonance Imaging (MRI) technique used in many hospitals, the superconducting super-collider project in Texas USA, the Large Hadron Collider (LHC) in CERN, the ITER project, the Experimental Advanced Superconducting Tokomak (EAST) in China and so on are all examples where superconductors are necessary materials. Another important application is the superconducting magnetic levitation system, which not only saves much of the electrical energy but also saves the space and weight of the vehicle and dramatically increases its speed [1]. Such important applications require deeper and exact understanding of the properties, e.g. mechanical and critical ones, of a superconductor.

The flat/steep band model [2, 3] for superconductivity has been proposed over many years in attempting to understand superconductivity from a chemistry point of view. The previous studies [4] indicate that the electronic structures of conventional elemental and complex superconductors show good agreement with this model. However, this model met a non-trivial challenge in the case of Calcium, although it finally proved to be the shortcoming of the first-principle methods instead of that of the flat/steep band model [5]. Nb_3Al/Nb_3Sn is a typical A15 superconductor, which though has been intensively studied, respectively, remains a challenge to the existing theories regarding their strain and scale-dependent critical properties such as the upper critical magnetic field and current density. On the other hand, Nb_3Al and Nb_3Sn have very important applications on the thermonuclear fusion experimental reactor as magnetic plasma-control materials for their overall merit factors. The results of earlier studies on the preparations, crystal structures, electronic structures, and various properties of A15 compounds have been systematically reviewed by many authors [6–9]. With respect to the electronic structures of A15 compounds, the earlier studies based on the first-principle methods [10–12] have revealed very flat bands around the Fermi level, which

Y. Ding · S. Deng (✉) · Y. Zhao
Key Laboratory of Magnetic Levitation Technologies and Maglev Trains, Ministry of Education of China, and Superconductivity and New Energy R&D Center, Southwest Jiaotong University, Chengdu 610031, China
e-mail: s.deng@swjtu.edu.cn

implies a van-Hove singularity in A15 system. However, the earlier experimental studies have never observed any Peierls-like structural instability in a wide temperature range. This fact itself indicates that either the flat band feature is over estimated as in the case of Calcium [5] or there are some other factors which counteract the effect of the "flat/bands" or both reasons may play a role. In this work, we study the electronic structure of Nb_3Al/Nb_3Sn using the state-of-the-art first-principle method.

2 Computational details

The electronic structure of Nb_3Al/Nb_3Sn has been calculated using the DFT theory as implemented in the Vienna ab initio simulation package (VASP) [13, 14]. The exchange–correlation potential has been calculated at the generalized gradient approximation (gga) level using the Perdew, Burke and Ernzerhof (PBE) formalism [15]. The $10 \times 10 \times 10$ and $20 \times 20 \times 20$ Monkhorst grids have been used to sample the first BZ for structural relaxation and self-consistent calculations, respectively. The cutoff energy for the plane wave expansion is chosen as 320 eV.

3 Results and discussion

3.1 Crystal structure

Nb_3Al/Nb_3Sn is a typical A15 compound with a space group $Pm\bar{3}n$. The unit cell structure of the compound is shown in Fig. 1. The structure belongs to a simple cubic crystal system and contains six Nb atoms and 2 Al or Sn atoms in one conventional crystallographic unit cell. Each unit cell contains six A-type atoms which lie on the surface of the unit cell and form chains along the axis directions, respectively. On each chain, the A-type atoms distribute equidistantly. The two B-type atoms occupy the corners and the center of a unit cell. The A-type and B-type atoms occupy the 6c and 2a Wyckoff positions of the space group, respectively.

3.2 Electronic structure

The electronic density of states (DOS) of Nb_3Al and Nb_3Sn is shown in Fig. 2. As can be seen from Fig. 2, except for tiny shifts and changes of the peak positions and heights, the overall agreement of the DOS curves of Nb_3Al and Nb_3Sn is very good. The DOS around the Fermi level has a sharp peak, which implies a dramatic change of physical properties if the Fermi level is changed by whatsoever means, e.g. doping or stress. By decomposing the

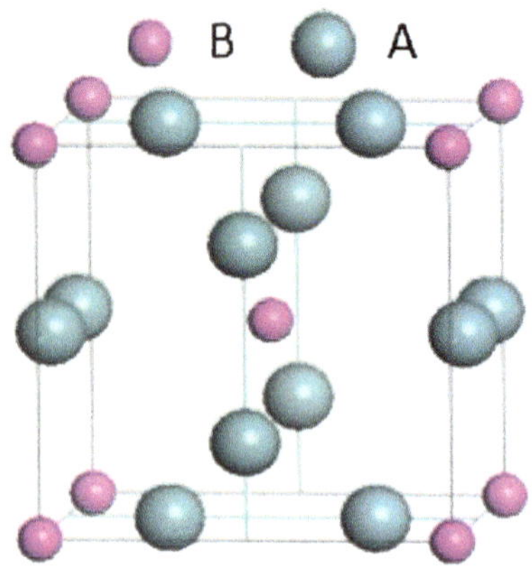

Fig. 1 The crystal structure of an A15 compound

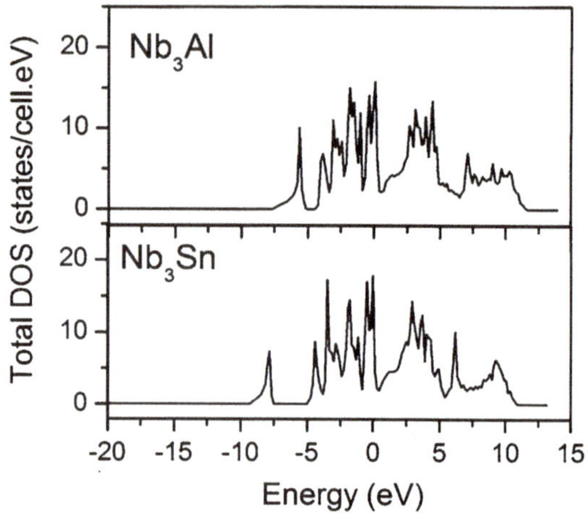

Fig. 2 Density of states of Nb_3Al and Nb_3Sn

electronic states into atomic centered states, we have found that the states around the Fermi level are essentially contributed by the Nb d-states. It is thus obvious that the superconducting properties are crucially decided by the Nb atoms. The energetically lowest peaks at −5.6 and −7.9 eV for Nb_3Al and Nb_3Sn, respectively, are due to the s-states of the B-type atoms. These results agree generally with those of the earlier studies [12, 16].

The first BZ for Nb_3Al and Nb_3Sn is shown in Fig. 3 where the symmetry lines and points are indicated. The coordinates of the symmetry points are (0, 0, 0), (1, 0, 0), (1, 1, 0), (1, 1, 1), for Γ, X, M, and R, respectively. The calculated band structures for Nb_3Al and Nb_3Sn are shown in Figs. 4 and 5, respectively. As for the DOS curves, the band structures of Nb_3Al and Nb_3Sn are quite similar. In both cases, the lowest bands are from the s, p states of the B-type atoms. The bands lying in the range of −3.2 to 5 eV for Nb_3Sn and those between −3 and 4.5 eV for Nb_3Al are contributed by the d-states of Nb. These bands split into the bonding and anti-bonding groups. As shown in Figs. 4 and

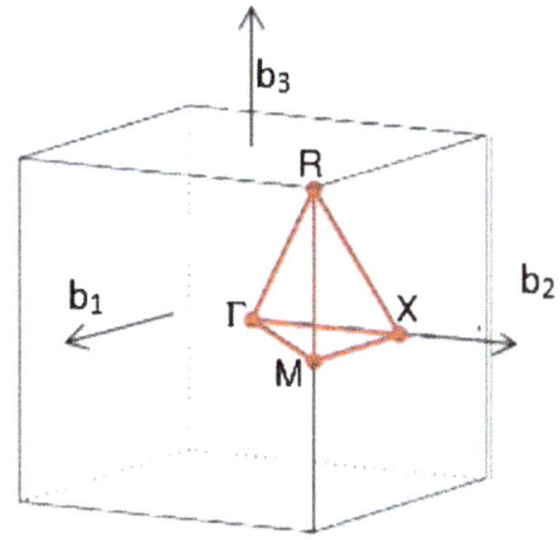

Fig. 3 The first BZ for Nb_3Al and Nb_3Sn structures

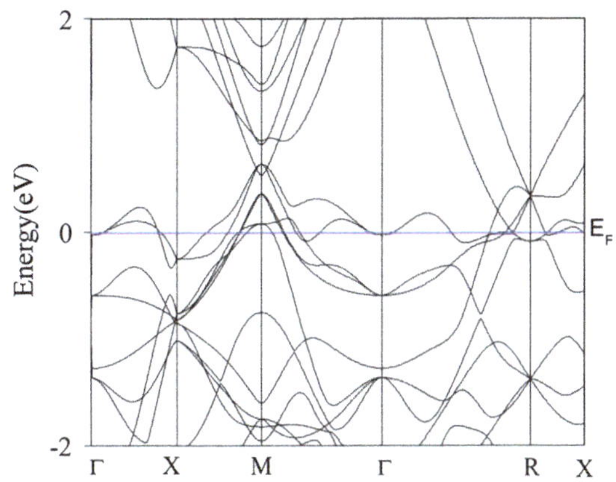

Fig. 4 Electronic band structure of Nb_3Al along symmetry directions

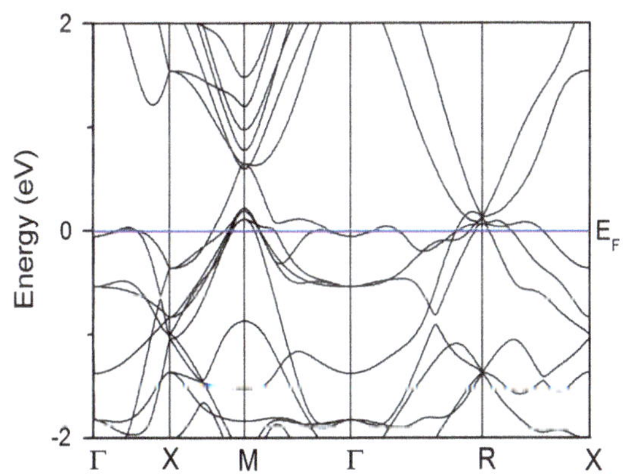

Fig. 5 Electronic band structure of Nb_3Sn along symmetry directions

5, the bonding bands are flatter than the anti-bonding bands.

As revealed in earlier studies [10–12], the bands around the Fermi level of Nb_3Al and Nb_3Sn appear rather flat which is, however, someway an artificial fact due to the chosen large energy scale in representing the band structure. Moreover, the earlier results are not consistent with each other with respect to the accurate position of the flat bands relative to the Fermi level [10–12]. Our investigations on the bands around the Fermi level indicate that the energy bands nearby Γ and R points are significantly flatter than some other bands, while those around the M point are obviously steeper. These steep bands play an important role in counteracting the possible structural instability caused by the high DOS (E_F). In this sense, Nb_3Al and Nb_3Sn fit more to the "flat/steep" band model than to the van-Hove singularity scenario. It is evident that more quantitative experimental data are needed to judge the theoretical results. In particular, the metric details of the Fermi surface are necessary to draw a solid conclusion.

4 Ending remarks

Our studies indicate that the importance of the steep band was not realized in earlier work. The steep band character in the A15 system was neglected due to the use of large energy scale. Our studies indicate clearly that the steep bands locate around the M point in the first BZ (Fig. 4). The obtained results reveal that Nb_3Al/Nb_3Sn fits more to the "flat/steep" band model than to the van-Hove scenario.

Acknowledgements This work is financially supported by the Science Foundation for International Cooperation of Sichuan Province (2014HH0016), the Fundamental Research Funds for the Central Universities (SWJTU2014: A0920502051113-10000), National Magnetic Confinement Fusion Science Program (2011GB112001).

References

1. See for example: http://www.superconductors.org/uses.htm
2. Deng S, Simon A, Köhier J (1998) Supercon-ductivity and chemical bonding in mercury. Angew Chem Int Ed 37(5):640–642
3. Deng S, Simon A, Köhier J (2005) Pairing mechanisms viewed from physics and chemistry. Struct Bond 114:103–141
4. Deng S, Simon A, Köhier J (2008) Calcium d states: chemical bonding of CaC6. Angew Chem Int Ed 47:6703–6706
5. Deng S, Simon A, Köhier J (2002) Flat-band steep band scenario and superconductivity the case of calcium. Solid State Sci 2(1):31–38
6. Weger M, Goldberg B (1973) Lattice and Electronic Properties of the β-Tungstens. Solid State Phys 28:1–177
7. Dew-Hughes D (1975) A-15 compounds: a review. Cryogenics 15(8):435–454
8. Godeke A (2006) A general scaling relation for the critical current density in Nb_3Sn. Supercond Sci Technol 19(10):100–116
9. Glowacki BA (1999) Niobium aluminide as a source of high-current superconductors. Intermetallics 7:117–140
10. Mattheiss LF, Weber W (1982) Electronic structure of cubic V_3Si and Nb_3Sn. Phys Rev B 25(4):2248–2269

11. Klein BM, Boyer LL, Papacon-stantopoulos DA (1978) Self-consistent augmented- plane-wave electronic-structure calculations for the A15 compounds V_3X and Nb_3X, X = Al, Ga, Si, Ge, and Sn. Phys Rev B 18(12):6411–64338
12. Pickett WE, Ho KM, Cohen LM (1979) Electronic properties of Nb3Ge and Nb3Al from self-consistent pseudotential. Phys Rev B 19(4):1734–1750
13. Kresse G, Furthmüller J (1996) Efficient iterative schemes for ab initio total-energy calculations using a plane-wave basis set. Phys Rev B 54(16):11169–11186
14. Kresse G, Furthmüller J (1996) Efficiency of abinitio total energy calculations for metals and semiconductors using a plane-wave basis set. Comput Mater Sci 6(1):15–50
15. Perdew JP, Burke K, Ernzerhof M (1996) Generalized gradient approximation made simple. Phys Rev Lett 77(3):865–3868
16. Paduani C (2007) Electronic Structure of the A_3B Compounds: A = Nb: B = Al, Ga, Ge, Sn and In. Braz J Phys 37(3B):1073–1079

14

Optimization of technical measures for improving high-temperature performance of asphalt–rubber mixture

Chuan Xiao · Tianqing Ling · Yanjun Qiu

Abstract Asphalt–rubber pavements often become damaged in high-temperature regions and appear rutted or wavy, and experience slippage. To improve the high-temperature performance of the asphalt–rubber mixture, technical measurements, such as, the optimal adjustment of gradation, technique of composite modification, and control of compaction were investigated. An optimal adjustment of aggregate gradation based on stone matrix asphalt improves the high-temperature stability of the asphalt–rubber mixture significantly. Through composite modification, the effect of asphalt–rubber modification was enhanced, and the dynamic stability and relative deformation indices of the asphalt–rubber mixture were improved significantly. Furthermore, compaction parameters had a significant influence on the high-temperature stability of the asphalt–rubber mixture. The rolling times for compacting the asphalt–rubber mixture should be controlled to within 18–20 round-trips at a molding temperature at 180 °C; if the rolling time is a 12 round-trip, the compaction temperature of the asphalt–rubber mixture should be controlled between 180 and 190 °C

Keywords Road engineering · Optimization · Laboratory test · Asphalt–rubber mixture · High-temperature performance

C. Xiao (✉) · Y. Qiu
School of Civil Engineering, Southwest Jiaotong University, P.O. Box 520, No. 111 First North Section, 2nd Ring Road, Chengdu 610031, Sichuan, China
e-mail: xcaaa6666@sina.com

T. Ling
School of Civil Engineering and Architecture, Chongqing Jiaotong University, Chongqing 400074, China

1 Introduction

Rapid growth of waste tires is a serious environmental problem because of their highly resistant chemical, biological, and physical properties. Many approaches have been considered to encourage the sustainable development. Using crumb rubber in asphalt, which initiated with the motivation to improve the binder properties, is one of the practical ways to tackle the increasing waste tires.

In general, the approaches used to incorporate crumb rubber modifier (CRM) in road paving materials are classified as the dry method and the wet method [2]. Wet method is applied in most of the rubberized asphalt projects in China, which entails adding the crumb rubber to the binder before mixing it with aggregate [3]. The behavior of asphalt–rubber with wet method depends on several factors, such as, the origin, fabrication process and grain size distribution of the crumb rubber, the type of base asphalt binder used in the mixture, and the temperature and time of the mixing process. Anderson et al. [4] investigated the rheological and physical properties of binders modified with rubber, for rubber contents below 20 % by weight. Huang et al. [5] and Shen and Amirkhanian [6] suggested the optimal preparation of asphalt–rubber according to comparative tests on material properties of asphalt binder.

In pavement destruction, asphalt–rubber has become increasingly attractive in the applications, such as, open graded friction course (OGFC), stress absorption membrane interlayer (SAMI), and super silent pavement (SSP) [2]. The asphalt–rubber pavements exhibit unique advantages in reducing pavement thickness, delaying reflection cracking, and decreasing traffic noise [7, 8]. However, an obvious problem in the application of the asphalt–rubber mixture is the lack of high-temperature stability used as structural layer, which could cause serious rutting under recycled vehicle loading. The indices, such as, viscosity, penetration, and softening point show that the asphalt–rubber shows excellent performance at high-temperature [9]. However, because of the interference of asphalt–rubber and aggregate during the compacting process and the low stiffness modulus and deformation characteristics of the asphalt–rubber mixture [8, 10], it is difficult to meet the desired demands when applying asphalt–rubber pavement in high-temperature regions. Furthermore, there is no unified technical specification for asphalt–rubber in China resulting in significant discrepancies in aggregate gradation, asphalt content, and mineral filler content when paving with asphalt–rubber mixture [11–13]. Improvements in the high-temperature performance of asphalt–rubber pavements are critical when they are applied in high-temperature regions and under heavy traffic conditions in China.

In this paper, we investigate the high-temperature stability of an asphalt–rubber mixture based on internal and external factors. At first, the optimal gradation was adjusted in the rutting tests with dynamic stability and relative deformation as evaluation indices. Then, the scheme of compound modification and optimization of the compaction parameters were analyzed to improve the high-temperature stability of the asphalt–rubber mixture. To obtain a reasonable scheme of compound modification, comparative tests of high-temperature performance were conducted between different binders and mixtures. The effects of rolling time and molding temperature on air void volume and the dynamic stability (DS) were investigated to determine optimal compaction parameters.

2 Test materials

2.1 Material properties of asphalt–rubber

SK 70# base asphalt and crumb rubber (30 mesh size) were used to produce asphalt–rubber for comparative tests of asphalt–rubber performance. The test methods followed "Standard Test Methods of Bitumen and Bituminous Mixtures for Highway Engineering" from the industry standard of China (JTG E20-2011) [14] with main performance indices listed in Table 1.

2.2 Material properties of aggregate and filler

The test methods followed "Test Methods of Aggregate for Highway Engineering" (JTG E42-2005) and the main indices of the aggregate and mineral filler in the asphalt–rubber mixtures are listed in Tables 2 and 3, respectively.

3 Test method and analysis

3.1 Optimization of aggregate gradation

3.1.1 Gradation-type selection

Based on the broad overview of a typical gradation type for an asphalt–rubber mixture, AR-AC-13 (based on Arizona standards [2, 12]), SMA-13 (traditional stone matrix asphalt [15]), and AC-13 (dense-graded asphalt mixture [15]) were chosen as research materials on which to conduct the rutting tests. Figure 1 shows the aggregate gradations of different mixtures.

The rutting tests on the different asphalt mixtures followed "Standard Test Methods of Bitumen and Bituminous Mixtures for Highway Engineering" (JTG E20-2011) with parameters listed in Table 4.

The rutting test results for the different asphalt mixtures are given in Table 5.

Table 5 shows that the preferential order of the three kinds of mixtures based on high-temperature performance is: SMA-13 > AR-AC-13 > AC-13. This occurs because of the different characteristics of the mixtures.

Because AC-13 is a "suspend-dense" structure mixture, there is interference between the asphalt–rubber binder and the aggregate during compaction. This type of mixture is difficult to compact completely with asphalt–rubber. This most likely explains why the high-temperature performance indices of AC-13 are the worst among the three kinds of asphalt mixtures.

AR-AC-13 has an aggregate gradation based on the Arizona standard with obvious gap gradation characteristics. By enhancing the high viscosity binder dosage and reducing the amount of fine aggregate, especially the filler, more voids appear in the mineral aggregate of AR-AC-13 and more

Table 1 Properties of asphalt–rubber

Performance index	Value	Technical standard [9]	Test method
180 °C rotation viscosity (Pa s)	2.8	2.5–5.0	T 0625
Softening point (°C)	67.6	>65	T 0606
Penetration (0.1 mm)	50.4	30–70	T 0604
Elastic recovery (%)	78.0	≥60	T 0662

Note The code of test method followed JTG E20-2011 [14]

Table 2 Properties of aggregate

Aggregate type	Apparent density ($g \cdot cm^{-3}$)	Bulk density ($g \cdot cm^{-3}$)	Crushed stone value (%)	Water-washing method <0.075 mm (%)	Flat and elongated particle in coarse aggregate (%)	Water absorption (%)	Sturdiness (%)	Sand equivalent (%)
10–15 mm gravel	2.928	2.834	11.2	0.2	7.8	1.84	0.3	–
5–10 mm gravel	2.912	2.826	11.6	0.2	7.1	1.88	0.3	–
Stone chip	2.725	2.725	–	–	–	–	1.2	78.4

Table 3 Properties of filler

Apparent density ($g \cdot cm^{-3}$)	Hydrophilic coefficient	Plasticity index (%)	Water content (%)	Heating stability	Pass percentage (%)			
2.710	0.6	2.2	0.43	Color did not change at 200 °C	<0.6 mm	<0.3 mm	<0.15 mm	<0.075 mm
					100.0	99.9	97.2	91.8

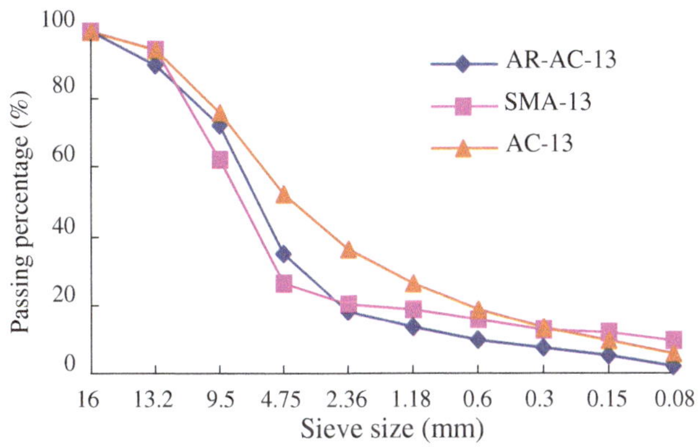

Fig. 1 Comparison of aggregate gradation

Table 4 Rutting test parameters

Test parameters	Value
Molding method	Wheel molding
Specimen size	300 × 300 × 50 mm
Test temperature	60 ± 1 °C
Pressure	0.7 ± 0.05 MPa
Loading distance	230 ± 10 mm
Loading speed	42 ± 1 times min^{-1}

significant features appear in the framework structure. However, rutting test results show that the typical "S" gradation type did not reach the expected target. One factor that contributes to the problem may be that with a reduction in filler, the asphalt mortar stiffness is reduced making the mixture prone to deformation. It is therefore difficult to achieve cohesion and stability in the AR-AC-13 mixture.

SMA-13 is also a gap gradation mixture, but compared with AR-AC-13, the asphalt–rubber mixture based on traditional SMA-13 has more fine aggregate and less asphalt binder proportion, and the deficiencies in AR-AC-13 can be overcome.

3.1.2 Gradation optimization

The rutting test results of SMA-13 with the asphalt–rubber were unable to meet heavy traffic demands [11]. We therefore selected SMA-13 (AR-SMA-13) for further adjustment of aggregate gradation. The optimal adjustment of the passing percentages through crucial sieves was studied to improve the high-temperature performance of AR-SMA-13. The key sieves for aggregate gradation were selected because: (1) the aggregate gradation should form a framework structure with excellent strength; (2) crumb rubber is coarse compared with conventional modifiers and

Table 5 Rutting test results for different asphalt mixtures

Mixture type	Asphalt content (%)	Dynamic stability (times mm^{-1})	Relative deformation (%)
AR-AC-13	7.0	1,721	4.3
SMA-13	5.8	3,219	2.9
AC-13	4.3	1,480	5.2

Note Dynamic stability is defined as the axle loading time when the mixture specimen generates a 1-mm deformation. Relative deformation is defined as the ratio between final deformation and original mixture specimen height [11]

it is necessary to decrease the filler proportion of AR-SMA-13 and increase voids in the mineral aggregate (VMA) so that there is enough filling space for the asphalt–rubber binder in the asphalt–rubber mixture; and (3) 2.36 mm is an important sieve size for aggregate gradation. The variation in the 2.36-mm passing percentage would not influence the mixture volume parameters significantly. It is therefore helpful to reduce the influence caused by variability in the other volume parameters.

As discussed above, the passing percentages through the 0.075- and 2.36-mm sieve sizes were selected as crucial sieves upon which to make adjustments. The changes in pass percentage are shown in Table 6.

Hot mix asphalt was designed according to the Marshall test [15] with results given in Table 7 (where VV is the volume of air voids and VFA is voids filled with asphalt).

Rutting tests [14] were conducted to determine the optimal gradation type for AR-SMA-13. The DS and relative deformation were chosen as evaluation indices. Table 8 shows the differences among the three types of mixtures from the rutting tests.

As shown in Table 8, the high-temperature stability of AR-SMA-13I reduced the filler proportion from 10 % to 8 %, yielded a better mixture than the other two types, and is the only mixture that meets the technical standards. Gradation adjustment to optimize the high-temperature performance of AR-SMA-13 is therefore feasible.

3.2 Compound modification

3.2.1 Preparation of compound-modified sample

Styrene–butadiene block copolymer (SBS) and viscosity-reducing additive (termed SAK) were chosen as modifiers to study compound modification on asphalt–rubber. Based on the different properties of SBS [16, 17] and SAK [18], different preparation programs were formulated for the two types of compound-modified asphalt–rubber:

(1) SBS-Rubber compound-modified asphalt (termed S-R asphalt): (a) heat base asphalt to 180 °C, add SBS (2 %) into base asphalt, and shear for 30 min using an emulsion shear apparatus at 180 °C and 3,500 rpm; (b) swell and develop for 30 min at 150 °C by manual mixing; (c) heat modified asphalt to 190–200 °C, add dry crumb rubber and then shear and develop for 45–60 min using an emulsion shear apparatus at 3,000 rpm.

(2) SAK-Rubber compound-modified asphalt (termed K-R asphalt): (a) heat base asphalt to 150 °C, add SAK (2.5 %) into base asphalt, and mix by hand; (b) heat modified asphalt to 190–200 °C, add dry crumb rubber, then shear and develop for 45–60 min at 3,000 rpm.

3.2.2 Asphalt binder tests

Pure asphalt–rubber, SBS asphalt, S-R asphalt, and K-R asphalt were selected to analyze the asphalt binder properties. The evaluation indices chosen were 180 °C rotation viscosity, penetration, softening point, and elastic recovery with results shown in Fig. 2.

Comparative test results for the different asphalt–rubber types show that the high-temperature performance of all three types of asphalt–rubber (pure, S-R, and K-R) was better than SBS asphalt. In terms of compound modification: (1) S-R asphalt exhibits a better high-temperature performance for all indices compared with pure asphalt–rubber; the 180 °C rotation viscosity, softening point, and elastic recovery increased by 10.7, 17.6, and 3.8 %,

Table 6 Adjustment of aggregate gradation

Gradation type	Pass percentage (%)								
	13.2 mm	9.5 mm	4.75 mm	2.36 mm	1.18 mm	0.6 mm	0.3 mm	0.15 mm	0.075 mm
Standard gradation [15]	95	62.5	27	20.5	19	16	13	12	10
Gradation I	95	62.5	27	18.5	17	14	11	10	8
Gradation II	95	62.5	27	16.5	15	12	9	8	6

Table 7 Results from the Marshall test

Gradation type	Asphalt content (%)	Bulk density ($g \cdot cm^{-3}$)	VV (%)	VMA (%)	VFA (%)	Marshall stability (kN)	Flow value (mm)
Standard gradation	5.8	2.435	4.2	17.1	75.4	7.94	2.63
Gradation I	6.0	2.444	4.3	17.5	75.4	8.02	2.31
Gradation II	6.1	2.437	4.1	17.6	76.7	8.21	2.57

Table 8 Comparison of different mixtures for high-temperature performance

Mixture type	Dynamic stability (times mm^{-1})		Relative deformation (%)	
	Test value	Standard	Test value	Standard
AR-SMA-13	3,219	≥3,500	4.3	≤3.1
AR-SMA-13I	3,688		2.9	
AR-SMA-13II	3,275		5.2	

respectively, while the 25 °C penetration decreased by 4.6 %. (2) SAK addition (viscosity-reducing additive) resulted in a K-R 180 °C rotation viscosity reduction of 60.7 % compared with pure asphalt. This helps enhance mixture compactness and provide structural strength. The 25 °C penetration of the K-R asphalt decreased by 7.1 %. Its softening point and elastic recovery increased by 28.3 and 1.3 %, respectively.

Compound modification in asphalt binder tests is therefore significant with the comprehensive high-temperature performance of the S-R asphalt being better than the other asphalt types.

3.2.3 *Asphalt mixture tests*

Rutting tests were conducted on the SBS asphalt mixture without fiber (SBS-SMA-13), asphalt–rubber mixture (AR-SMA-13), SBS-AR compound-modified mixture (SBS-AR-SMA-13), and SAK-AR compound-modified mixture (SAK-AR-SMA-13) with results shown in Fig. 3.

The following is concluded from the graphical illustrations in Fig. 3:

(1) Based on the DS and relative deformation, the preferential order of the four kinds of mixtures was: SBS-AR-SMA-13 > SAK-AR-SMA-13 ≈ SBS-SMA-13 > AR-SMA-13.

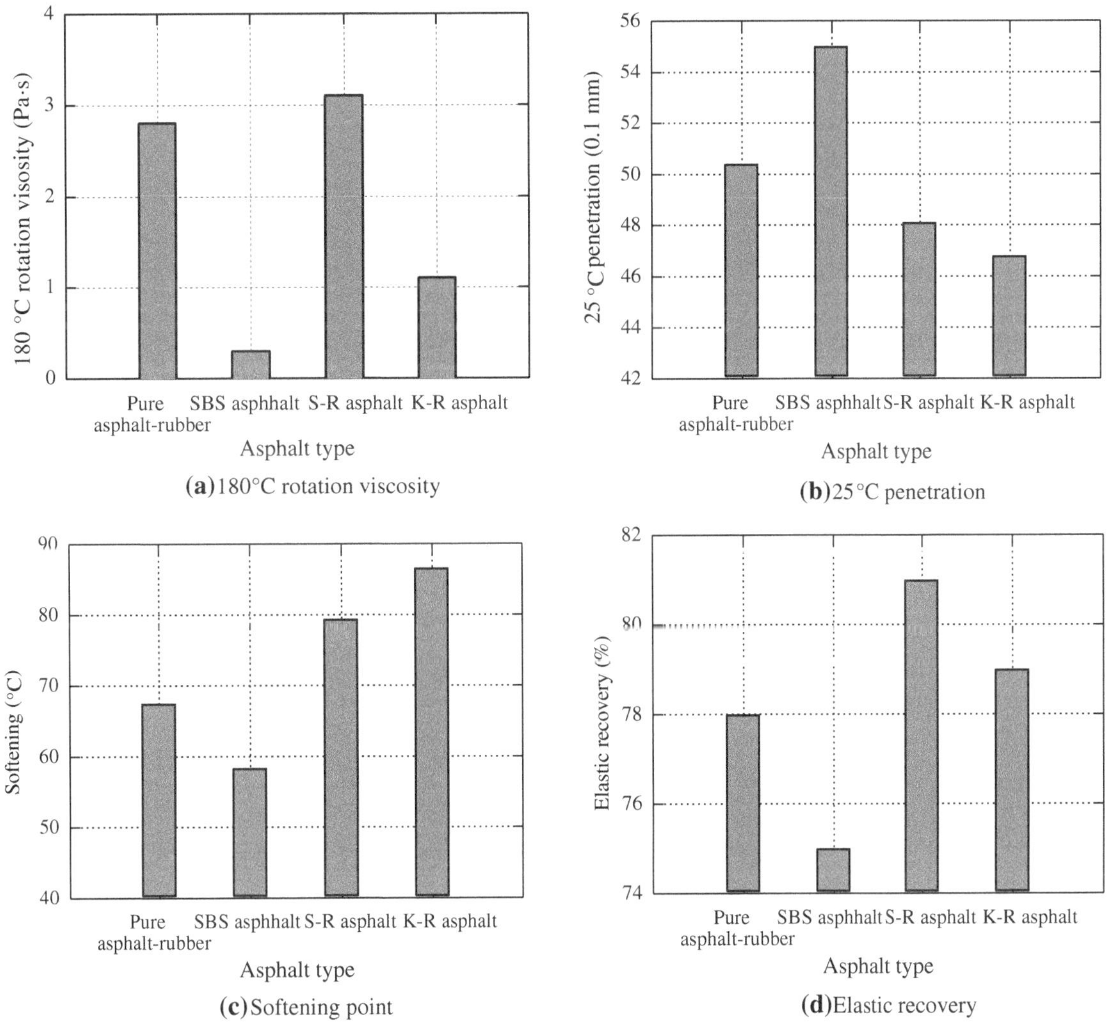

Fig. 2 Comparison of asphalt binder test results

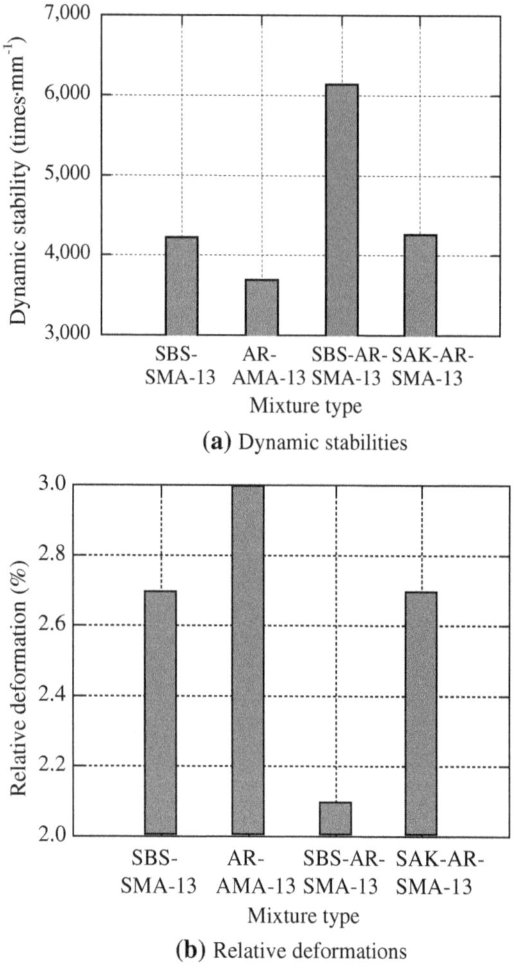

Fig. 3 Comparison of rutting test results

(2) Compared with AR-SMA-13, the DS of SBS-AR-SMA-13 increased by 66.3 % and the relative deformation increased by 30.0 %. The DS of SAK-AR-SMA-13 increased by 15.5 % and its relative deformation increased by 10.0 %. Therefore, compound modification can improve high-temperature stability.

(3) The high-temperature viscosity of SBS asphalt was lower than that of pure asphalt–rubber, but SBS-SMA-13 had good rutting resistance. Therefore, for different types of asphalt with different mechanisms, the viscosity index is unilateral at times. We now need to integrate factors comprehensively to evaluate the high-temperature performance of asphalt and its mixtures.

3.3 Effect of molding parameters

During the rutting test, the molding temperature and rolling times were closely related to the compactness and stability of the mixture. As demonstrated in the current standard (JTG E20-2011), 12 round-trips are recommended for the rolling time for molding rutting test specimens. The compactness of the baseline asphalt mixture meets the specification demand after 12 round-trips of wheel-rolling. The standard also recommends a temperature of approximately 140–170 °C to mold rutting test specimens. Because of the properties of high viscosity asphalt–rubber, the compactness and high-temperature stability of the asphalt–rubber mixture with recommended molding parameters are far below standard values [19].

We selected rolling time and forming temperature as molding parameters to analyze the compaction effect on high-temperature performance of the asphalt–rubber mixture.

3.3.1 Rolling times for compaction

Different rolling times (10, 12, 14, 16, 18, 20, 22, and 24 round-trips) were selected for molding asphalt–rubber mixture specimens at uniform temperature (180 °C). From Fig. 4, we conclude that:

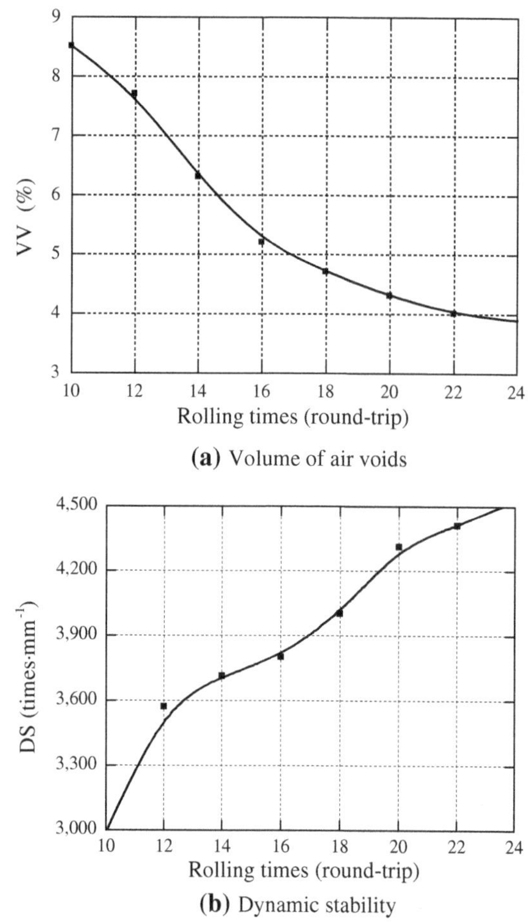

Fig. 4 Relationship between VV and DS with rolling time

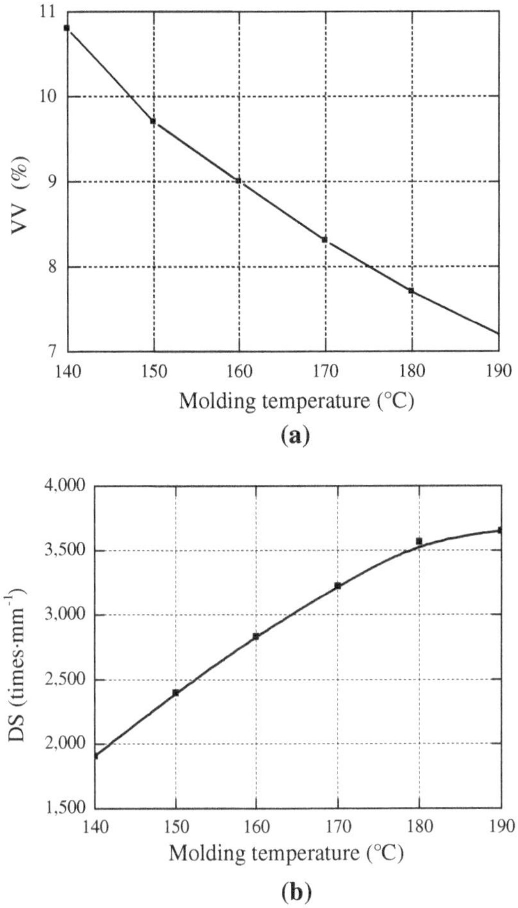

Fig. 5 Relationship of VV and DS with molding temperature

(1) The VV decreases with increase in rolling time, but the rate of change decreases gradually. If the asphalt–rubber mixture is compacted with 12 round-trips by the wheel-rolling method, its VV is 7.7 % (larger than the objective of 3 %–5 % [11]). This occurs because the viscosity of the asphalt–rubber binder is high, there is a thick covering of the aggregate with asphalt mortar, and the asphalt–rubber mixture is therefore harder to compact.

(2) The compactness of the asphalt–rubber mixture increases with increase in rolling time and therefore the DS increases significantly. For the asphalt–rubber mixture, an increase in rolling time contributes to better structural strength and stability. To enhance mixture compactness and achieve high-temperature stability, the rolling times of the asphalt–rubber mixture should be controlled strictly. However, if the compactness were too high, it would result in interference between the binder and aggregate and bleeding in the asphalt pavement. So that the mixture VV reaches its objective value (3 %–5 % [11]) and so that there is no interference in compaction, the rolling times for molding asphalt–rubber mixtures should be controlled between 18 and 20 round-trips.

3.3.2 Molding temperature

Different molding temperatures (140, 150, 160, 170, 180, and 190 °C) were chosen to mold rutting specimens at uniform rolling times (12 round-trips). As shown in Fig. 5, the molding temperature was closely related to the VV and DS. The VV decreased with increase in molding temperature. As the molding temperature increased, the high-temperature stability of the asphalt–rubber mixture improved significantly. The VV of the mixture molded at 140 °C was 1.5 times that molded at 190 °C and the DS was 52.1 % of those molded at 190 °C. Figure 5b indicates that with decline in temperature, the downtrend of the DS was more significant. The compaction temperature must therefore be controlled strictly to ensure good performance of the asphalt–rubber mixture. If the rolling time to mold rutting specimens is set at 12 round-trips, the temperature must be controlled at 180–190 °C to meet technical standards (DS $\geq$ 3,500 times mm^{-1}).

4 Conclusion

(1) The high-temperature stability of the mixtures varied as: AR-SMA-13 > AR-AC-13 > AC-13. The high-temperature performance of AR-SMA-13 can be improved by adjustment of the SMA-13 gradation. This results in a decrease of the 0.075-mm passing percentage from 10 to 8 % and that of the 2.36-mm passing percentage from 20.5 to 18.5 %.

(2) The effects of compound modification in asphalt–rubber are significant. The comprehensive high-temperature performance of S-R asphalt is better than the other types of asphalt. Compared with pure asphalt–rubber, the K-R asphalt with SAK improved the high-temperature performance indices, such as, the softening point, penetration, and elastic recovery. Its viscosity reduced significantly and therefore enhances mixture compactness to yield structural strength.

(3) The high-temperature performance of the four mixtures was: SBS-AR-SMA-13 > SAK-AR-SMA-13 $\approx$ SBS-SMA-13 > AR-SMA-13. The high-temperature stability can therefore be improved by compound modification, especially SBS compound modification in asphalt–rubber.

(4) Compaction parameters, such as, molding temperature and rolling times were closely related to the high-temperature stability of the asphalt–rubber mixture. With increase in rolling time, the compactness and

dynamic stability of the asphalt–rubber mixtures increased gradually and the rolling times for molding the asphalt–rubber mixtures should be controlled for 18–20 round-trips at uniform temperature (180 °C). With decrease in compaction temperature, the compactness and dynamic stability of the asphalt–rubber mixture decreased by degrees. If setting rolling times of 12 round-trips were used as the uniform case (as for the SBS asphalt mixture), the compaction temperature must be controlled at 180–190 °C to meet technical standards.

References

1. Yang ZF, Li MJ, Wang XD (2005) The history and status of rubber powder used in road-building. J Mater Civ Eng 22(7):19–22 (in Chinese)
2. Bahia HU, Davies R (1994) Effect of crumb rubber modifier (CRM) on performance related properties of asphalt binders. J Assoc Asphalt Paving Technol 63:414–438
3. Sun ZW, Chen B (2007) Guide for technical application of asphalt rubber. China Communications Press, Beijing (in Chinese)
4. Anderson J, Pampulim V, Saim R, et al (2000) Asphalt rubber laboratory properties related to type and process technology of crumb rubber. In: Asphalt Rubber 2000 Proceedings, Portugal, pp 371–382
5. Huang P, Lv WM, Zhang FQ et al (2001) Research on performance and technology of the rubber powder modified asphalt mixture. China J Highw Transp 14(12):2–5 (in Chinese)
6. Shen J, Amirkhanian S (2005) The influence of crumb rubber modifier (CRM) microstructures on the high temperature properties of CRM binders. Int J Pavement Eng 6(4):265–271
7. Bennert T, Hanson D, Maher A et al (2005) Influence of pavement surface type on tire/pavement generated noise. J Test Eval 33(2):94–100
8. Wang XD, Li MJ, Lu KJ et al (2008) The applied technology of the crumb rubber in the asphalt and mixture. China Communications Press, Beijing, pp 111–124 (in Chinese)
9. Ye ZG, Kong XM, Yu JY et al (2003) Investigation on crumb rubber modified asphalt. J Wuhan Univ Technol 25(1):11–14 (in Chinese)
10. Ling TQ, Xiao C, Xia W et al (2010) Characteristics of asphalt–rubber mortar and aggregate gradation optimization based on high temperature condition. J Civ Archit Environ Eng 32(5):47–52 (in Chinese)
11. Ministry of Communications Highway Research Institute (2008) Guide for design and construction of asphalt rubber and mixtures. China Communications Press, Beijing (in Chinese)
12. Arizona Department of Transportation Standard Specifications for Road & Bridge Construction, Phoenix, AZ, 2000
13. Pasquini E, Canestrari F, Cardone F et al (2011) Performance evaluation of gap graded asphalt rubber mixtures. Constr Build Mater 25(4):2014–2022
14. P.R. China Ministry of Communications (2011) Standard test methods of bitumen and bituminous mixtures for highway engineering (JTG E20-2011). China Communications Press, Beijing (in Chinese)
15. P.R. China Ministry of Communications (2004) Technical specifications for construction of highway asphalt pavement (JTG F40-2004). China Communications Press, Beijing (in Chinese)
16. Lu XH, Isacsson U, Ekblad J (1999) Phase separation of SBS polymer modified bitumens. J Mater Civ Eng 11(1):51–57
17. Wang T, Cai HM, Zhang YZ (2008) Study on mechanism of SBS modified asphalt. Pet Asphalt 6(22):10–14 (in Chinese)
18. Gao XF, Liu LP, Liu HF et al (2009) Performance evaluation of warm-asphalt mixture with SAK. Highw Eng 34(6):51–54 (in Chinese)
19. Xiao C (2009) Research on high temperature performance and construction technology of asphalt–rubber and mixture. Dissertation, Congqing Jiaotong University, Chongqing (in Chinese)
20. Huang WD, Wei W, Huang Y et al (2010) Influencing factor research on high-temperature performance of asphalt rubber mixture. J Tongji Univ 38(7):1023–1028 (in Chinese)

15

The preparation process and feature of the topological insulator Bi_2Te_3

Peng Chen · Dajin Zhou · Pingyuan Li · Yajing Cui · Yongliang Chen

Abstract Topological insulators are insulating in the bulk but have metallic surface states. Its unique physicochemical properties can find numerous applications in electronics, spintronics, photonics, the energy sciences, and the signal control of transportation. We report an experimental approach to synthesize the high-quality single crystal of topological insulator Bi_2Te_3 by using self-flux method. We obtained the optimal preparation conditions by adjusting the parameters of heat treatment, and successfully prepared the single-crystal Bi_2Te_3 sample. The as-grown samples have a surface with bright metallic luster and are soft and fragile. Furthermore, Bi_2Te_3 has the obvious layer structure from SEM results. The data of X-ray diffraction and scanning electron microscope show that Bi_2Te_3 single crystal grows along the *c*-axis with the order of $Te^{(1)}$–Bi–$Te^{(2)}$–Bi–$Te^{(1)}$ and crystallizes in the hexagonal system with space group of R/3 m. The ρ–T curve shows that ρ decreases with temperature, showing metallic behavior over the whole temperature range.

Keywords Bi_2Te_3 · Single crystal · Topological insulator

1 Introduction

Topological insulators (TIs) are a new class of quantum states of matter with topologically protected conducting surface states, arising from the topology of the bulk electronic band structure [1–3]. They are insulating in the bulk but possess metallic surface states protected by time-reversal symmetry [4]. The surface structure is composed of single Dirac cones, and belongs to the Dirac particles system [5]. The carrier can be transmitted in the surface without scattering and energy loss. The topological surface states are gapless, and ambipolar conduction in the gate voltage tuned devices is taken as a signature of TI surface transport [6]. Because of these features, TI materials have great application prospects in the field of transistors, storage devices, magnetic sensors, and energy efficient products. Three-dimensional TIs including Sb_2Te_3, Bi_2Se_3, and Bi_2Te_3 are also excellent thermoelectrics, which have been intensively investigated for optimized thermoelectric performance in the recent decades [7, 8]. The theories and experiments have successively confirmed that they have large bulk band gap and single Dirac cone surface states [9–12]. They have quickly become a hot spot in the study of topological insulator material. The topological characteristic of Bi_2Te_3 was predicted by first-principles calculation and analysis of the model in Ref. [10] and observed by the angle-resolved photoemission spectroscopy (ARPES) [12, 13]. The Fermi surface (FS) is the shape of the snowflake-like. The Bi_2Te_3 single crystal was predicted to be a bulk insulator, and the electron carriers were observed in experiment to arise from crystal imperfections, specifically vacancies and defects [12]. It is found that increasing Se concentration effectively suppresses the bulk carrier transport and induces semiconducting behavior in $Bi_2(Te_{1-x}Se_x)_3$. An ambipolar field effect from the gapless surface states was observed in $Bi_2(Se_{0.8}Te_{0.2})_3$ [14].

The single crystal along the *c*-axis with $Te^{(1)}$–Bi–$Te^{(2)}$–Bi–$Te^{(1)}$ cycle stacked growth, belongs to the hexagonal

P. Chen · D. Zhou · P. Li · Y. Cui · Y. Chen (✉)
Key Laboratory of Magnetic Levitation Technologies and Maglev Trains, Ministry of Education of China, and Superconductivity and New Energy R&D Center, Southwest Jiaotong University, Chengdu 610031, China
e-mail: llchen@home.swjtu.edu.cn

system, and space group is R/3 m. Between Te and Bi layer is a covalent bond, and Van-der Waals bond between Te and Te layer. So the single Bi_2Te_3 is cleavage easily along the basal plane. Near the melting point temperature Bi_2Te_3 deviates from the strict ratio of chemical components. It presents a little excess Bi that Bi accounted for 40.065 % of atomic weight ratio and Te accounted for 59.935 % [15].

Bi_2Te_3 can greatly improve the speed of computer chips and efficiency. Using existing semiconductor technology, this kind of material can allow electrons to move on the surface without energy consumption at room temperature, so the application of this chip to modern transportation will lead to a leap.

Early by directional solidification method and the method of zone melting method prepared Bi_2Te_3 single crystal or columnar crystals; commercial production is also using these methods. Bi_2Te_3 as thermoelectric material was prepared by hot pressing method [16], the SPS (spark plasma sintering) method [17], hot extrusion method [18], solvent method [19], and so on. The samples with irregular polyhedron did not show the structural characteristics and contained an impurity phase [15]. Recently, the topological property of Bi_2Te_3 has been confirmed [12, 13]. In order to better study the bizarre properties of the TIs, we need the high-quality single-crystal material. So we have done many times trials according to the structure characteristics of Bi_2Te_3 single crystal with the self-flux method. To try and improve the process of proportioning, sample preparation, sintering, testing, etc., we summarize an optimization method of preparation of Bi_2Te_3 single crystal of high-quality material.

2 Experimental

High-purity elemental Bi (99.999 %) and Te (99.999 %) were used for the Bi_2Te_3 crystal growth. Bi and Te mixed with a molar ratio of 2:3 were fully grinded in glove box filled with argon at least 2 h, enclosed into evacuated quartz tubes, and then in heat treatment with the process of Fig. 1. We design three groups of Bi_2Te_3 samples (see Table 1). The first group (H-750–H-850) is to change the highest temperature in the range of 750–850 °C, and keep heating time, quenching temperature, cooling rate unchanged; the second group (Q500–Q580) is to change quenching temperature from 500 to 580 °C, and keep other conditions unchanged; the third group(C-4–C-6) is only to change cooling rate.

The crystal structure was studied by powder X-ray diffraction (XRD) using an X'Pert MRD diffractometer with Cu $K\alpha$ radiation. All observed reflections were indexed.

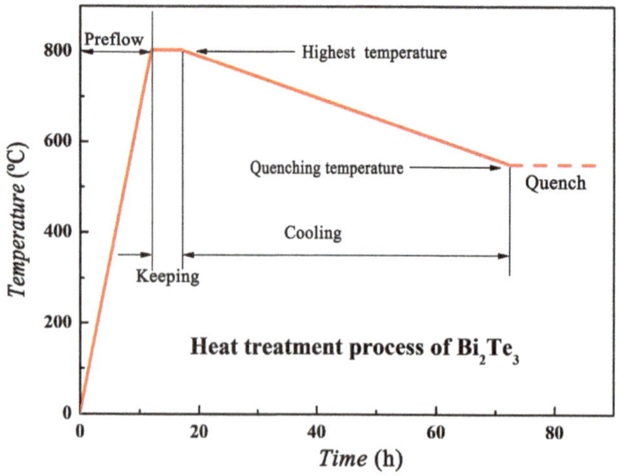

Fig. 1 Heat treatment process of Bi_2Te_3

Table 1 The grouped control variables in the heat treatment process (*H* highest, *Q* Quenching, and *C* Cooling)

No.	Highest temperature (°C)	Keeping time (h)	Quenching temperature (°C)	Cooling rate (°C/h)
H-750	**750**	5	550	5
H-780	**780**	5	550	5
H-800	**800**	5	550	5
H-820	**820**	5	550	5
H-850	**850**	5	550	5
Q-500	800	5	**500**	5
Q-550	800	5	**550**	5
Q-560	800	5	**560**	5
Q-570	800	5	**570**	5
Q-580	800	5	**580**	5
C-4	800	5	550	**4**
C-5	800	5	550	**5**
C-6	800	5	550	**6**

Bold values indicate differences of the heat treatment processes

Lattice constants were determined from LeBail refinements. Microstructure and composition of the sample were analyzed using a field emission scanning electron microscope (FESEM) equipped with an energy dispersive X-ray analysis (EDX). Resistivity measurements were performed with a physical property measurement system (PPMS, Quantum Design).

3 Results and discussions

Figure 2a, c, e show the XRD data for all the samples. Most of XRD peaks correspond to (00*L*) reflections, (006),

(0015), (0018), and (0021) distributed in the about 17.5°, 44.5°, 54.5°, and 64.5°, respectively. Calculating the cell parameters of every sample based on each of the main peak of sample, we draw out the curve of cell parameters *c* with temperature (see Fig. 2b, d, f). Bi_2Te_3 single crystal grows along the *c*-axis, so we only consider the value of the *c*. Figure 2b, d show that the value of *c* increases gradually with the highest temperature and the quenching temperature gradually; Fig. 2f shows that *c* value decreases with the increase of cooling rate.

According to the standard PDF card, we can know that the Bi_2Te_3 lattice parameters are $a = b = 4.3852$ and $c = 30.4830$ Å [20]. In Fig. 2b, d, f, $c = 30.48$ Å corresponds to the highest temperature 810 °C, quenching temperature 560 °C, and the cooling rate 4.4 °C/h.

We have got the sample of Bi_2Te_3 with the highest temperature 810 °C, quenching temperature 560 °C, and the cooling rate 4.4 °C/h. Figure 3 shows XRD pattern of the powder and single crystal as-grown samples Bi_2Te_3. After the Bi_2Te_3 standard pattern contrast, the vast majority

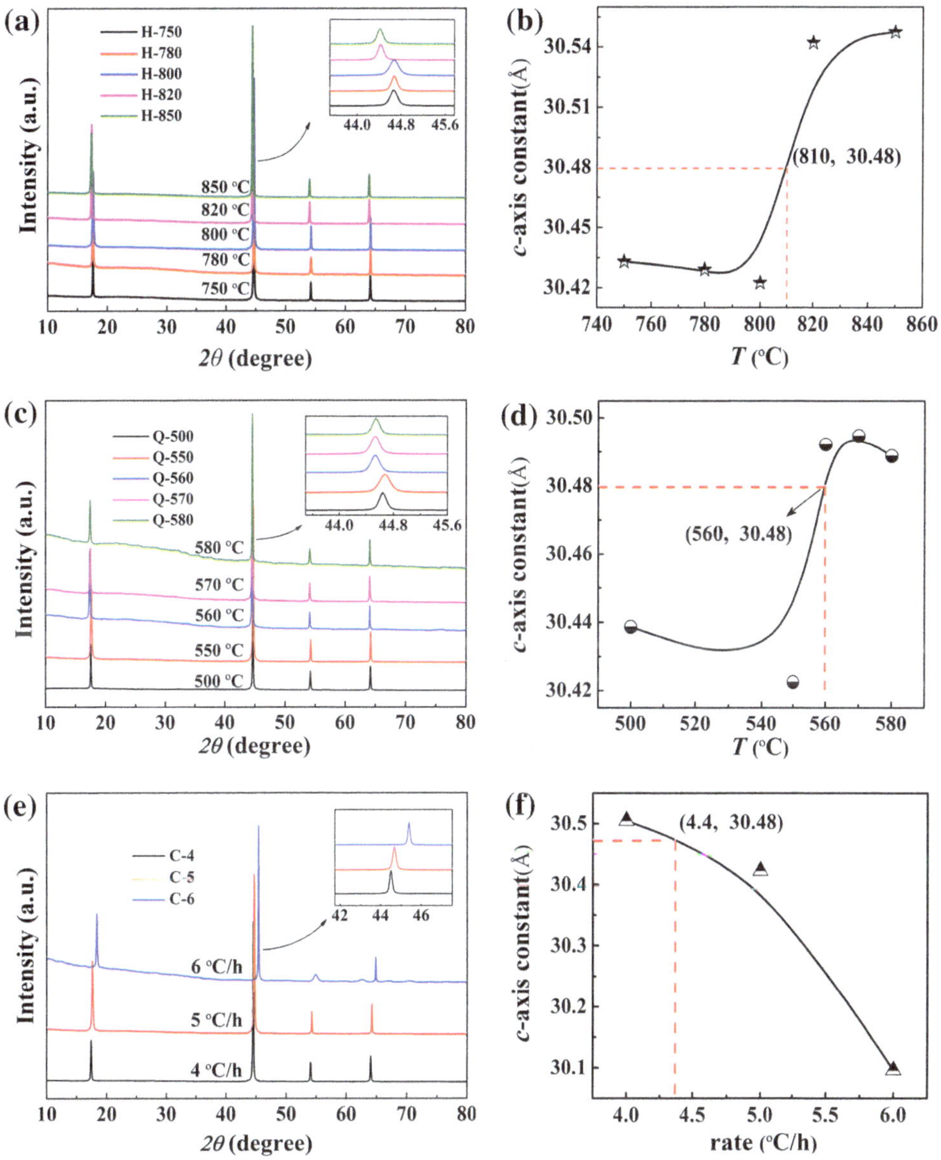

Fig. 2 X-ray diffraction curves of the H-750–H-850 (**a**), Q-500–Q-580 (**c**), and C-4–C-6 (**e**) and curves of cell parameters *c* with the highest temperature (**b**), quenching temperature (**d**), and cooling rate (**f**)

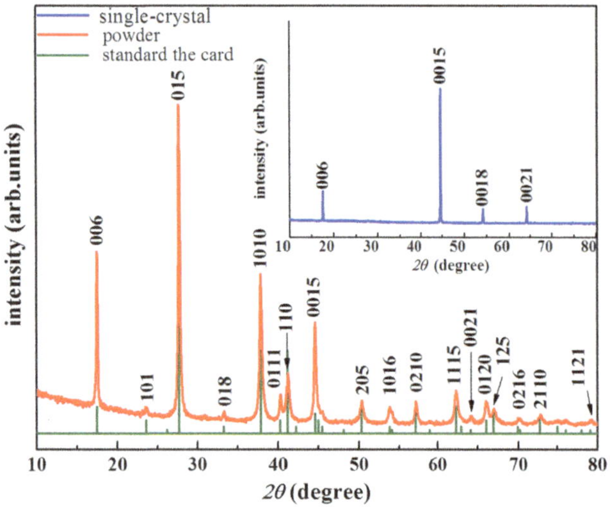

Fig. 3 Bi_2Te_3 powder X-ray diffraction pattern, inset is single crystal X-ray diffraction pattern

Table 2 Five strong peaks value with its corresponding interplanar distance *d*

Peak	006	015	1,010	110	0,015
d	5.07422	3.22059	2.37772	2.19265	2.03361

of the main peak marked in the picture can better match with standard pattern, indicating that the high-purity Bi_2Te_3 sample was obtained. The single-crystal sample only has (00*L*) peaks (*L* is 3*n*). It grows along the *c*-axis with the order of $Te^{(1)}$–Bi–$Te^{(2)}$–Bi–$Te^{(1)}$ and crystallizes in the hexagonal system with space group of R/3 m. We calculated lattice parameters by taking the five strong peaks' interplanar distances *d* (Table 2) from the XRD pattern of Bi_2Te_3 powder. The lattice parameters are $a = b = 4.38456$ and $c = 30.50063$ Å, which are consistent basically with the standard PDF card ($a = b = 4.3852$ and $c = 30.4830$ Å). However, the *c* values of the samples H-800 and Q-550 are 30.42248 and 30.42369 Å, respectively, which have large difference with standard values.

It can be seen from Fig. 4a that the sample surface has a bright metallic luster; there is an obvious hierarchical structure, and the surface is relatively flat. Figure 4b is EDX spectrum picture of Bi_2Te_3, Bi(L) and Te (M) in weight percentage of 52.82 % and 47.18 %, and the atomic percentage of 40.61 % and 59.39 %, respectively. In this sample, the atom ratio of Bi and Te was close to 2:3, Bi slightly excess. These results also deviated in component the stoichiometric ratio. It is consistent with the Bi–Te alloy phase diagram.

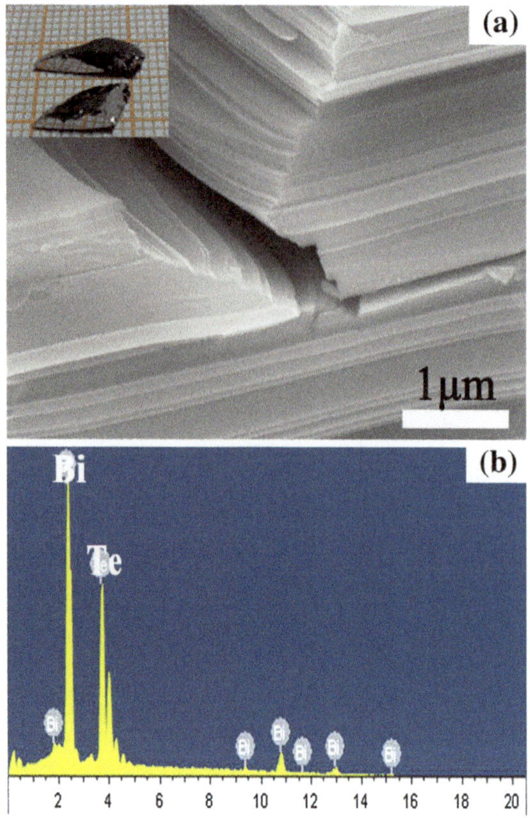

Fig. 4 SEM photographs of the single crystal of Bi_2Te_3 in 25,000 times and inset shows its physical picture. The conditions are that the highest temperature is 810 °C, quenching temperature is 560 °C, and the cooling rate is 4.4 °C/h (**a**) and EDX spectrum picture of the Bi_2Te_3 single crystal (**b**)

Figure 5a shows the resistance dependence of temperature curve of the Bi_2Te_3 with the magnetic field parallel to the *c*-axis, and it can be seen that the sample shows metallic behavior over the whole temperature range. The values of resistances increase gradually with the temperature and the intensity of the applied magnetic field. The entire curves have the same variation trend. We did not observe any induced insulators conduction or abnormal behavior due to the additional magnetic field. The tiny fluctuation of the resistance is found near 50 K. Figure 5b is the curve of the resistances dependence of the magnetic field parallel to the *c*-axis at $T = 5$ K. The values of resistances increase gradually with the applied magnetic field. At the high field, the curve of *R* versus *H* is linear. However, it deviates from the straight line at the part of the low filed. The linear variation law of the magnetic resistance may be associated with the surface state of TI. At low field the magnetic resistance deviates from the linearity, which may be associated with the weak anti-localization (WAL).

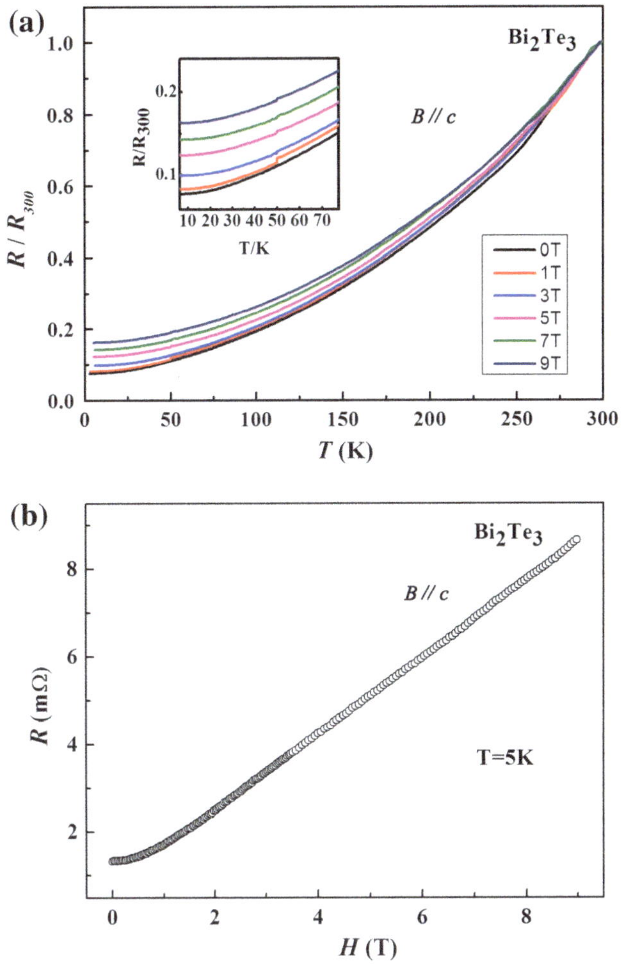

Fig. 5 **a** The *R–T* curve (0–9 T) of the Bi_2Te_3. The resistances at 300 K are 0.14331, 0.14224, 0.14527, 0.14495, 0.14729, and 0.14737 Ω, respectively. **b** The *R–H* curve ($T = 5$ K) of the Bi_2Te_3

4 Conclusions

We prepared the high-quality single-crystal synthesis of TI Bi_2Te_3 with the self-flux method. By controlling single variable and adjusting the parameters of heat treatment, we determined the optimal conditions for the highest temperature 810 °C, the quenching temperature 560 °C, and the cooling rate 4.4 °C/h. Bi_2Te_3 single crystal prepared under this condition has smooth surface with bright metallic luster and obvious layered structure, showing metallic behavior over the temperature range 7–300 K. At low temperature, the resistance increases as additional magnetic field increase, and shows linear magnetic resistance behavior in the high field.

Acknowledgments This work was supported by the National Magnetic Confinement Fusion Science Program (Grant No. 2011GB112001), the Program of International S&T Cooperation (Grant No. 2013DFA 51050), the National Natural Science Foundation of China (Grant No. 11104224, 11004162, 51271155), the Research Fund for the Doctoral Program of Higher Education of China (Grant No. 20110184120029), the Fundamental Research Funds for the Central Universities (Grant No. 2682013ZT16, SWJTU11ZT31, 2682013CX004, SWJTU11BR063), and the Science Foundation of Sichuan Province (Grant No. 2011JY0031, 2011JY0130).

References

1. Kane CL, Mele EJ (2005) Z_2 topological order and the quantum spin Hall effect. Phys Rev Lett 95:146802
2. Bernevig BA, Zhang SC (2006) Quantum spin Hall effect. Phys Rev Lett 96:106802
3. Qu DX, Hor YS, Cava RJ (2012) Quantum oscillations in magnetothermopower measurements of the topological insulator Bi_2Te_3. Phys Rev Lett 109:246602
4. Wang J, Chang CZ, Li HD et al (2012) Interplay between topological insulators and superconductors. Phys Rev B 85:045415
5. Bernevig BA, Hughes TL, Zhang SC (2006) Quantum spin Hall effect and topological phase transition in HgTe quantum wells. Science 314:1757
6. Kong DS, Chen YL, Cha JJ, Zhang QF et al (2011) Ambipolar field effect in the ternary topological insulator $(Bi_xSb_{1-x})_2Te_3$ by composition tuning. Nat Nanotechnol 6:705–709
7. Scherrer H, Scherrer S (1996) In: Rowe DM (ed) Handbook of thermoelectrics. CRC Press, New York, pp 211–237
8. Hor YS, Richardella A, Roushan P et al (2009) P-type Bi_2Se_3 for topological insulator and low-temperature thermoelectric applications. Phys Rev B 79:195208
9. Qi XL, Hughes TL, Zhang SC (2008) Topological field theory of time-reversal invariant insulators. Phys Rev B 78:195424
10. Hsieh D, Qian D, Wray L et al (2008) A topological Dirac insulator in a quantum spin Hall phase. Nature 45:970–974
11. Zhang HJ, Liu CX, Qi XL et al (2009) Topological insulators in Bi_2Se_3, Bi_2Te_3 and Sb_2Te_3 with a single Dirac cone on the surface. Nat Phys 5:438–442
12. Chen YL, Analytis JG, Chu JH et al (2009) Experimental realization of a three-dimensional topological insulator Bi_2Te_3. Science 325:178–181
13. Hsieh D, Xia Y, Qian D et al (2009) Observation of time-reversal-protected single-Dirac-cone topological-insulator states in Bi_2Te_3 and Sb_2Te_3. Phys Rev Lett 103:146401
14. Wang ZH, Qiu RLJ, Lee CH et al (2013) Ambipolar surface conduction in ternary topological insulator $Bi_2(Te_{1-x}Se_x)_3$ nanoribbons. ACS Nano 7(3):2126–2131
15. Ni HL, Zhao XB (2007) Synthesization and properties of Bi_2Te_3 based nanocomposite thermoelectric materials. Zhejiang University
16. Kim TS, Kim IS, Kim TK et al (2002) Thermoelectric properties of p-type 25% Bi_2Te_3+75%Sb_2Te_3 alloys manufactured by rapid solidification and hot pressing. Mater Sci Eng B 90:42–46
17. Keawprak N, Sun ZM, Hashimoto H et al (2005) Effect of sintering temperature on thermoelectric properties of pulse discharge sintered $(Bi_{0.24}Sb_{0.76})(2)Te_3$ alloy. J Alloy Compd 397:236–244
18. Seo J, Lee C, Park K (1998) Effect of extrusion temperature and dopant on thermoelectric properties for hot-extruded p-type Te-doped $Bi_{0.5}Sb_{1.5}Te_3$ and n-type SbI_3-doped $Bi_2Te_{2.85}Se_{0.15}$. Mater Sci Eng B 54:135–140
19. Zhang HT, Luo XG, Wang CH et al (2004) Characterization of nanocrystalline bismuth telluride (Bi_2Te_3) synthesized by a hydrothermal method. J Cryst Growth 265:558–562
20. National Bureau of Standards (US) Monograph 25 (1964) Sec 3, 16

16

Control strategy of maglev vehicles based on particle swarm algorithm

Hui Wang · Gang Shen · Jinsong Zhou

Abstract Taking a single magnet levitation system as the object, a nonlinear numerical model of the vehicle–guideway coupling system was established to study the levitation control strategies. According to the similarity in dynamics, the single magnet-guideway coupling system was simplified into a magnet-suspended track system, and the corresponding hardware-in-loop test rig was set up using dSPACE. A full-state-feedback controller was developed using the levitation gap signal and the current signal, and controller parameters were optimized by particle swarm algorithm. The results from the simulation and the test rig show that, the proposed control method can keep the system stable by calculating the controller output with the full-state information of the coupling system, Step responses from the test rig show that the controller can stabilize the system within 0.15 s with a 2 % overshot, and performs well even in the condition of violent external disturbances. Unlike the linear quadratic optimal method, the particle swarm algorithm carries out the optimization with the nonlinear controlled object included, and its optimized results make the system responses much better.

Keywords Maglev control · Vehicle–guideway coupling vibration · Particle swarm algorithm · Full-state feedback

H. Wang (✉) · G. Shen · J. Zhou
Institute of Railway and Urban Mass Transit, Tongji University, Shanghai 201804, China
e-mail: wh053@163.com

G. Shen
e-mail: elsg@sh163.net

J. Zhou
e-mail: jinsong_zhou@126.com

1 Introduction

Levitation control, as one of the most important techniques for electromagnetic suspension (EMS) vehicles, has been always drawing the worldwide experts' attentions and favors [1, 2]. Due to the guideway flexibility, the magnet couples with the elastic guideway in dynamics when levitating on it. The vehicle and the guideway may vibrate intensively if the controller is designed inappropriately, which will affect the stable operation of maglev vehicles [3]. Increase of the guideway mass and its stiffness is the common engineering application to avoid the phenomena, but this will result in a substantial increase in the construction cost of maglev line, and severely restrain the promotion of maglev transportation [4, 5]. Plenty of academics have been focusing on the problem of vehicle–guideway coupling vibration in recent years and some progresses have been made. Lee et al. [6] established a model of guideway beam with the modal superposition method, and studied the influence of some factors on the levitation gap and the dynamic magnification factor, such as the vehicle speed, irregularity, guideway deflection ratio, span length, span continuity, and damping ratio. Yau [7, 8] investigated the influence of girder settlement on the system dynamic characteristics when the maglev vehicle moves on a series of guideways. To suppress the self-excited vibration, Zhou and Hansen [9] proposed a concept of a virtual tuned mass damper (TMD) that used an electromagnetic force to emulate the force of a real TMD acting on the girder. Wang et al. [10, 11] investigated the disadvantages that the traditional maglev line structure brings to solve the vehicle–guideway coupling vibration, and proposed a new structure of maglev line system. However, the maglev vehicle–guideway coupling system is composed of a magnet and elastic guideways. In the levitation controller, if the feedback variables cannot reflect completely the system dynamic information, the acceptable dynamic properties of system are hardly obtained. The authors' recent studies [12, 13] also indicated that the introduction of guideway kinematic information into the controllers is conductive to the system stability, and they developed the full-state-feedback controllers using the linear quadratic regulator (LQR) algorithm.

For the levitation control system, the controlled object is nonlinear. When the LQR algorithm is used, however, the controller parameters are calculated based on linearized models, which is different from the objective fact. Consequently, when the system deviates far away from the equilibrium point, the system performance may go worse. Furthermore, the index function in the LQR algorithm has to be expressed in the quadratic form, which brings great restrictions. Therefore, in this work, taking the nonlinear vehicle–guideway coupling system as the controlled object, we develop a controller with the full states feedback method and calculate the output with the full dynamic information. Then, the controller parameters are optimized with the particle swarm optimization (PSO) algorithm, and the nonlinear controlled object is included in the optimization process, which distinguishes it from the traditional linear optimization algorithm. In addition, a test rig for the elastic-track single magnetic levitation system is built, on which the effectiveness of the controlled strategy is verified.

2 Nonlinear controlled object

Due to the application of independent controllers and mechanical decoupling devices, the single magnet levitation system can be treated as the basic unit of maglev vehicles, and its stable levitation is the basis for the system to achieve stable operation. Hence, it is feasible to investigate the dynamic properties of maglev vehicles by studying the single magnet levitation system. The schematic diagram of a single magnet-guideway coupling system is shown in Fig. 1, in which the levitation magnet is treated as a lumped mass as its length is rather smaller than the guideway span.

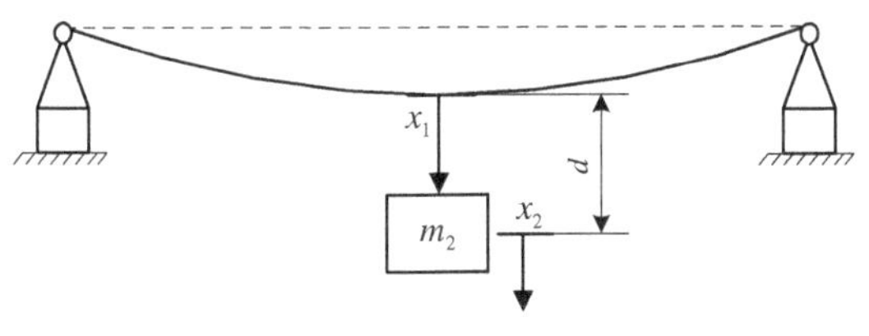

Fig. 1 Vehicle–guideway coupling model of the single magnet

In terms of the Newton's second law, the magnet dynamics equation can be written as

$$m_2\ddot{x}_2 = -F + f_d, \tag{1}$$

where m_2 is the magnet mass, F denotes the force between the track and the supporting beam, and f_d is the constant load acting on the magnet.

The voltage across the electromagnet U can be expressed as

$$U = R \cdot I + \frac{\mu_0 n^2 A_N}{2d} \cdot \dot{I} - \frac{\mu_0 n^2 A_N I}{2d^2} \cdot \dot{d}, \tag{2}$$

where R is the coil resistance, n is the number of coil turns, I is the current in the coil, A_N is the magnet working area, d is the levitation gap, and μ_0 is the vacuum permeability.

The electromagnetic force between the magnet and the guideway can be expressed as [14]

$$F = \frac{\mu_0 n^2 A_N}{4}\left(\frac{I}{d}\right)^2. \tag{3}$$

To analyze the dynamic properties of the vehicle–guideway coupling system, the guideway is generally treated as a simply supported Bernoulli–Euler beam [15]. According to the modal superposition theory [16], its dynamic characteristics can be described by the product of the shape function and the generalized coordinate, as follows:

$$X_1 = \sum_{i=1}^{n} \Delta_i(x) q_i(t), \tag{4}$$

where q_i and Δ_i are respectively the ith modal coordinate and the order shape function of the simply supported beam. In the normal coordinate system, the guideway vibration differential equation can be expressed as

$$\ddot{q}_i + 2\xi_i p_i \dot{q}_i + p_i^2 q_i = F\Delta_i, \tag{5}$$

where p_i and ξ_i denote the ith modal frequency and damping ratio, respectively.

The guideway span is very large in engineering practice, and it is unrealistic to build a 1:1 maglev vehicle–guideway coupling vibration test rig in laboratory. In addition, it can only guarantee the similarity in the first natural frequency when a small proportional continuous beam is built to simulate a real guideway, but almost impossible in the high-order mode. According to the vibration theory of the continuous beam, the damping effect becomes much stronger, and more vibration nodes exist at high-order modes. Therefore, compared with the lower-order modes, the high-order modes contribute less to the integrated dynamic performances. Therefore, it is reasonable to describe the guideway dynamic behaviors using its first-order mode, as shown by Eq. (6):

$$\ddot{q}_1 + 2\xi_1 p_1 \dot{q}_1 + p_1^2 q_1 = F\Delta_1 \tag{6}$$

If multiplying $\Delta_1 (\Delta_1 \neq 0)$ to the both sides of Eq. (6) and setting $x_1 = q_1\Delta_1$, we can derive the following equation to describe the vibration characteristics of the guideway midpoint:

$$\ddot{x}_1 + 2\xi_1 p_1 \dot{x}_1 + p_1^2 x_1 = F\Delta_1^2. \tag{7}$$

Making a further transformation to Eq. (7), we can obtain

$$m_1\ddot{x}_1 + c\dot{x}_1 + k_t x_1 = F, \tag{8}$$

where

$$m_1 = \frac{1}{\Delta_1^2}, c = \frac{2\xi_1 p_1}{\Delta_1^2}, \text{ and } k_t = \frac{p_1^2}{\Delta_1^2}.$$

Obviously, from Eq. (8), the single magnet-guideway coupling system can be simplified to a system as shown in Fig. 2. In this figure, the suspended track system, composed of the lumped mass and the spring-damping element, is used to simulate the vibration of the specified location on the guideway.

3 Design of the control system

3.1 Control strategy

The full-state-feedback control strategy, by calculating the controller output with the full information of the coupling system, is supposed to keep the system soundly stable. However, not all the state variables are measurable in engineering practices [17]. Therefore, taking the current signal and the levitation gap signal as the outputs, we try to derive full system states through the state observer and develop a full-state controller. If we suppose that $\boldsymbol{x}_o$ and $\boldsymbol{y}_o$ are the system state and output gained by state observer, and $\boldsymbol{G}$ is the feedback matrix of the output errors, the schematic of control system can be illustrated by Fig. 3.

In Fig. 3, $\boldsymbol{y}_r = [d \quad I]^T$,

$$\boldsymbol{A}_o = \begin{bmatrix} 0 & 1 & 0 & 0 & 0 \\ \frac{P_s - k_t}{m_1} & -\frac{c}{m_1} & -\frac{P_s}{m_1} & 0 & \frac{P_I}{m_1} \\ 0 & 0 & 0 & 1 & 0 \\ -\frac{P_s}{m_2} & 0 & \frac{P_s}{m_2} & 0 & -\frac{P_I}{m_1} \\ 0 & (\eta - 1)\frac{P_s}{P_I} & 0 & (1-\eta)\frac{P_s}{P_I} & -\frac{R}{L} \end{bmatrix},$$

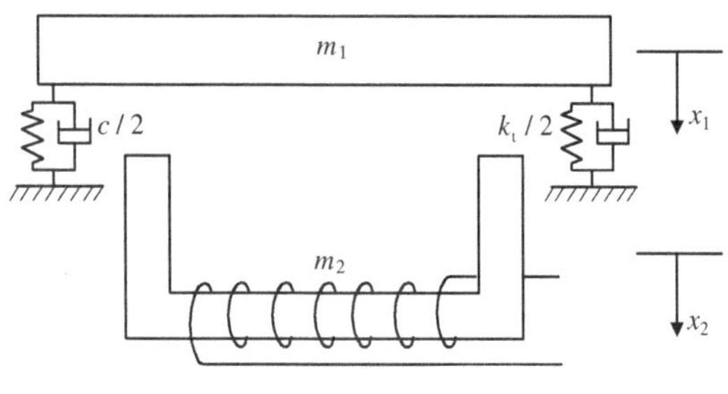

Fig. 2 Simplification of vehicle-guideway coupling model

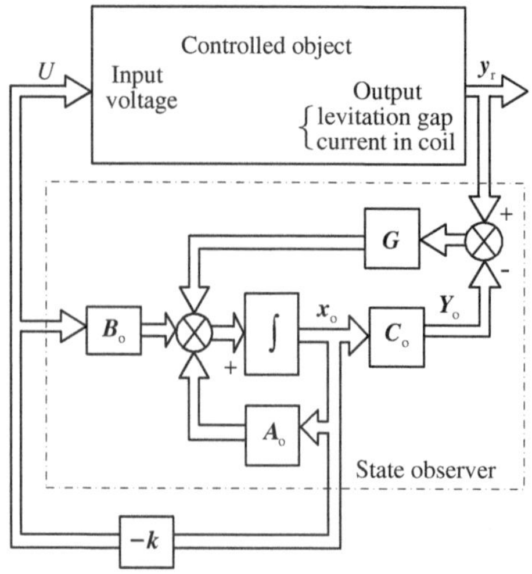

Fig. 3 Schematic of maglev control system

$$\boldsymbol{B}_o = \begin{bmatrix} 0 \\ 0 \\ 0 \\ 0 \\ \frac{1}{L} \end{bmatrix}, \boldsymbol{C}_o = \begin{bmatrix} -1 & 0 \\ 0 & 0 \\ 1 & 0 \\ 0 & 0 \\ 0 & 1 \end{bmatrix}^T,$$

where P_s and P_I are the gap coefficient and current coefficient, respectively; η is the magnetic flux leakage; and L is the electrical inductance.

The control strategy can be expressed by the following equation

$$U = -\boldsymbol{k}\hat{\boldsymbol{x}}, \tag{9}$$

where $\boldsymbol{k} = [K_1 \quad K_2 \quad K_3 \quad K_4 \quad K_5]$, and it is the controller feedback coefficient vector; $\hat{\boldsymbol{x}} = [\hat{x}_1 \quad \hat{\dot{x}}_1 \quad \hat{x}_2 \quad \hat{\dot{x}}_2 \quad \hat{I}]^T$, and its elements are successively the track vertical displacement, the track velocity, the magnet displacement, the magnet velocity, and the current in the magnet coil.

3.2 Optimization of controller parameters

The vector $\boldsymbol{k}$ can be calculated using the LQR algorithm, but this kind of algorithm is based on some linearized models. Furthermore, the index function for the LQR algorithm has to be expressed in the quadratic form, which brings great restrictions. Therefore, we take the nonlinear vehicle–guideway coupling model as a part of the closed system and optimize the controller parameters using the PSO algorithm.

The PSO algorithm is a stylized representation of the movement of organisms in a bird flock or fish school. It works by having a population (called a swarm) of candidate solutions (called particles). These particles are moved around in the search-space according to a few simple formulae. The movements of particles are guided by their own best known positions in the search-space, as well as the entire swarm's best known positions [18].

If we suppose that the solution space is d-dimension and the number of particles is n, the present position and historical optimal position of the ith particle can be respectively expressed by two vectors: $\boldsymbol{z}_i = [z_{i1} \;\; z_{i2} \;\; \cdots \;\; z_{id}]$, $\boldsymbol{p}_i = [p_{i1} \;\; p_{i2} \;\; \cdots \;\; p_{id}]$. By setting the swarm's best position $\boldsymbol{g} = [g_1 \;\; g_2 \;\; \cdots \;\; g_d]$, we can update their positions and velocities according to the following two equations [19]:

$$v_{ij}(t+1) = wv_{ij}(t) + c_1 r_1(t)[p_{ij}(t) - z_{ij}(t)] + c_2 r_2(t)(g_j(t) - z_{ij}(t)), \tag{10}$$

$$z_{ij}(t+1) = z_{ij}(t) + v_{ij}(t+1), \tag{11}$$

where the subscript i is the particle identification and j the element identification; r_1 and r_2 are the random numbers within the scope $[0, 1]$; w is the inertia weight; c_1 and c_2 are the learning factors and they determine the influences of experiential information from the particle itself and others, respectively. A bigger w accelerates the iteration convergence speed in the early stage, but weakens the local searching ability and reduces the solution accuracy. On the contrary, a smaller w will result in a narrow searching space. In order to get a proper compromise between the searching ability and accuracy, the value of w in this paper decreases linearly with the increase of iterations, as shown by

$$w = w_{\max} - \frac{w_{\max} - w_{\min}}{i_{\max}} \cdot i. \tag{12}$$

where $i_{\max}$ represents the maximum iterations, and i is the present iteration number. According to Eq. (12), w decreases within the scope $(w_{\min}, w_{\max}]$.

4 Test rig for the single magnet static-levitation system

In order to verify the effectiveness of control strategies, a test rig for the elastic-track single magnetic levitation system is built based on dSPACE. The main components and connection block diagram are shown in Fig. 4.

In Fig. 4, the host PC is mainly used to develop the control strategies and provide a graphical user interface platform. The dSPACE used in this paper is a real-time control system based on the DS1104 controller board and equipped with several 16-bit A/D and D/A converters. For the single magnet levitation table, as shown in Fig. 5, the magnet is bolted at one end of the swing arm, and the other end of the arm is pined to the base, so that the movement of magnet can be regarded as a linear movement in the vertical direction when the arm swings within a small angle range. The track (the steel plate in Fig. 5) is connected to the base via the springs, and it can vibrate vertically. The elastic vibration of guideway beam in practice is simplified as the track's vibration. By alternating different springs, we can change the track natural frequency. The main parameters of the test rig are shown in Table 1.

5 Simulation and test on the test rig

A nonlinear model of the vehicle–guideway coupling system was established in MATLAB/Simulink according to the schematic of the closed-loop system (Fig. 3). To facilitate the comparison between the simulation results and the test rig results, the parameter values in the model are set according to Table 1. In the PSO algorithm, the fitness function is used as the criteria to judge the personal best positions and the swarm's best position. The following factors are taken into account when we set the fitness function: the levitation gap, the magnet acceleration, and the voltage across the magnet, which are, respectively, related with the driving safety, the ride comfort, and the required control energy. Therefore, the fitness function J can be written as

$$J = \int_0^{\infty} \left(q_1 \hat{d}^2 + q_2 \ddot{x}_2^2 + rU^2\right) \mathrm{d}t, \tag{13}$$

where q_1, q_2, and r are the corresponding weighted coefficients. The J expressed by Eq. (13) can be also treated as

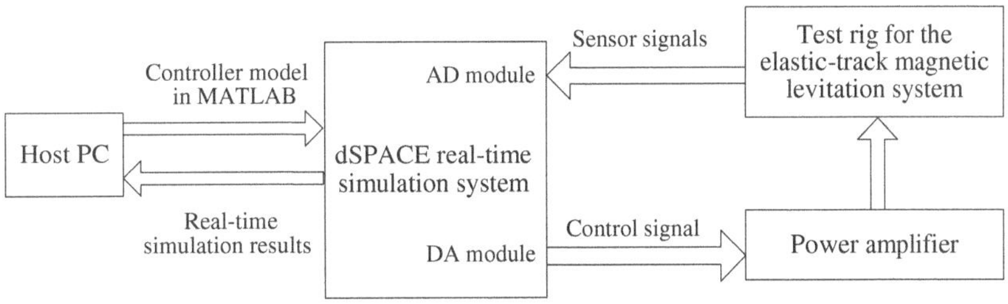

Fig. 4 Diagram of test rig for the elastic-track magnetic levitation control system

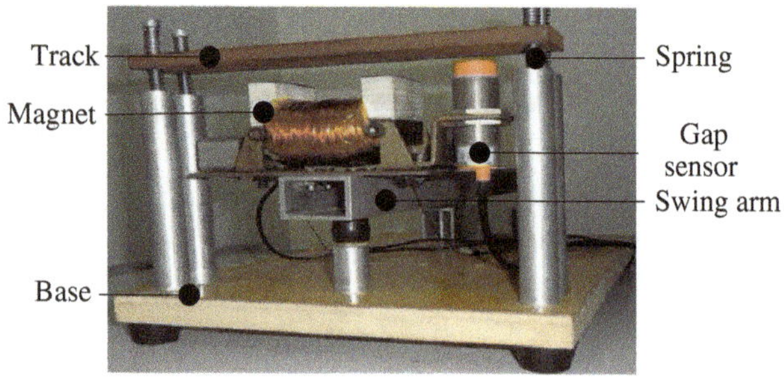

Fig. 5 Test rig for the elastic-track magnetic suspension system

Table 1 Parameters of the test rig

Physical quantities	Values
Mass of the magnet, m_2	1.8 kg
Mass of the track, m_1	0.8 kg
Electric resistance of the magnet, R	12.5 Ω
Electric inductance of the magnet, L (gap: 10 mm)	0.3 H
Magnetic flux leakage, η (gap: 10 mm)	0.65
Stiffness of the springs, k_t	30 kN/m
Current sensor	ACS712ELCTR-05
Gap sensor	LXC-M15P2

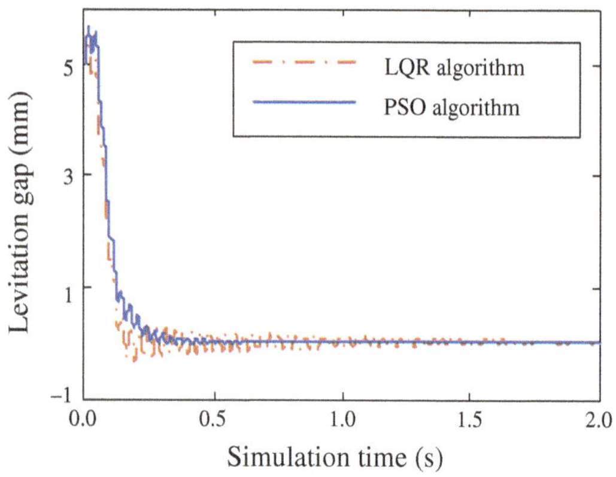

Fig. 6 Levitation gap of the magnet

the index function for the LQR algorithm. Taking $q_1 = 1$, $q_2 = 1 \times 10^6$, and $r = 1$, we derive two $\boldsymbol{k}$'s with the LQR algorithm and the PSO algorithm, respectively. By setting that the initial vertical position of the magnet deviates 5 mm from the equilibrium point and the simulation time is 2 s (the same values will be taken in the following time-domain simulations unless otherwise specified), we have calculated the time-domain system responses and the results are shown in Figs. 6 and 7. In the process of PSO algorithm, the particle number n is set to 40, the inertia weight w scope is (0.6, 1.2), and the maximum number of iteration is 100.

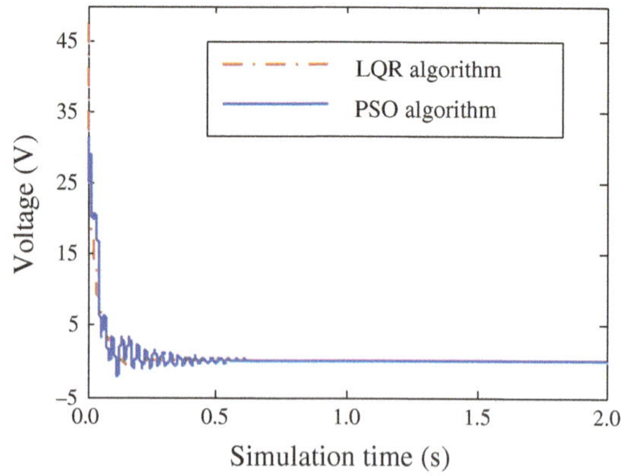

Fig. 7 Voltage across the magnet coil

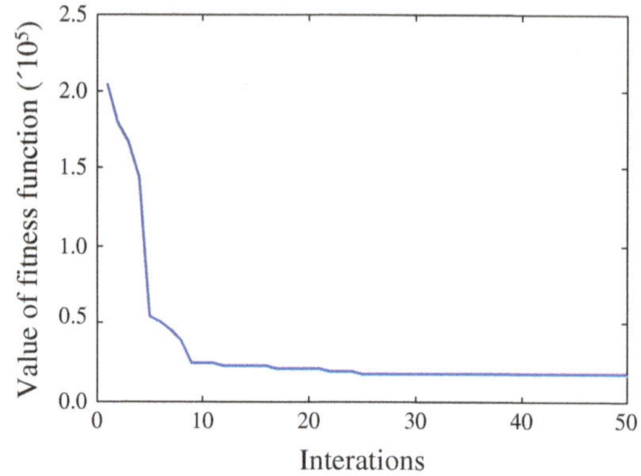

Fig. 8 PSO optimization process

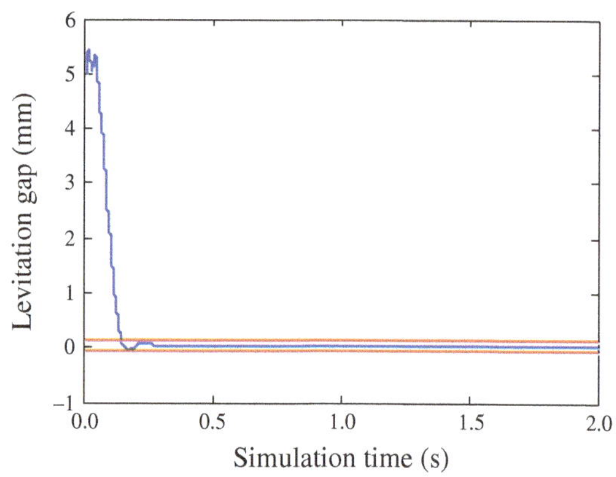

Fig. 9 Levitation gap of the magnet

From Figs. 6 and 7, compared with the LQR algorithm, the PSO algorithm reduces both the number of oscillations and the vibration amplitude. The dynamic performance of the system applying the PSO algorithm is much better, and the output of the controller is cut down obviously, which means the requirement of the power source is reduced.

In addition, the fitness function in the POS algorithm can be described in a much more flexible form. To obtain a

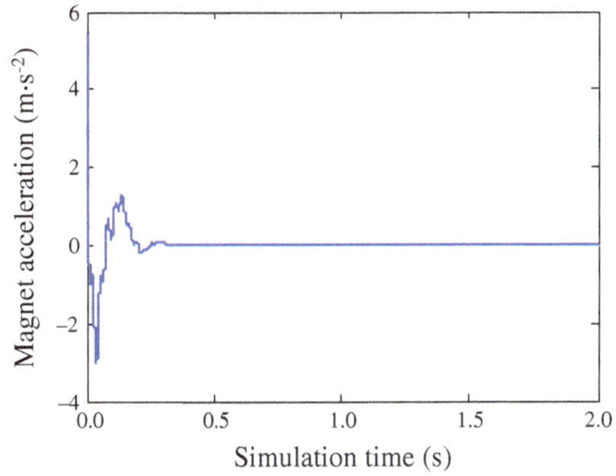

Fig. 10 Magnet acceleration

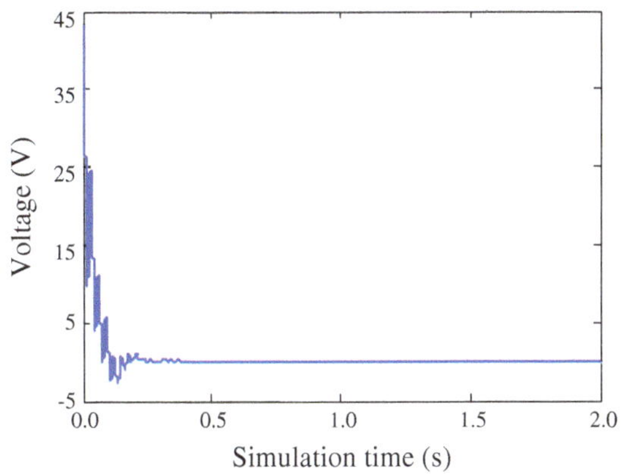

Fig. 11 Voltage across the magnet coil

better system response, we use the following equation as the fitness function:

$$J = \int_0^{\infty} \left(q_1 \hat{d}^2 + q_2 \hat{\ddot{x}}_2^2 + rU^2 \right) \mathrm{d}t + f_\mathrm{p} + f_\mathrm{v}, \tag{14}$$

where

$$f_\mathrm{p} = \begin{cases} 0, & |V_\mathrm{max}| < 45 \\ 500(V_\mathrm{max} - 45), & |V_\mathrm{max}| \geq 45 \end{cases}$$

is the function of maximum voltage V_max, $f_\mathrm{v} = 5\,n_\mathrm{o}$, and n_o denotes the number of oscillations.

By setting the particle number to 40, the inertia weight to (0.6, 1.2], and the maximum iteration to 50, we calculated the fitness function value, and the result is shown in Fig. 8. With the optimized controller parameters, the time-domain system responses are also calculated, as shown in Figs. 9, 10, 11. From the figures, the maximum overshoot is about 2.0 % and it takes about 0.14 s for the system to reach the steady state if the stable region is set to 2 %.

The effectiveness of the control strategy is tested on the hardware-in-loop test rig. Because of the existence of high-frequency noise, the gap signal and current signal from sensors are filtered. A simple first-order low-pass filter is adopted, and the transfer function is

$$H_\mathrm{f} = \frac{200}{s + 200}. \tag{15}$$

The time-domain system responses measured from the test rig are shown in Fig. 12. In the initial condition, the magnet deviates 3.3 mm from the equilibrium point, and the track is struck at the sixth and nineth seconds, respectively. From Fig. 12, it takes about 0.15 s for the system to reach the steady state, and the stable current in the coil is 1.78 A. After suffering two times of external violent shocks, the system can quickly return to the equilibrium position. The test results verify the effectiveness of the control method.

6 Conclusions

A nonlinear numerical model of vehicle–guideway coupling system is established. Based on the similarity in dynamic properties, the single magnet vehicle–guideway

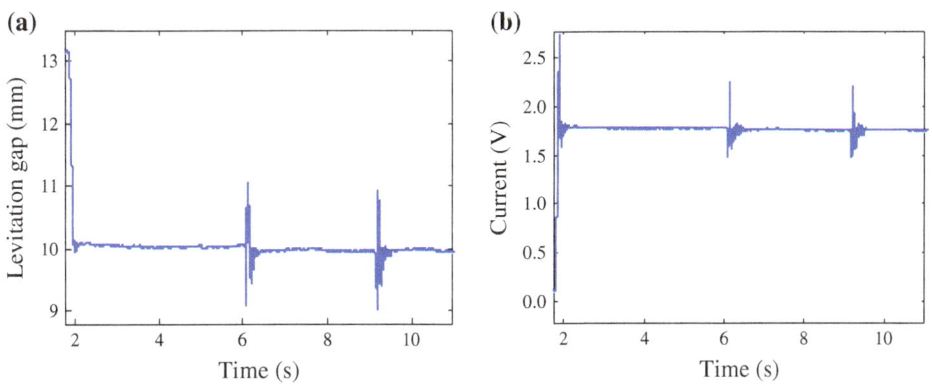

Fig. 12 Test data measured from test rig. **a** Levitation gap of magnet. **b** Current in the magnet coil

coupling system is simplified and the corresponding small-scale hard-in-loop test rig is built. A full-state-feedback controller was developed using the levitation gap signal and the current signal. The controller parameters are calculated using the LQR algorithm and the PSO algorithm, respectively. The results from the simulation and test rig show that, the control method proposed in this paper, by calculating the controller output with the full information of the coupling system, is effective to keep the system stable. Compared with the LQR algorithm, the PSO algorithm carries out the optimization directly taking the non-linear system as the object, and its optimized results make the system responses much better.

References

1. Li XN, Tong LS (2011) Technology research of medium and low speed maglev train. Electr Loc Mass Transp Veh 2(34):1–4 (in Chinese)
2. Lee HW, Kim KC, Lee J (2006) Review of maglev train technologies. IEEE Trans Magn 42(7):1917–1925
3. Zhou DF, Hansen CH, Li J et al (2010) Review of coupled vibration problems in EMS maglev vehicles. Int J Acoust Vib 15(1):10–23
4. Zhai WM, Zhao CF (2005) Dynamics of maglev vehicle/guideway system(I)—magnet/rail interaction and system stability. Chin J Mech Eng 41(7):1–10 (in Chinese)
5. Tao X (2008) Research on the low-speed magnetic levitation rail beam key design parameters. Tongji University, Dissertaion (in Chinese)
6. Lee JS, Kwon SD, Kim MY et al (2009) A parametric study on the dynamics of urban transit maglev vehicle running on flexible guideway bridges. J Sound Vib 328:301–317
7. Yau JD (2009) Vibration control of maglev vehicles traveling over a flexible guideway. J Sound Vib 321:184–200
8. Yau JD (2009) Response of a maglev vehicle moving on a series of guideways with differential settlement. J Sound Vib 324: 816–831
9. Zhou DF, Hansen CH, Li J (2011) Suppression of maglev vehicle–girder self-excited vibration using a virtual tuned mass damper. J Sound Vib 330:883–901
10. Wang H, Zhong XB, Shen G (2013) A new maglev line system design and control strategy. J Tongji Univ 41(7):1112–1118 (in Chinese)
11. Wang H, Zhong XB, Shen G (2013) Experimental study on the control strategy of maglev vehicle considering the track vibration. Paper presented at the 23rd Internal Symposium on Dynamics of Vehicles on Roads and Tracks, Qingdao 19–23 Aug 2013
12. Wang H, Zhong XB, Shen G et al (2013) Comparison between simulation and test results of an observer-controlled maglev vehicle on elastic guideway. Sonderdruck Schriftenreihe der Georg-Simon-Ohm-Hochschule Nürnberg 54:1–11
13. Wang H, Zhong XB, Shen G (2013) Analysis and experimental study on the maglev vehicle–guideway interaction based on the full-state feedback theory. J Vib Control.
14. Meisinger R (1975) Control systems for flexible maglev vehicles riding over flexible guideways. Veh Dyn Syst 4:200–202
15. Han HS, Yim BH, Lee NJ et al (2009) Effects of the guideway's vibrational characteristics on the dynamics of a maglev vehicle. Veh Syst Dyn 47(3):309–324
16. Zhang Z, Wang FQ (1991) Vibration analysis. Southeast University Press, Nanjing (in Chinese)
17. Katsuhiko O (2002) Modern control engineering. Tsinghua University Press, Beijing
18. Gaing ZL (2004) A partical swarm optimization approach for optimum design of PID control in AVR system. IEEE Trans Energy Conv 19(2):384–391
19. Shi YL, Hou CZ, Su HB (2008) Auto-disturbance-rejection controller design based on particle swarm optimization algorithm. J Syst Simul 20(2):433–436 (in Chinese)

17

Magnetic levitation performance of high-temperature superconductor over three magnetic hills of permanent magnet guideway with iron shims of different thicknesses

Yuming Gong · Gang Liang · Lifeng Zhao · Yong Zhang · Yong Zhao · Xuyong Chen

Abstract Superconducting magnetic levitation performance, including levitation force and guidance force, is important for the application of high-temperature superconducting maglev. Both of them are not only affected by different arrays of superconductors and magnets, but also by the thickness of the iron shim between permanent magnets. In order to obtain the best levitation performance, the magnetic field distribution, levitation force, and guidance force of a new type of three magnetic hills of permanent magnet guideway with iron shim of different thicknesses (4, 6, and 8 mm) are discussed in this paper. Simulation analysis and experiment results show that the guideway with iron shim of 8 mm thickness possesses the strongest magnetic field and levitation performance when the suspension gap is larger than 10 mm. However, with the decreasing of suspension gap, the guideway with iron shim of 4 mm thickness possesses the best levitation performance. The phenomena can be attributed to the density distribution of flux and magnetization of iron shim.

Keywords High-temperature superconductor · Three magnetic hills of permanent magnet guideway · Thickness of iron shim · Levitation force · Guidance force

Y. Gong · G. Liang · L. Zhao (✉) · Y. Zhang · Y. Zhao
Superconductivity and New Energy R&D Center, Southwest Jiaotong University, Chengdu 610031, China
e-mail: zhaolf@swjtu.edu.cn

X. Chen
Beijing Xinli Machinery Co, Ltd, Beijing 100039, China

1 Introduction

Since high-temperature superconducting (HTS) materials were found in 1986, people have paid much attention to the practical application of the permanent magnet–HTS levitation system. With the continuous development of superconducting technology [1–3], HTS maglev has been widely used in superconducting flywheel energy storage system [4], bearing [5, 6], HTS motors [7, 8], HTS cable [9], HTS maglev vehicle [10, 11], etc. On December 31, 2000, the world's first manned HTS maglev vehicle was successfully developed in Southwest Jiaotong University [11]. Subsequently, Russia and Germany developed their HTS maglev test vehicles in 2004 [12].

For HTS maglev vehicles, levitation and guidance force are very important parameters [13–19]. The improvement of levitation force can effectively enhance the carrying capacity of the maglev system. Guidance force directly determines the stability of the HTS maglev vehicles. Traditionally, permanent magnet guideway often possesses single magnetic hill [20], which cannot ensure the stability of HTS maglev vehicle running at high speed. In the meantime, researchers have paid much attention to the design of permanent magnet guideway, especially the permanent magnet arrangement in different ways [21–24]. However, little attention was paid to the iron shim inside guideway.

In this paper, the levitation and guidance force of high-temperature superconductor over the redesigned three magnetic hills of permanent magnet guideway with iron shim of different thicknesses are studied through simulation and experiments. The relationship between levitation performance and the thickness of iron shim obtained may provide a valuable reference for the optimum design of permanent magnet guideway for HTS maglev system.

2 Simulation analysis

2.1 Structure of permanent magnet guideway

The guideway consists of four permanent magnets with section size of 40 mm × 80 mm, three iron shims, and two other irons used as clips to fix the whole structure, as shown in Fig. 1. The trademark of the permanent magnet is N45. The arrows in Fig. 1 indicate the direction of magnetization of magnets. Thus, three magnetic hills' distribution over the permanent magnet guideway can be obtained.

2.2 Magnetic field simulation with Ansoft

By simulating magnetic field with Ansoft Maxwell, we calculated the distribution of magnetic induction intensity (B), as shown in Fig. 2. However, the actual magnetic induction intensity on the guideway surface is the same as that 2 mm above guideway in simulation [21], so the reference system was moved up 2 mm.

2.3 Magnetic field analysis

The thicknesses of iron shim are set as 4, 6, and 8 mm; and their guideways are marked with *a, b,* and *c*, respectively. The simulation results of B_Z (vertical component of B) at different heights above the three guideways are shown in Fig. 3.

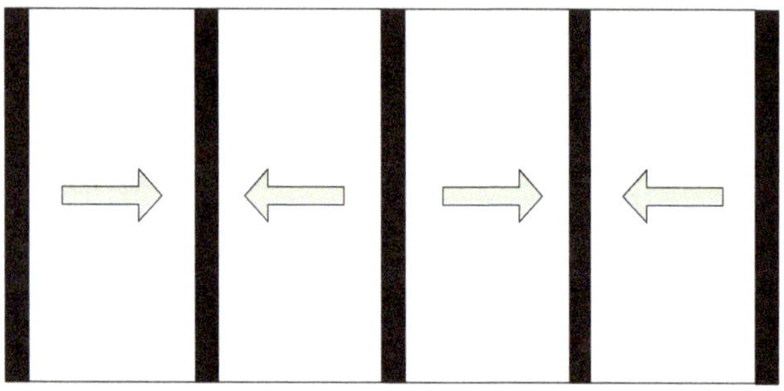

Fig. 1 The structure of permanent magnet guideway

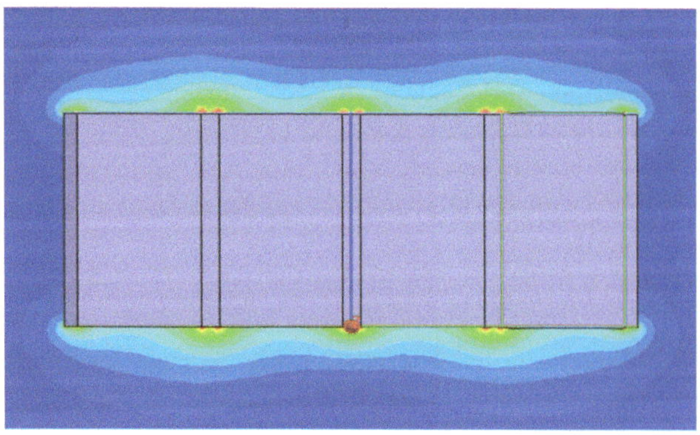

Fig. 2 Distribution diagram of magnetic induction intensity

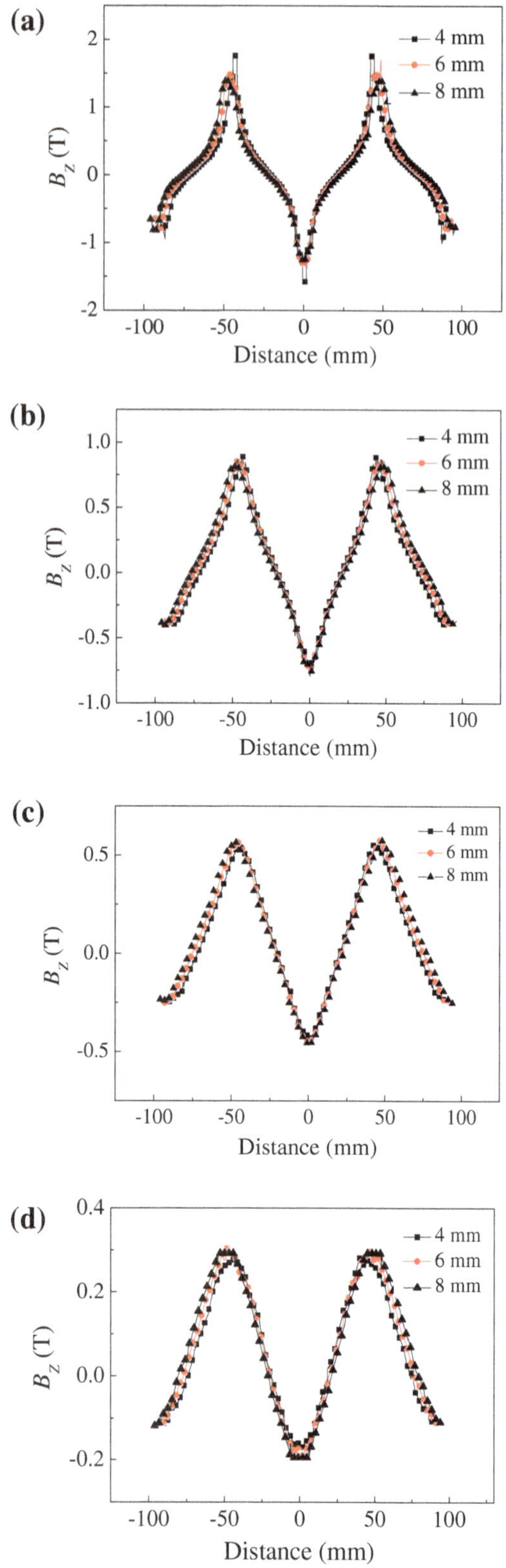

Fig. 3 B_Z distribution at different heights above guideways. **a** 0 mm. **b** 5 mm. **c** 10 mm. **d** 20 m

When the height is less than 5 mm above guideway, the B_Z of the guideway *a* with 4 mm iron shim is the largest, and it decreases with the height increasing. When the height is larger than 5 mm, with the increase of height, B_Z above guideway *a* is the least, while that above guideway *c* turns to be the largest. As we know, larger density of flux causes larger B_Z, and this should be the reason that the B_Z of the iron shim in 4 mm thickness is the largest when the height is less than 5 mm. However, the iron shim is magnetized by permanent magnets and presents flux. Thus the thicker iron shim should provide more flux. Furthermore, according to the magnetization curve of the iron shim, a thinner iron shim produces less flux than a thicker one in its saturation magnetization zone. This may cause B_Z above guideway *c* to be the largest when the height is larger than 5 mm.

Figures 4 and 5 show the distribution of B_Z above the middle and side iron shims, respectively, which indicate similar characteristics of B_Z distribution for iron shims in different place. Nevertheless, it also can be noticed that the B_Z of side iron shims is a little larger than that of the middle one.

As the levitation force is associated to the B_Z, we use B_Z to estimate the levitation force of the HTS bulks and magnet levitation system. We take the levitation force F_L simply as proportional to B_Z [14], namely$F_L \propto B_Z$.

3 Experiment

3.1 Tests of levitation force

Three YBCO superconducting bulks were placed above the three magnetic hills, as shown in Fig. 6. Superconducting bulks were located in a container and cooled with liquid nitrogen.

One of the real guideways applied in experiments is shown in Fig. 7.

The HTS bulks placed inside container were cooled at different heights (15, 20, 25, and 30 mm) and pressed to 5 mm above guideway, and then returned to the original locations, respectively. The container was moved down and up at a speed of 50 mm/min. The variations of the average levitation force with distance for HTS bulks are shown in Fig. 8.

According to the experimental results, no matter what the cooling height is, the levitation force of guideway *a* is less than that of the others. Analyzing the experimental data in detail, we found that the levitation force of

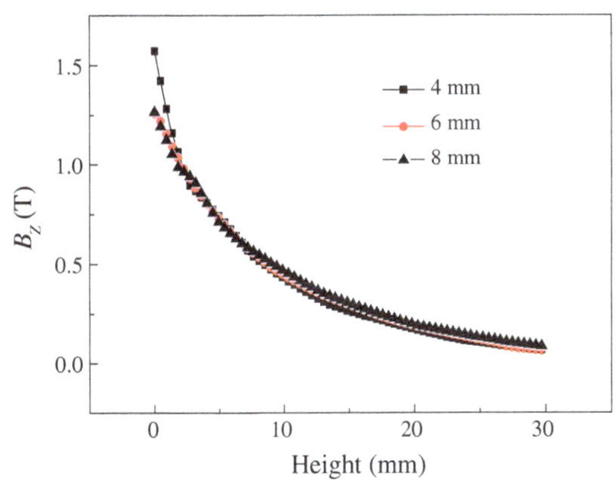

Fig. 4 B_Z distribution above middle iron shim

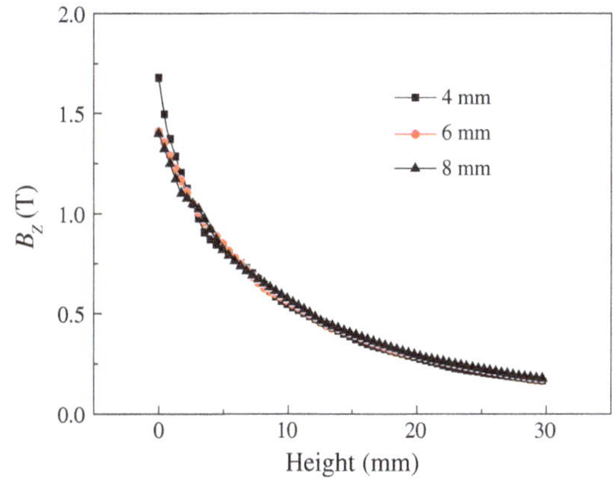

Fig. 5 B_Z distribution above side iron shim

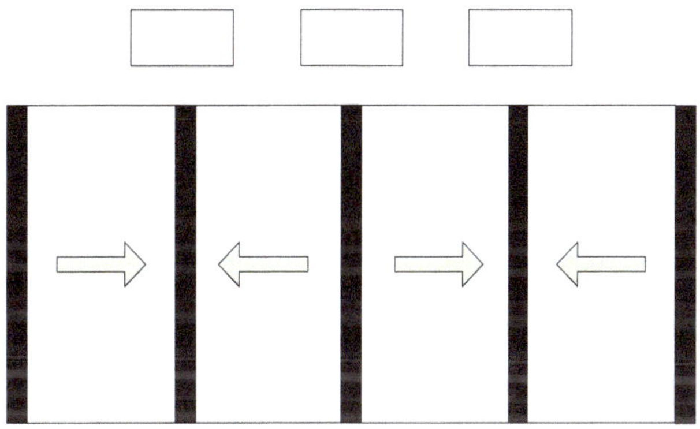

Fig. 6 HTS bulks above guideway

Fig. 7 Photography of experiment guideway *c*

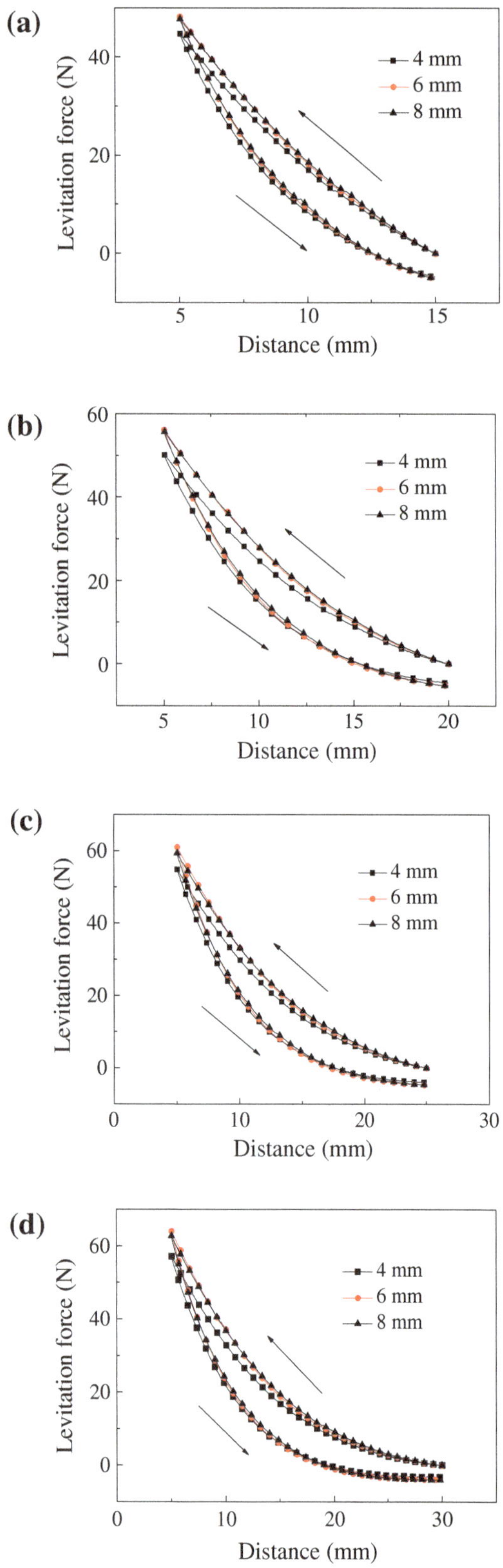

Fig. 8 Levitation force over different guideways with different cooling heights. **a** 15 mm. **b** 20 mm. **c** 25 mm. **d** 30 mm

Table 1 The actual B_Z distribution above middle iron shim

Height (mm)	B_Z (G)		
	Guideway *a*	Guideway *b*	Guideway *c*
0	14,022	12,911	10,746
5	7,366	7,453	7,357
10	4,445	4,564	4,630
15	2,703	2,905	3,136
20	1,811	1,985	2,088
25	1,064	1,268	1,373
30	707	803	936

Table 2 The actual B_Z distribution above side iron shims

Height (mm)	B_Z (G)		
	Guideway *a*	Guideway *b*	Guideway *c*
0	14,815	13,252	11,476
5	8,024	8,286	8,023
10	4,916	5,081	5,159
15	3,324	3,484	3,619
20	2,303	2,474	2,568
25	1,775	1,867	1,981
30	1,297	1,362	1,469

guideway *b* is a little larger than that of guideway *c* when the distance is less than 7 mm. With the increase of distance, however, the levitation force of guideway *b* become less than that of guideway *c*.

We further measured the vertical distribution of the B_Z in the center of middle iron shim and side iron shim from surface to the height of 30 mm. The results are shown in Tables 1 and 2.

B_Z distributions and their corresponding levitation force imply that higher B_Z causes larger levitation force. Comparing the distribution with levitation force, we take the relationship between them as $F_L \propto B_Z$ on the whole. It is obvious that the extra B_Z contributed by magnetized iron shim plays a key role in the process.

3.2 Tests of guidance force

HTS bulk over the permanent magnet guideway possesses guidance force when it is cooled with liquid nitrogen and captures a certain number of fluxes [16]. Guidance force has much to do with the field cooling height [16, 17], so the test height was set as the field cooling height of 15, 20, and 25 mm, respectively. The field cooling superconductor bulks were pushed along the horizontal direction by the

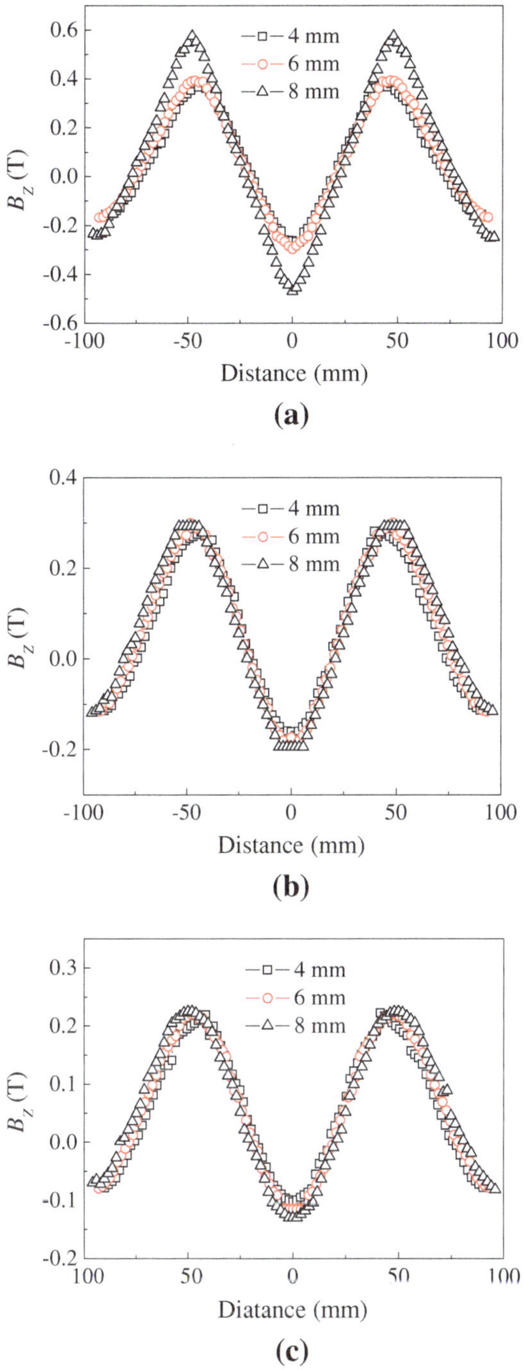

Fig. 9 Distribution of B_Z above guideways with different field cooling heights. **a** 15 mm. **b** 20 mm. **c** 25 mm

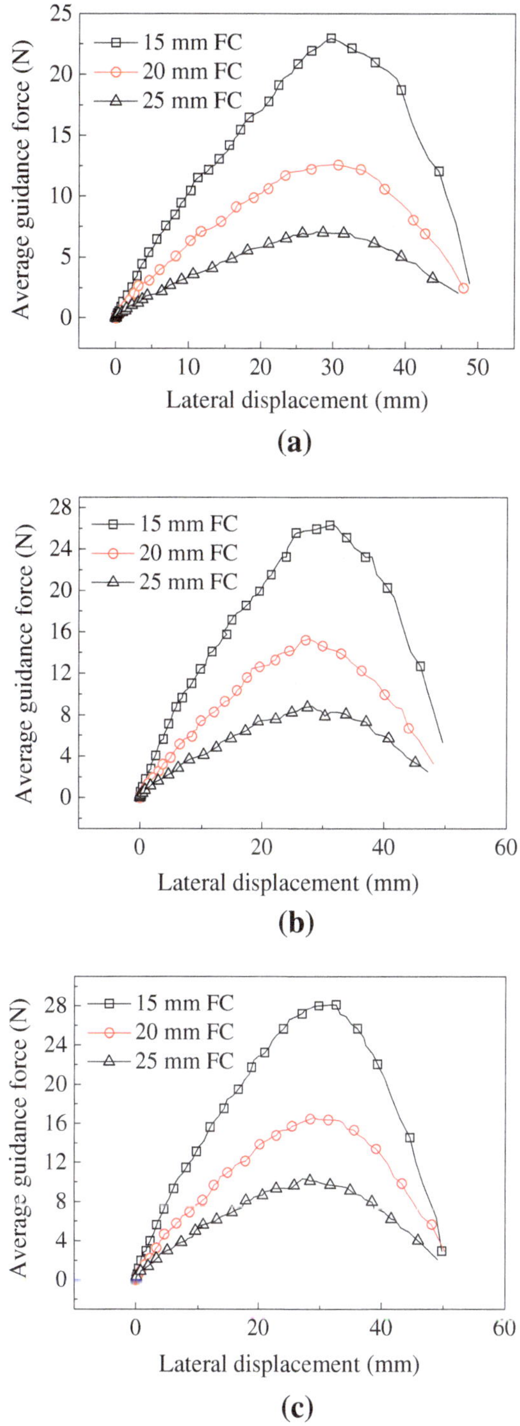

Fig. 10 Average guidance force of superconductor with different field cooling heights over the same guideway. **a** guideway *a*. **b** guideway *b*. **c** guideway *c*

stepper motor connected with a force sensor. The guideways *a*, *b*, and *c* were also used for the tests.

As we know, the guidance force is related to the magnetic flux captured by high-temperature superconductor, when it achieves the superconducting state [16, 17]. It can be seen from Fig. 9 that the B_Z of *c* is the largest among the three guideways. The higher the density of magnetic flux, the bigger the B_Z, and hence the larger guidance force. This rule is also reflected in Fig. 10.

Figure 10 shows the average guidance force of superconductor with different field cooling heights over the same guideway. The position where the superconductor

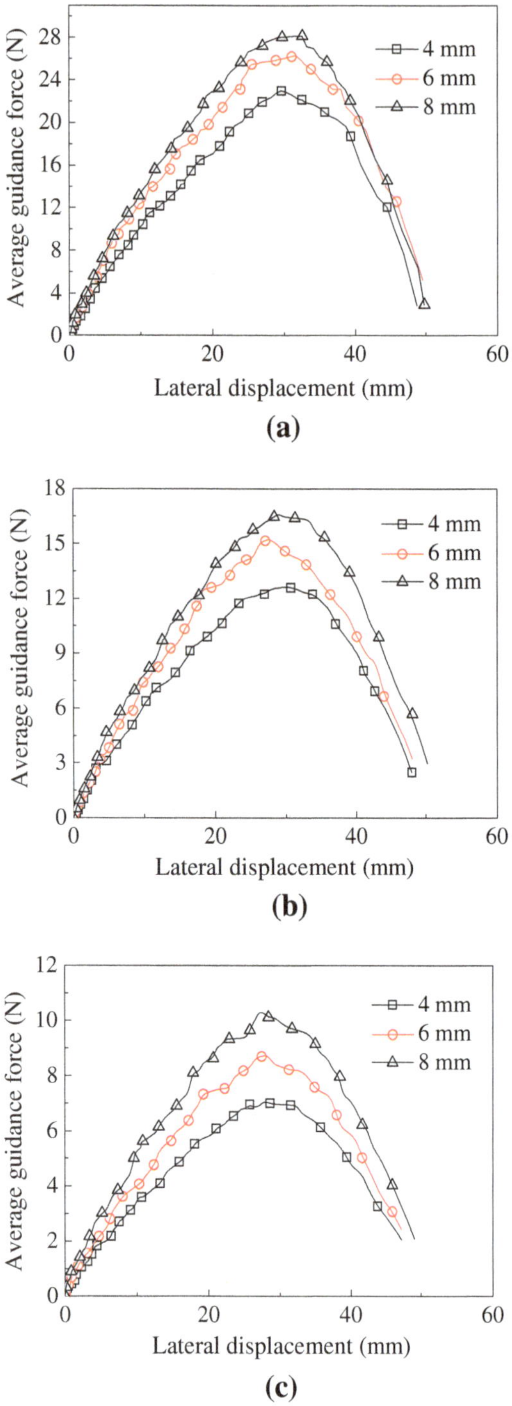

Fig. 11 Average guidance force of superconductor with different field cooling heights over different guideways. **a** 15 mm. **b** 20 mm. **c** 25 mm

was cooled is regarded as zero in displacement. We can see that when the displacement is less than 20 mm, namely half the size of a single permanent magnet, the guidance force is almost linear with the lateral displacement. When it is larger than 20 mm, with the displacement increasing, the guidance force gradually tends to saturation and decline at last. Otherwise, with the decrease of field cooling height, the performance of guidance force gets better, which is similar to that described in literature [16, 17] for the single magnetic hill guideway and Halbach twin hills permanent magnet array.

The average guidance forces of superconductor bulks for different field cooling heights over different guideways are shown Fig. 11. We can see that, for the same field cooling height of 15 and 20 mm, the thicker iron shim causes a larger guidance force. Moreover, the decrease of cooling height enlarges guidance force obviously. On the other hand, the decrease of the field cooling height will reduce the performance of HTS levitation force [16].

The main reason for the phenomenon observed above is due to the increase of captured flux with the iron shim getting thicker or the decrease of cooling height. The macro equivalent current density J, which is composed of eddy currents to maintain flux quantum beam in the numerous flux pinning centers, was enhanced with increase of the initial captured flux. The guidance force of HTS bulk will thus increase according to the guidance force calculation formula [16]:

$$F_G = \int_V J_Y B_Z \mathrm{d}V, \tag{1}$$

where F_G is guidance force, J_Y stands for the horizontal component of J, and V is the volume of the HTS bulk.

4 Conclusion

In this paper, simulation of the vertical component of magnetic induction intensity (B_Z) is combined with experiments on levitation performance to discuss the variation of guidance force and levitation force with the thickness of iron shim and cooling height. Narrower iron shim may cause stronger B_Z on the surface of guideway, but does not in the space. On the contrary,wider iron shim provides stronger B_Z at higher height above guideway. Actually, the working height of superconducting bulks is usually not on the surface of guideway. Thus, it is very important to choose a proper thickness of iron shim to obtain the strongest B_Z and the best levitation and guidance performance according to the different working height. This is meaningful for the application of HTS magnet levitation system.

Acknowledgments This work was supported by the National Magnetic Confinement Fusion Science Program (2011GB112001), the Program of International S&T Cooperation (S2013ZR0595), the National Natural Science Foundation of China (No. 51271155), the Fundamental Research Funds for the Central Universities (SWJTU11CX073, SWJTU11ZT16 and SWJTU11ZT31), and the Science Foundation of Sichuan Province (2011JY0031 and 2011JY0130).

References

1. Yang TX (2008) Status of research in high temperature superconductivity technology in china. J China Acad Electron Inf Technol 3(2):122–127
2. Jiang ZJ (2009) Development of high temperature superconductor applications. Cryog Supercond 37(10):17–21
3. Jin JX (2008) High temperature superconductivity in the past twenty years part 2—towards to practical applications. J Electron Sci Technol China 6(2):237–254
4. Suzuki T, Suzuki H, YasakaY, et al (1993) Adv Supercond (6): 1937
5. Fukuyama H, EndoS, TakizawaT, et al (1993) Adv Supercond (6): 1341
6. Deng ZG (2009) Research and development status of high temperature superconducting magnetic bearings. Trans China Electrotech Soc 24(9):1–8
7. Zheng LH (2007) Developments of high temperature superconducting motors. Electric Mach Control Appl 34(3):1–6
8. Xu YY (2013) A survey of the development of the low and high Tc superconducting motors/generators. Chin J Low Temp Phys 35(1):43–51
9. He Q (2013) A summarization of R & D on HTS cable. Cryogen Superconduct 41(5):31–35
10. Wang JS, Wang SY (2005) Synthesis of bulk superconductors and their properties on permanent guideway. Frontiers Superconduct Mater 4:885–912
11. Wang JS, Wang SY (2002) The first man-loading high temperature superconducting maglev test vehicle in the world. Physica 378–381:809–814
12. Schultz L, De Haax O, Verges P et al (2005) Superconductively levitated transport system - the SupraTrans project. Appl Supercond 15(2):2301–2305
13. Deng ZG (2008) Two kinds of potential application of superconducting bulk magnet for high temperature superconducting maglev vehicle. Rare Metal Mater Eng 37(A04):350–354
14. Ren ZY (2002) Studying about relation between levitation force of YBaCuO Bulk over NdFeB guideway and magnetic field and magnetic field gradient. Chin J Low Temp Phys 24(4):293–297
15. Ren ZY (2000) Levitation force of multi-block YBaCuO high temperature superconductors over a permanent magnetic railway. Cryogen Superconduct 28(2):17–21
16. Ren ZY (2004) Experimental investigation and numerical calculation on high temperature superconducting maglev based on a permanent magnetic guideway. Dissertation, southwest Jiaotong University
17. Qin YJ (2009) Study on the guidance force characteristics of bulk high-T_c superconductor over permanent magnetic guideway by experiment and simulation. Dissertation, southwest Jiaotong University
18. Wang XR (2003) Research on guidance force of high temperature superconductor YBCO bulks over the permanent magnet guideway. Dissertation, southwest Jiaotong University
19. Qin GP (2010) Study of the relationship between levitation force and the arrangement of high temperature superconductivity bulks over guideway. Chin J Low Temp Phys 32(4):257–261
20. Halbach K (1980) Nucl Instrum Methods 169:1–10
21. Liang X (2012) Simulation and optimization of permanent magnet guideway arrangement based on ansoft. Cryogen Superconduct 41(3):49–52
22. Guo F (2008) Structural parameter optimization design for HTS permanent maglev orbit. Rare Metal Mater Engineer 37(A04):385–388
23. Wang LL (2007) The levitation performance of YBCO above three permanent magnet guideways. Cryogen Superconduct 35(4): 300–303
24. Dang QH (2011) PMG optimization design based on HTS bulk magnetic stiffness characteristic. Cryogen Superconduct 39(7):32–35

18

Relative efficiency appraisal of discrete choice modeling algorithms using small-scale maximum likelihood estimator through empirically tailored computing environment

Hyuk-Jae Roh · Prasanta K. Sahu · Ata M. Khan · Satish Sharma

Abstract Discrete choice models are widely used in multiple sectors such as transportation, health, energy, and marketing, etc., where the model estimation is usually carried out by using commercial software. Nonetheless, tailored computer codes offer modellers greater flexibility and control of unique modelling situation. Aligned with empirically tailored computing environment, this research discusses the relative performance of six different algorithms of a discrete choice model using three key performance measures: convergence time, number of iterations, and iteration time. The computer codes are developed by using Visual Basic Application (VBA). Maximum likelihood function (MLF) is formulated and the mathematical relationships of gradient and Hessian matrix are analytically derived to carry out the estimation process. The estimated parameter values clearly suggest that convergence criterion and initial guessing of parameters are the two critical factors in determining the overall estimation performance of a custom-built discrete choice model.

H.-J. Roh (✉)
City of Regina, Regina, SK S4P 3C8, Canada
e-mail: hyukjae.roh@gmail.com

P. K. Sahu
National Institute of Construction Management and Research, Pune 411045, India
e-mail: prasantsahu222@gmail.com

A. M. Khan
Department of Civil and Environmental Engineering, Carleton University, Ottawa, ON K1S 5B6, Canada
e-mail: ata.khan@carleton.ca

S. Sharma
Environmental Systems Engineering, University of Regina, Regina, SK S4S 0A2, Canada
e-mail: Satish.Sharma@uregina.ca

Keywords Estimation algorithms · Visual basic application · Convergence criterion · Binary logit · Maximum likelihood

1 Introduction

It is common practice to carry out mathematical modeling for transportation planning, design, and operation by using readily available commercial software packages. Generally, the estimation processes are embedded in the software while developing the software package. This leads to constraints for users due to lack of information on the underlying analytical procedure for the model parameter estimation. This practice also applies to discrete choice model parameter estimation. There has been a general consensus that transportation modellers do not need to put extra efforts to come up with tailored computer codes dedicated solely for their unique modelling need. This may work in most cases; however, it may fail to work for very specific cases. In this context, the lack of flexibility in the computing environment could be exemplified through a situation, where the researchers choose 'non-linear-in-parameters' specification. Commercial packages cannot handle such a specification, and thus, the researcher might be forced to write a specific computer code dedicated only for that work [1, 2].

Advances in statistical computing language are readily available and could be applied as modeling base to all estimation procedures in a systematic way. A number of suitable computing languages for model estimation have improved significantly over the past years from, when reliance was almost exclusively on commercial software. As the statistical computing language of modeling has evolved significantly, more sophisticated modeling procedures could now be tailored by researchers in accordance with the level of transport modeling analysis. In an effort to relax limitations of software, researchers are strongly encouraged to use their own modeling computer codes [3]. On the basis of a clear understanding of analytical procedures that are repeatedly taking place during the estimation processes, researchers would be able to use different estimation algorithms and empirically modify/improve some portions of estimation processes. Eventually, with diverse numerical experiments, this repetitive task will help in proposing a new advanced/improved estimation procedure based on certain performance criterion.

Additionally, transportation researchers have not paid much attention on how to go beyond the limitation of using commercial packages in terms of carrying out entire modelling procedures with tailored computing environments [4]. This purpose is pursued in this research with a small scale discrete choice model, namely the binary logit model. The main objective of this research is to implement all analytical procedures involved in discrete choice modelling with visual basic application (VBA) and to present results of numerical experiments solely through possible tailored modeling. The other objectives of this research are (a) to identify the factors affecting the model estimation performance and (b) to suggest potential research topics related to custom-built modeling.

2 Derivation of multinomial logit

This section presents a binary logit choice function derived from the multinomial logit (MNL) choice function. The binary logit model is used as the target choice function for experimentation in this research. The derivation procedure adopted here helps in understanding not only the model's limitations but also the conceptual background based on which the model is built. Consequently, researchers can differentiate MNL from other types of discrete choice models with utmost clarity and may try with a variety of experimental model estimation trials. Based on the research by Train [3] and Ben-Akiva and Lermans [5], the algebraic manipulation to get a final logit choice function is described here in details. MNL model, which is a prototype model of discrete choice models, is formulated based on the random utility concept. The probability that any alternative in C_n is chosen by decision makers can be formulated as in Eq. 1. C_n is actual choice set that the decision maker (n) can consider in the given choice situation.

$$P_n(i) = P(U_{ni} > U_{nj}, \quad \forall j \in C_n, i \neq j), \tag{1}$$

where $P_n(i)$ is the probability of choosing a alternative i by decision maker n; U_{ni} and U_{nj} are representative utility of the alternatives i and j of the decision n, $\forall j \in C_n$, $i \neq j$ indicates comparison of all alternatives except for the same alternative. Based on assumption, the utility can be divided into systematic and random parts (shown in Eq. 2).

$$P_n(i) = P\left(S_{ni} + R_{ni} > S_{nj} + R_{nj}, \forall j \in C_n, i \neq j\right), \tag{2}$$

$$= P\left(R_{nj} < S_{ni} - S_{nj} + R_{ni}, \forall j \in C_n, i \neq j\right), \tag{3}$$

where S_{ni} is the systematic (or deterministic, observable, constant) part of the representative utility of alternative i when faced by decision maker n and is observable to a researcher, and R_{ni} is the random (unobservable, error) part of the representative utility of alternative i when faced by decision maker n and is unobservable to a researcher. Usually, various discrete choice models are derived from Eq. (3), which depends on the mathematical distribution (e.g., Gumbel Type I, and Weibull, etc.)that the random component ($R_{ni,}$ R_{nj}) follows. Therefore, using density function of Gumbel Type I (or Weibull), the density for each random parts relation is shown in Eq. 4 and the cumulative distribution is presented in Eq. 5.

$$f(R_{na}) = e^{-R_{na}} e^{-e^{-R_{na}}}, \tag{4}$$

$$F(R_{na}) = e^{-e^{-R_{na}}}. \tag{5}$$

In Eq. (3), if R_{ni} is not known, then the choice probability of $P_n(i)$ is a cumulative distribution, whose value is determined at S_{ni}– S_{nj} + R_{ni}, as given in Eq. (5).

$$F\left(R_{nj}\right) = e^{-e^{-(S_{ni}-S_{nj}+R_{ni})}}, \tag{6}$$

where R_{nj} is distributed independently for all alternatives, $i \neq j$ and it is logical that the Eq. 3 means that the choice probability is the product of the individual value of cumulative distribution obtained using Eq. 6. Equation 7 presents this function.

$$P_n(i)|R_{ni} = \prod_{j \neq i} e^{-e^{-(S_{ni}-S_{nj}+R_{ni})}}. \tag{7}$$

As mentioned earlier, R_{ni} is unknown and is assumed to have a distribution. By integration of $P_n(i)|R_{ni}$, $P_n(i)$ will be determined as shown in Eq. 8.

$$P_n(i) = \int \left(\prod_{j \neq i} e^{-e^{-(S_{ni}-S_{nj}+R_{ni})}}\right)(e^{-R_{ni}} e^{-e^{-R_{ni}}}) \mathrm{d}R_{ni}. \tag{8}$$

Some more algebraic interpolations need to be made to reach the final form of logit choice function. The remained processes are given below:

$$P_n(i) = \int_{R_{ni}=-\infty}^{\infty} \left(\prod_{j \neq i} e^{-e^{-(S_{ni}-S_{nj}+R_{ni})}}\right)\left(e^{-R_{ni}} e^{-e^{-R_{ni}}}\right) \mathrm{d}R_{ni}, \tag{9}$$

$$= \int_{R_{ni}=-\infty}^{\infty} \left(\prod_{j \neq i} e^{-e^{-(S_{ni}-S_{nj}+R_{ni})}}\right)\left(e^{-R_{ni}} e^{-e^{-(S_{ni}-S_{ni}+R_{ni})}}\right) \mathrm{d}R_{ni}, \tag{10}$$

$$= \int_{R_{ni}=-\infty}^{\infty} \left(\prod_{j} e^{-e^{-(S_{ni}-S_{nj}+R_{ni})}}\right)(e^{-R_{ni}}) \mathrm{d}R_{ni}, \tag{11}$$

$$= \int_{R_{ni}=-\infty}^{\infty} \left(e^{-e^{-(S_{ni}-S_{n1}+R_{ni})}} e^{-e^{-(S_{ni}-S_{n2}+R_{ni})}} \cdots e^{-e^{-(S_{ni}-S_{ni}+R_{ni})}} \right) (e^{-R_{ni}}) \mathrm{d}R_{ni}, \tag{12}$$

$$= \int_{R_{ni}=-\infty}^{\infty} \left(e^{-e^{-(S_{ni}-S_{n1}+R_{ni})} - e^{-(S_{ni}-S_{n2}+R_{ni})} \ldots - e^{-(S_{ni}-S_{ni}+R_{ni})}} \right) (e^{-R_{ni}}) \mathrm{d}R_{ni}. \tag{13}$$

The Eq. (13) can be simplified as follows:

$$P_n(i) = \int_{R_{ni}=-\infty}^{\infty} \left(e^{-\sum_j e^{-(S_{ni}-S_{nj}+R_{ni})}} \right) (e^{-R_{ni}}) \mathrm{d}R_{ni}, \tag{14}$$

$$= \int_{R_{ni}=-\infty}^{\infty} \left(e^{-e^{-R_{ni}} \sum_j e^{-(S_{ni}-S_{nj})}} \right) (e^{-R_{ni}}) \mathrm{d}R_{ni}. \tag{15}$$

Let $e^{-R_{ni}} = Z$, and after differentiating with respect to R_{ni}, it becomes $-e^{-R_{ni}}\mathrm{d}R_{ni} = \mathrm{d}Z$. It is noted that if $R_{ni} \to \infty$ then, $Z \to \infty$, if $R_{ni} \to \infty$ then $Z \to 0$. Replacing these terms in Eq. 15, the final logit choice function is obtained as below (See Eq. 18).

$$P_n(i) = \int_{Z=0}^{\infty} e^{-Z\sum_j e^{-(S_{ni}-S_{nj})}} \mathrm{d}Z, \tag{16}$$

$$= \left[\frac{e^{-Z\sum_j e^{-(S_{ni}-S_{nj})}}}{-\sum_j e^{-(S_{ni}-S_{nj})}} \right]_0^{\infty}, \tag{17}$$

$$= \frac{e^{S_{ni}}}{\sum_j e^{S_{nj}}}. \tag{18}$$

The final form of probabilistic MNL model consists of two parts. The denominator represents the total utility of all the alternatives considered in the choice set and theutility of a particular alternative is given in the numerator. The binary logit model is utilized as numerical experimental base in this research, which can be simplified from Eq. 18 by assuming that only two alternatives are available in the mode choice market. Assuming two alternatives (auto and transit) are available; the Eq. 18 can be simplified into the binary logit choice formulation as shown in Eq. 19, where S_{na} and S_{nt} are the systematic part of the representative utility of alternative auto and transit, respectively.

$$P_n(autoortransit) = \frac{e^{S_{na}} (or\ e^{S_{nt}})}{e^{S_{na}} + e^{S_{nt}}}. \tag{19}$$

3 Methodology

3.1 Data, utility, likelihood, and log-likelihood function

Numerical experiments, in this research, are carried out using simple database that are successfully applied by Ben-Akiva and Lermans [5] for model parameter estimation of a given model specification. The reason behind this is, that a proven data base provides guarantee for successful convergence of the experimental model run. Additionally, the value of parameter estimates generated through experimental runs can be compared to true values and hence it may be concluded that all experimental runs are conducted on the right estimation track. Data plugged in model specification within tailored computing environments allow research to keep track of computational efforts: such as log-likelihood values generated for each observation until model convergence is achieved.

The model specification exemplified in a Ref. [5] is adopted in this research in order to compare models in the exact same conditions. Two variables are included in model specification. The first one is a generic variable belonging to both modes (i.e., auto and transit) and the other one is a mode-specific constant for auto mode (since the transit mode is designated as the reference mode). In other words, T_A and T_T are included as service-level attributes in the utility function for these two modes. Consequently, the final model specifications are shown as below in Eqs. 20 and 21

$$V_{An} = \beta_1 + \beta_2 T_A, \tag{20}$$

$$V_{Tn} = \beta_2 T_T. \tag{21}$$

where V_{An} and V_{Tn} are the utility functions of auto and Transit. β_1 and β_2 are model parameters to be estimated in the computation process.

In the next step, we considered the estimation method, which is one of core parts in custom-built modeling to get an in-depth understanding about the data involvement in discrete choice analysis. We applied the maximum likelihood function throughout the research: developing the computer code, deriving formulas for elements of Hessian matrix, and Gradient vector etc. Newton–Raphson estimation algorithm is employed mainly because of its excellence in convergence characteristics, and, more importantly, its higher level of accuracy for the parameter estimated. Data used for this research is assumed to be selected at random from a population and the likelihood function of the entire sample is the product of the likelihoods of each observation. Therefore the final form of the function specified for this research is given in Eq. 22. We used the log-likelihood function for the ease of computation and this is given in Eq. 23. The binary logit choice function tailored formulations for the two modes (auto, transit) are given in Eq. 24.

$$L(\beta_1, \beta_2) = \prod_{n=1}^{21} P_n(A)^{y_{An}} P_n(T)^{y_{Tn}}, \tag{22}$$

$$LL(\beta_1, \beta_2) = \sum_{n=1}^{21} [y_{An} lnP_n(A) + y_{Tn} lnP_n(T)], \tag{23}$$

$$P_n(A) = \frac{e^{V_{An}}}{e^{V_{An}} + e^{V_{An}}}, \quad P_n(T) = \frac{e^{V_{Tn}}}{e^{V_{An}} + e^{V_{An}}}. \tag{24}$$

Replacing Eqs. 20, 21 and 24 in 23, the final form of log-likelihood function specified for this experimental study is presented in Eq. 25. This equation is used as a base equation for obtaining a variety of formulas with which a tailored-computer code is developed.

$$LL(\beta_1,\beta_2)=\sum_{n=1}^{21}\left[y_{An}\ln\left(\frac{e^{\beta_1+\beta_2T_A}}{e^{\beta_1+\beta_2T_A}+e^{\beta_2T_T}}\right)+y_{Tn}\ln\left(\frac{e^{\beta_2T_A}}{e^{\beta_1+\beta_2T_A}+e^{\beta_2T_T}}\right)\right]. \quad (25)$$

.

3.2 Derivation of gradient and Hessian formulas

The gradient of the log-likelihood function, which is a vector of the first derivatives of Eq. 25, plays an important role in determining the direction for parameter estimation. We derived the first-order partial derivatives of the Eq. 25; knowing the importance of the estimation procedure for developing the computer code. The most critical part of developing a computer program for estimation is to understand the relationship between the data and derived formulations used for estimation. Particularly, in discrete choice model, the functional form of the model looks simple; however mapping a relationship between the formulae and data set from the operational perspective is vital to define. In addition, the expansion of the given log-likelihood function is computationally extensive and costly [4]. In this section, we have presented only the final formulation of the gradient vector, which has been derived by using partial derivatives in analytical terms. The elements included in a gradient vector can be described as below in the mathematical form by Eqs. 26 and 27.

$$\frac{\partial LL(\beta_1,\beta_2)}{\partial\beta_1}=\sum_{n=1}^{21}[y_{An}P_n(T)-y_{Tn}P_n(A)], \quad (26)$$

$$\frac{\partial LL(\beta_1,\beta_2)}{\partial\beta_2}=\sum_{n=1}^{21}[y_{An}P_n(T)(T_{An}-T_{Tn})-y_{Tn}P_n(A)(T_{Tn}-T_{An})]. \quad (27)$$

We calculated the real gradient values through the estimation process and the value is constantly updated by new values until iteration process is terminated with final parameters estimates. A gradient vector of log-likelihood function can be summarized in a vector (column vector) form as shown on Eq. 28. In the context of this research, the column vector includes the two explanatory variables.

$$G_t=\frac{\partial LL(\beta_1,\beta_2)}{\partial\beta_k}=\begin{bmatrix}\frac{\partial LL(\beta_1,\beta_2)}{\partial\beta_1}\\ \frac{\partial LL(\beta_1,\beta_2)}{\partial\beta_2}\end{bmatrix}=\begin{bmatrix}\sum_{n=1}^{21}[y_{An}P_n(T)-y_{Tn}P_n(A)]\\ \sum_{n=1}^{21}[y_{An}P_n(T)(T_{An}-T_{Tn})-y_{Tn}P_n(A)(T_{Tn}-T_{An})]\end{bmatrix}. \quad (28)$$

Finally, the mathematical expression of a gradient vector can be translated into computer code based on the interpretation implied in the above formula expansion and its numerical value is continuously changed during iterations until the process is terminated.

In order to employ Newton–Raphson estimation algorithm as an iteration method in custom-built discrete choice modeling, it is needed to derive the second-order partial derivatives of the log-likelihood formulation for the given choice situation. Therefore it becomes obvious for the researchers (or modellers) to derive analytically the second-order partial derivatives of the log-likelihood function [4], which contains full information associated with sample used in the analysis. Indeed, the process involved in the analysis is extremely troublesome and computationally intensive; however, there is no commercial software that offers the same level of understanding to researchers. Also, the experience gained practically in a way that all procedures associated with model estimation are mimicked analytically. In this context, the Newton–Raphson algorithm combined with the real Hessian matrix shows the best estimation performance in discrete mode choice modeling compared to any other algorithm operated by approximate Hessian matrix such as BHHH, BHHH-2, BFGS to name a few [3]. Based on the knowledge of the Hessian matrix symmetry, only the off-diagonal elements of Hessian matrix are treated mathematically. The following part of this section is mainly focused on the description of the formulas resulted from the second-order partial derivatives of the log-likelihood function for the off-diagonal elements of the Hessian matrix. For the purpose of simplicity, off-diagonal elements included in the Hessian matrix is formulated in the forms of relations given by Eqs. 29,30 and 31 by calculating the second-order derivatives with respect to $\beta_1\beta_1$, $\beta_2\beta_2$ (or $\beta_1\beta_1$ and $\beta_2\beta_2$). The formulations are shown below.

$$\frac{\partial^2 LL(\beta_1,\beta_2)}{\partial\beta_1\partial\beta_1}=\sum_{n=1}^{21}[-y_{An}P_n(T)P_n(A)-y_{Tn}P_n(A)P_n(T)], \quad (29)$$

$$\frac{\partial^2 LL(\beta_1,\beta_2)}{\partial\beta_1\partial\beta_2(or\partial\beta_1\partial\beta_2)}=\sum_{n=1}^{21}[y_{An}P_n(T)P_n(A)(T_{Tn}-T_{An})-y_{Tn}P_n(T)P_n(A)(T_{An}-T_{Tn})], \quad (30)$$

$$\frac{\partial^2 LL(\beta_1,\beta_2)}{\partial\beta_2\partial\beta_2} = \sum\nolimits_{n=1}^{21}\left[-y_{An}P_n(T)P_n(A)(T_{An}-T_{Tn})^2 - y_{Tn}P_n(T)P_n(A)(T_{An}-T_{Tn})^2\right] \quad (31)$$

.

The three elements consist of Hessian matrix is given in Eq. 32.

$$H_t = \begin{bmatrix} \frac{\partial^2 LL(\beta_1,\beta_2)}{\partial\beta_1\partial\beta_1} & \frac{\partial^2 LL(\beta_1,\beta_2)}{\partial\beta_1\partial\beta_2} \\ \frac{\partial^2 LL(\beta_1,\beta_2)}{\partial\beta_2\partial\beta_1} & \frac{\partial^2 LL(\beta_1,\beta_2)}{\partial\beta_2\partial\beta_2} \end{bmatrix}. \quad (32)$$

Finally, all prerequisites that researchers need before the custom-built discrete choice modeling process are prepared.

3.3 Visual basic application (VBA)

Computer codes are written for estimation algorithms and the maximum likelihood method that are explained in the previous section. The VBA package in Microsoft Excel was utilized for this purpose. It is notable to know that there are two critical parts for researchers to be concerned about in developing computer codes. The first is to know how the log-likelihood function computer codes could be realized and the second is to know how the estimation algorithms could be included in the language of computer codes. All estimation runs are conducted using a personal computer (Intel Pentium 4, CPU 2.80 GHz, 512 MB Ram).

4 Iteration algorithms and estimation factors

In off-the-shelf software application for modeling, the software primarily enables a researcher to find the optimal set of parameter estimates under a given estimation algorithm, where, estimation factors are prefixed by the vendor. However in custom-built modeling environments, finding parameter estimates typically depends on iteration method predefined earlier in chosen estimation algorithm. Furthermore, to make the estimation operational, a researcher will be directed to input initial estimation factors which would affect the overall estimation performance at the initial stage of estimation. These factors include step size and convergence criterion. Six algorithms are applied to experimentally estimate parameters with a given model specification in the binary logit choice model framework. It is noticed during the course of this research, that the iteration method involved in each six algorithms mutates each other than updating process of approximate Hessian. Therefore, we discussed in detail the role of estimation factors including the estimation method based on the Newton–Raphson (NR) algorithm in the following paragraphs.

Newton–Raphson algorithm uses the real Hessian matrix in its estimation routine by employing true Hessian values generated from the second-derivative of log-likelihood function. The computations for obtaining real Hessian matrix are expensive and costly. Moreover, majority of work associated with the analytical derivation of log-likelihood function are extensive, nevertheless it is very useful [4]. In this research, the mathematical relationships for the real Hessian matrix are formulated by means of direct analytical derivatives from the given log-likelihood function (Eq. 25) and the final form of matrix is provided through the Eqs. 29–31. Translating estimation procedures into computer codes involves: 1) understanding the fundamental iteration rule for Newton–Raphson algorithm, and 2) separating the functionality of estimation procedures into independent, interchangeable modules. The estimation method of Newton–Raphson employed in this research can be derived by taking the second order Taylor's approximation of $LL(\beta_{t+1})$ around $LL(\beta_t)$ and the details are as follows. The Taylor's series expansion of $f(x)$ around $f(x_0)$ can be written as below [4].

$$f(x) = \frac{f(x_0)}{(0)!} + \frac{f^{(1)}(x_0)}{(1)!}(x-x_0) + \frac{f^{(2)}(x_0)}{(2)!}(x-x_0)^2 + \cdots + \frac{f^{(n)}(x_0)}{(n)!}(x-x_0)^n. \quad (33)$$

The above function can be shortened by taking the second-order approximation of Taylor's series as shown below:

$$LL(\beta_{t+1}) \cong \frac{LL(\beta_t)}{(0)!} + \frac{LL^{(1)}(\beta_t)}{(1)!}(\beta_{t+1}-\beta_t) + \frac{LL^{(2)}(\beta_t)}{(2)!}(\beta_{t+1}-\beta_t)^2, \quad (34)$$

$$= LL(\beta_t) + G_t(\beta_{t+1}-\beta_t) + \frac{1}{2}H_t(\beta_{t+1}-\beta_t)^2. \quad (35)$$

It should be noted that $f(x)$ is $LL(\beta_{t+1})$ and $f(x_0)$ is $LL(\beta_t)$ in Eq. (34). Based on the Eq. (35), the maximization value of β_{t+1} can be found by setting $\frac{\partial LL(\beta_{t+1})}{\partial\beta_{t+1}} = 0$. The first derivatives with respect to β_{t+1} results in

$$\frac{\partial LL(\beta_{t+1})}{\partial\beta_{t+1}} = G_t + H_t(\beta_{t+1}-\beta_t). \quad (36)$$

Consequently, by solving the Eq. 36 for the value of β_{t+1}, the iteration rule of Newton–Raphson is represented by the following formula:

$$\beta_{t+1} = \beta_t + \lambda(-H_{tReal}^{NR})^{-1} G_t, \tag{37}$$

where β_t is the parameter estimates after t iterations, β_{t+1} is the parameter estimates after $t + 1$ iterations, λ is the step size, $(-H_{tReal}^{NR})^{-1}$ is the inverse of the negative real Hessian matrix in iteration t, G_t is a vector of gradient in iteration t, $(-H_{tReal}^{NR})^{-1} G_t$ is the search direction of the function in iteration process. Stated more simply, β_{t+1} which is $(t + 1)$ th parameter estimates is equal to β_t which is tth parameter estimates plus negative inverse Hessian multiplied by gradient. A way of the estimation method of other five algorithms considered in this research can be found in detail in the previous study conducted by Roh and Khan [4], Train [3], and Green [6].

In this study, both H_{tReal}^{NR} and G_t are the average value of the Hessian and gradient. The Hessian and gradient are calculated at each step and continuously updated till the convergence. Based on the fact that the log-likelihood function of logit probability is always globally concave [7], the following paragraph explains how parameters are searched during iteration process from the initial guessing points to final convergence of parameter estimates.

The "the direction to follow" concept in the context of iteration method is shown in Fig. 1. G_t is the slope of the log-likelihood function and H_{tReal}^{NR} shows the change in slope of the log-likelihood function. In cases of global concave function, H_{tReal}^{NR} is always negative and thus $(H_{tReal}^{NR})^{-1}$ is also negative. The negative of this negative Hessian, $-(H_{tReal}^{NR})^{-1} G_t$ is positive. Therefore, the direction of $-(H_{tReal}^{NR})^{-1} G_t$ depends entirely on the sign of G_t. If the G_t is positive, β_t moves to the right direction (Fig. 1a), otherwise, β_t moves to the left direction (Fig. 1b) according to the iteration rule seen in Eq. 37. Its step size is given by a researcher at the initial stage of iteration and its magnitude is given by the term, $\lambda(-H_{tReal}^{NR})^{-1} G_t$, included in the iteration rule. In either case, β_t is moved to the best value that maximizes the log-likelihood function.

The step size implied by $\lambda(-H_{tReal}^{NR})^{-1} G_t$ is decreased, if the curvature represented by $-(H_{tReal}^{NR})^{-1}$ has a large value in its magnitude (Fig. 2a); otherwise, the step size is increased (Fig. 2b).

For understanding "step size" in the context of iteration method, it is useful to know the maximization of $LL(\beta)$:$(LL(\beta) = A + B\beta + C\beta^2)$, and it's convergence to

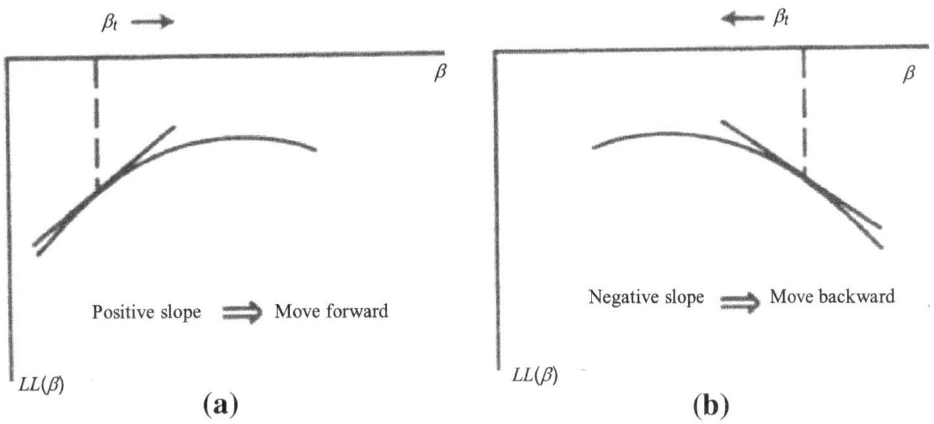

Fig. 1 Direction of step follows the slope [3]

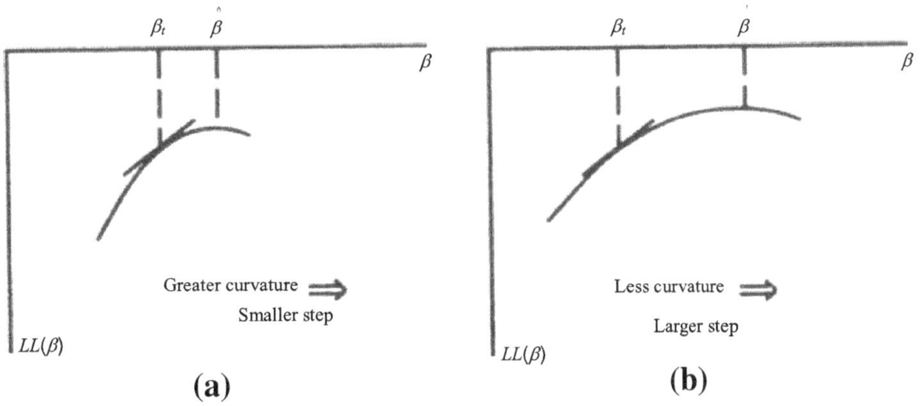

Fig. 2 Step size is inversely related to curvature [3]

maximum. In the case of quadratic function, the optimum value can be obtained with one iteration [3]. However, the problem is that most $LL(\beta)$ are not quadratic in nature and they require more than one iteration to attain the optimum. This problem can be solved by choosing an appropriate initial step size, and hence minimizing the iteration number and convergence time. The step size determination is explained in the subsequent paragraph.

4.1 Case–I: $LL(\beta)$ is not a quadratic function

Figure 3 shows two different functions represented by solid and dashed lines. The solid line represents the actual $LL(\beta)$ and the quadratic form is represented by the dashed line. After one iteration ($\lambda = 1$), if the given function is quadratic, then β_{t+1} reaches to the maximum point. But, the actual $LL(\beta)$ is not quadratic, thus β_{t+1} goes beyond the maximum point and as a result of which, $LL(\beta_{t+1}) < LL(\beta_t)$. To ensure that $LL(\beta_{t+1}) > LL(\beta_t)$, and to guarantee an increasing $LL(\beta)$, one can decrease λ to 1/2. After this operation, it is logical to compare the values between $LL(\beta_{t+1})$ and $LL(\beta_t)$. If $LL(\beta_{t+1}) < LL(\beta_t)$, then λ may be further decreased to 1/4 The λ value will be decreased to smaller values such as 1/8 and 1/16 repeatedly until $LL(\beta_{t+1}) > LL(\beta_t)$, is satisfied at certain λ.

4.2 Case–II: $LL(\beta)$ is a quadratic function

The second case is completely opposite to the first case. As shown in Fig. 4, the solid line is for actual $LL(\beta)$ and dashed line is for quadratic. At first iteration with $\lambda = 1$, β_{t+1}, reaches to the maximum for the quadratic case. But, the actual $LL(\beta)$ is not quadratic and the maximum point is far away and λ should be adjusted by taking a large step in order to satisfy $LL(\beta_{t+1}) > LL(\beta_t)$. In Fig. 4, it can be seen that $\lambda = 2$ is the best step size to be chosen for estimation process.

Subsequently, VBA tailored computing codes developed earlier are utilized for the six different estimation algorithms to compare their convergence behaviour in terms of performance measures. The following section presents and discusses the results from the experiments.

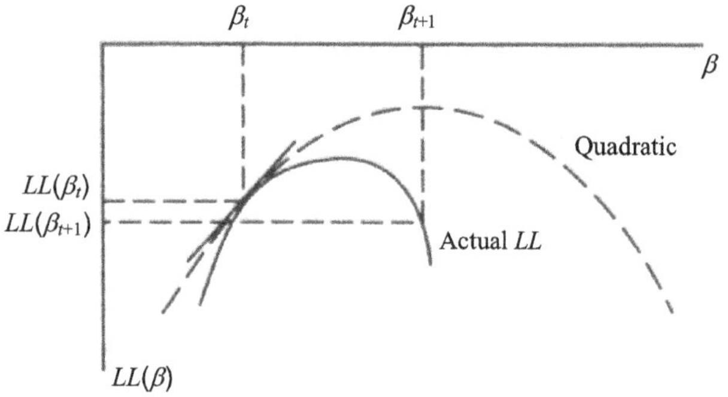

Fig. 3 Finding a proper step size in the first case [3]

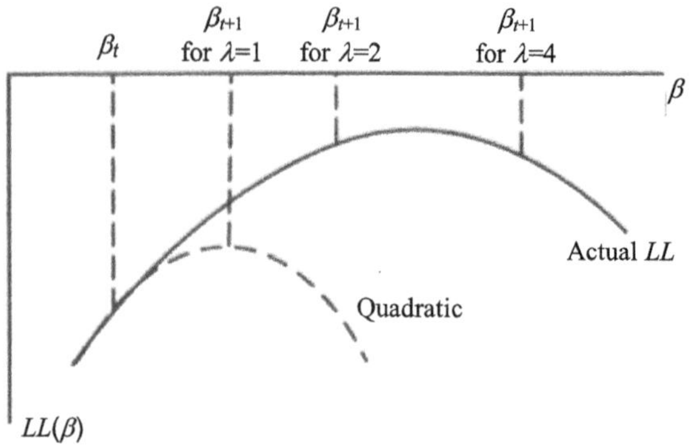

Fig. 4 Finding a proper step size in the second case [3]

5 Experimental results

The computer codes developed for this research worked successfully within the simulated discrete choice model estimation environments. The values of parameter estimates of the model are compared with true value [5] and it is found that the values of parameter estimates are closer in magnitude across six algorithms. This means that the computer codes developed for this empirical estimation worked on right estimation track. The parameter estimates for six algorithms are presented in Table 1 with different step sizes, which λ shows the best performance in a given convergence criterion: $CR = 10^{-4}$. It can also be noted that all estimation procedures involved in discrete choice modelling are completely translated into tailored computer estimation codes. The purpose of this research is to get in-depth understanding of calculation mechanism underlying in discrete choice modeling, not to develop a model. Therefore, statistical explanations of parameters are not within the scope of this research.

The relative performances of algorithms are compared by using the three performance measures: number of iteration, iteration time, and convergence time. In the first case, we chose different convergence criteria and the second case is involved with different initial guesses of the parameter values. After applying various step sizes along with the iteration rule, the experimental runs that converged successfully are selected and summarized.

Before presenting experiment results, overall convergence characteristics of six algorithms is presented using 3D convergence surface as shown in Fig. 5. Only a run showing the best performance for each algorithm is depicted in 3D convergence surface. Variation of log-likelihood values for all observations in each iteration are continuously saved in server until the estimation process is terminated. Unique

Table 1 Parameters estimated from six algorithms ($CR = 10^{-4}$)

Variables	NR ($\lambda = 1$)	BHHH ($\lambda = 1/2$)	BHHH-2 ($\lambda = 1/2$)	SA ($\lambda = 16$)	DFP ($\lambda = 16$)	BFGS ($\lambda = 8$)
β_1	−0.237575	−0.237462	−0.237428	−0.237588	−0.237575	−0.237576
β_2	−3.186590	−3.186410	−3.186355	−3.186671	−3.186590	−3.186590

NR Newton–Raphson, *BHHH* Berndt, Hall, Hall and Hausman, *SA* Steepest Ascent, *DFP* Davidon, Fletcher, and Powell, *BFGS* Broydon, Fletcher, Goldfarb, and Shanno

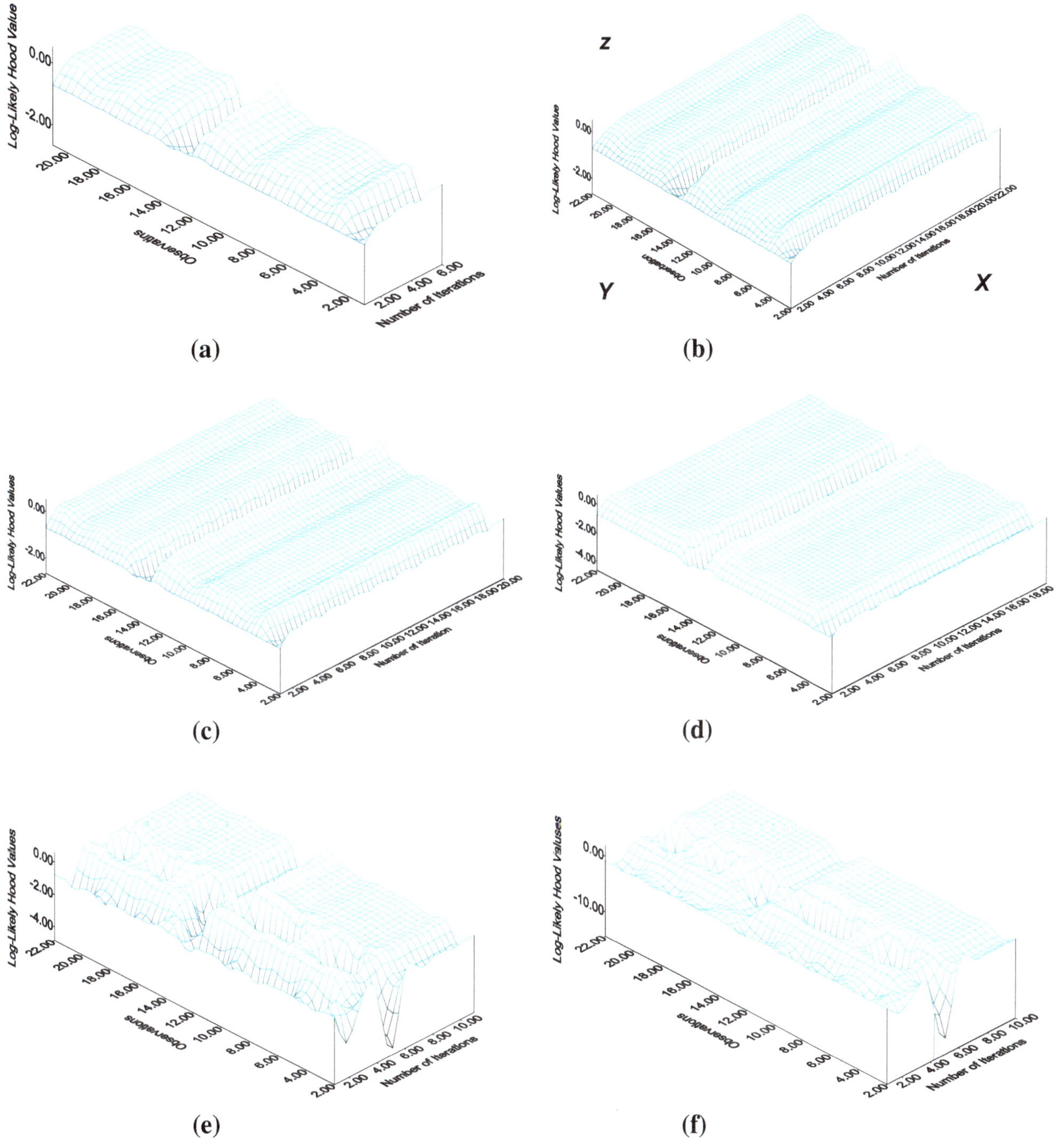

Fig. 5 Estimation surface converged for six algorithms with $CR = 10^{-4}$, **a** Newton–Raphson ($\lambda = 1$), **b** BHHH ($\lambda = 1/2$), **c** BHHH-2 ($\lambda = 1/2$), **d** Steepest Ascent ($\lambda = 16$), **e** DFP ($\lambda = 16$), **f** BFGS ($\lambda = 8$)

convergence behaviour of estimation algorithms determined by Hessian updating mechanism prefixed in each algorithm can be clearly demonstrated. Log-likelihood values are presented on *Z*-axis; the number of iterations is on *X*-axis, and observation numbers for estimation sample are on *Y*-axis (Fig. 5b). In Fig. 5a, log-likelihood value is calculated for each observation from the first iteration and it showed the same value (−0.693147181). This value resulted from the assumption that all parameter values are zero in the initial stage of estimation in the model specification. NR, as shown in Fig. 5a, shows the most simple convergence behaviour in terms of small number of iteration (*X*-axis), meaning that short convergence time. The log-likelihood value for NR was stared with −14.55609079 and converged with the value of −6.166042212 at 6th run. Algorithms performance for DFP and BFGS are shown in Fig. 5e, f respectively. It can be easily seen, that their convergence pattern are similar in terms of both the number of iterations and convergence time. BHHH and BHHH-2 are presented in Fig. 5b, c respectively and are also very similar in their convergence behaviour. These similarities are the result of using almost same iterative process in terms of updating approximate Hessian matrix. SN shown in Fig. 5d, showed similar performance with BHHH and BHHH-2.

5.1 Numerical experiment 1

The two different levels of convergence criteria used in this experiment are used from two earlier studies [5, 8, 9]. The iteration is completed with $\beta_{t+1,k}$ as the solution, when the prefixed stopping precision is satisfied. The mathematical expressions used to represent convergence criteria are shown below. Numerical experiment1 generally uses $\beta_0 = [\beta_{01} = 0,\ \beta_{02} = 0]'$, as initial guessing of starting points.

$$\left[\frac{1}{k}\sum_{k=1}^{k}(\beta_{t+1,k} - \beta_{t,k})^2\right]^{1/2} < CR = 10^{-4}, \tag{38}$$

$$\left[\frac{1}{k}\sum_{k=1}^{k}(\beta_{t+1,k} - \beta_{t,k})^2\right]^{1/2} < CR = 10^{-6}. \tag{39}$$

5.1.1 Experimental run with $CR = 10^{-4}$

The result from the experimental runs (trials with convergence) based on convergence criteria ($CR = 10^{-4}$) for six different algorithms are summarized graphically in Figs. 6 and 7. From these figures, it is conclusive that the NR algorithm is the most appropriate in terms of all three performance measures despite its mathematical difficulty to compute the real Hessian of the log-likelihood function. The performance results based on for one iteration time are found less than 1 s because of a simple model specification. We tried with varied step sizes for all six algorithms to obtain the maximum value for the log-likelihood function. The step size can be considered at random during the initial stage of estimation. NR with step size $\lambda = 1$ showed better performance (converged after 6 iterations) than the other algorithms as far as the number of iteration criteria is concerned.

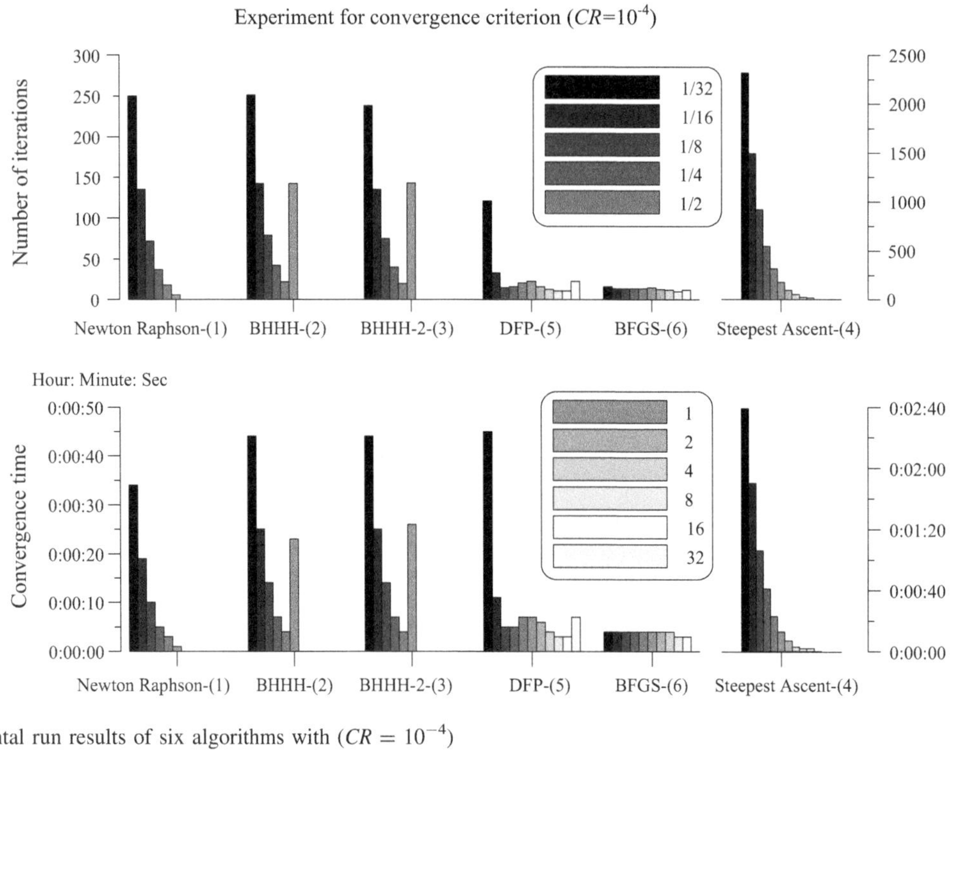

Fig. 6 Experimental run results of six algorithms with ($CR = 10^{-4}$)

The convergence time can be used as a sign of the operational efficiency of an algorithm because it shows critically the relative superiority in convergence speed of an algorithm. In the case of number of iterations, there are step size differences in the same algorithm; also differences between the six algorithms (Fig. 6). The SA algorithm showed the worst performance. The application of the convergence time showed notable differences between some algorithms. The NR algorithm showed the best performance with respect to the convergence time for all step sizes, followed by the BFGS, DFP and BHHH, BHHH-2, and SA algorithm in order. Particularly, with step size λ

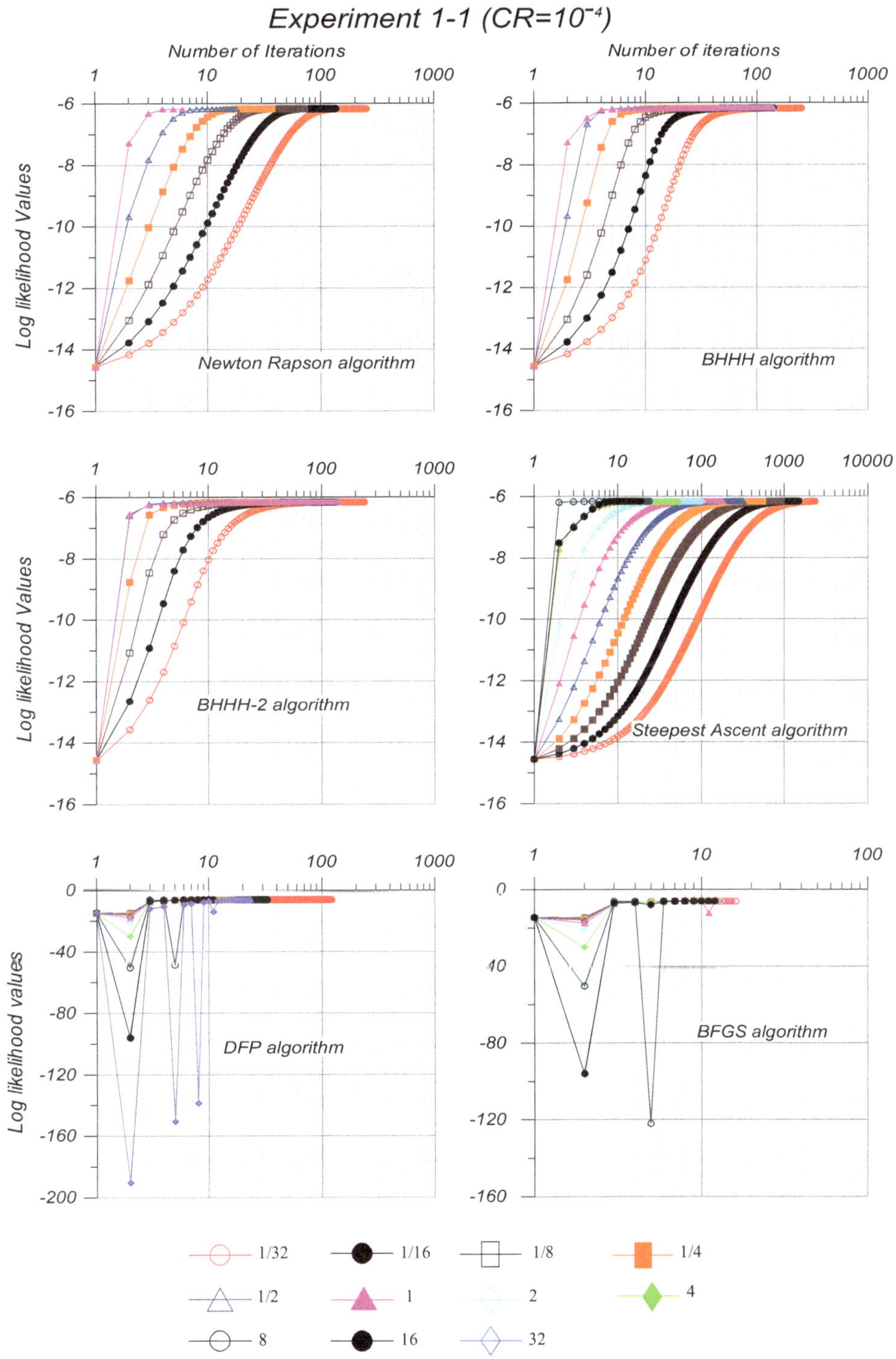

Fig. 7 Illustration of variation of Log-likelihood values in iteration process

equal to 1, the NR algorithm needs just 1 s to finish calculation. Conversely, SA algorithm takes 2 min 39 s with the step size $\lambda = 1/32$.

5.1.2 Experimental run with $CR = 10^{-6}$

In experiment for the evaluation of performance with different convergence criteria ($CR = 10^{-6}$), a different precision level of stopping convergence criteria having the same mathematical expression as shown in Eq. 39 is used. The comparison results are presented graphically in Fig. 8. For all algorithms, the estimation performance has deteriorated as compared to experiment conducted with $CR = 10^{-4}$ in terms of all two measures. These deteriorations took place for all algorithms, step sizes, and performance measures. For example, in the case of SA algorithm ($\lambda = 1/32$), the experiment convergence time $CR = 10^{-6}$ increased significantly (from 0:02:39 to 0:09:43), also the difference of the number of iterations is significant (from 2,320 to 7,033). The pattern of deteriorations is generally the same for all six algorithms. However, NR algorithm required only seven iterations for step size 1. In the run using $CR = 10^{-6}$ with a step size one, one iteration number is increased as compared to the run using $CR = 10^{-4}$ with a step size one. The convergence behaviour is identical to experiment using $CR = 10^{-4}$ except the fact that slight differences occur for all two performance measures in terms of numerical values.

5.2 Numerical experiment 2

The second numerical experiment is intended for the examination of variations in algorithm performance, which may arise by considering various starting points for the parameters. This experiment is initiated to understand how to guess the starting points, and analyze the prospective of the different estimation performances. Nonetheless, searching a new suitable method for a good initial guess of the starting points is not within the scope of this experiment. In this study, the focus is to know that, whether it can make a difference in estimating performance by changing initial starting points, particularly in custom built environment. The outcome of this experiment will help in suggesting a new methodical approach of guessing starting points for a better operational performance. This may be considered as future research of this study. This experiment is designed to in line of this new interesting topic. In this investigation, another vector of parameter is introduced: $\beta_0 = [\beta_{01} = -0.1, \beta_{02} = -0.1]'$, in which two parameters are changed from 0 to -0.1 so as to reflect the case, where an analyst has prior knowledge of the parameter estimates and a negative ($-$) sign can be assigned to the two parameters. The convergence criterion remains same as it is given in subsect. 5.1.1. Detailed results for all algorithms are presented in Fig. 9. Overall, a little improvements were observed in terms of all performance but, in a few cases, deterioration has also happened. Taking

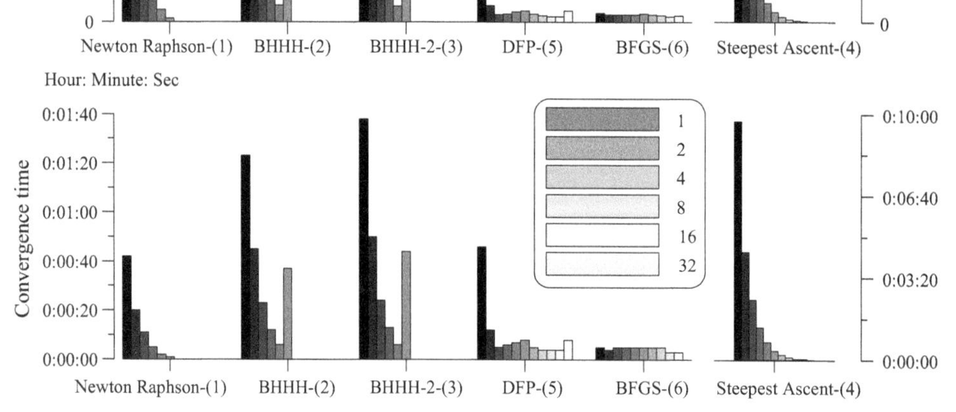

Fig. 8 Experimental run results of six algorithms with $CR = 10^{-6}$

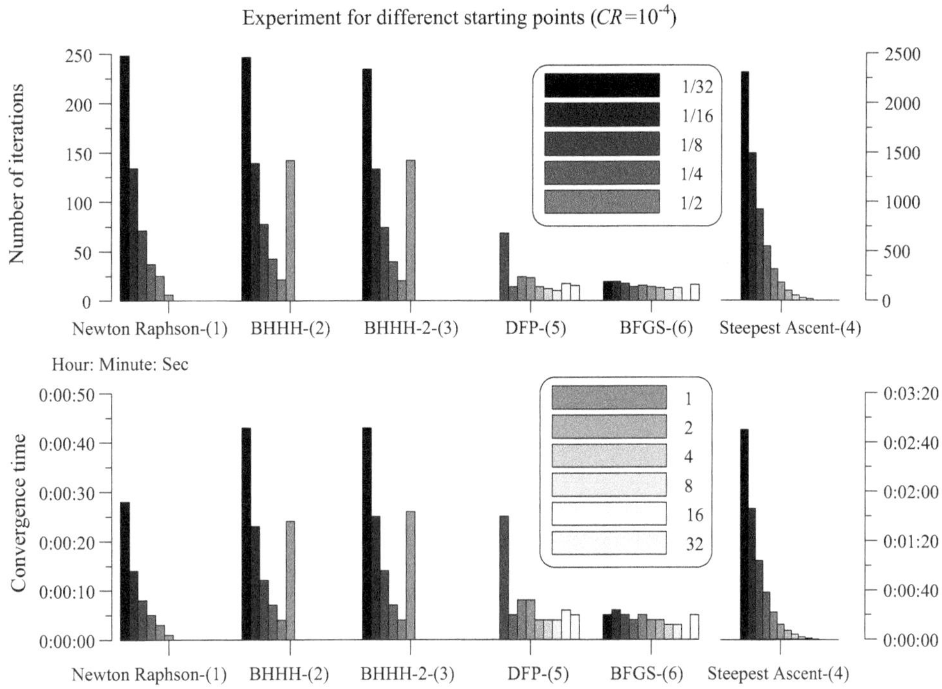

Fig. 9 Experimental run results of six algorithms with different starting points

the example of DFP algorithm, a run with step sizes 1/32 and 1/16 did not converge with new starting points, however the same run converged in experiment with ($CR = 10^{-4}$ and 10^{-6}). The rationale for this dissimilar convergence behaviour is not definite in this study.

6 Conclusion and discussion

Discrete choice analysis has been of interest to researchers in many diverse disciplines for consumer's choice behaviour prediction. Particularly, in transportation domain, there have been many applications of discrete choice modeling techniques to predict the behaviour of users consuming transportation systems or services. However, almost all research associated with discrete choice modeling focuses only on the interpretation of the estimated results or on the efforts to find policy implications in model parameters to support decision making process. Furthermore, almost all types of discrete choice models are carried out using commercial statistical packages. In the earlier generation of discrete choice modeling, it was common practice to follow instructions of a selected specific computer package. However, as the choice model specification becomes more complicated and custom-built modeling is increasingly in demand, researchers working in the modeling domain are challenged to go beyond the first generation modeling practices. Nonetheless, there is a lack of literature available to research on estimation algorithms as well as important components in it. Although, some literatures on this may be available in econometric field [8–10], they are not directly related to the concerns of transportation modellers. Also, these studies have a different theoretical base, which puts some constraints to write the computer codes for transportation modeling application.

Several factors that dominate estimation performance are detected through this experimental model estimation. These are: (a) convergence criterion; (b) initial guessing of starting points. In this research a convergence criterion of $\left[\frac{1}{k}\sum_{k=1}^{k}(\beta_{t+1,k}-\beta_{t,k})^2\right]^{1/2}$, is used, which is generally accepted convergence criterion in many literatures [8, 9]. Based on the results, some important aspects might be considered in more elaborated research associated with custom-built modeling. Possibly, potential research extensions are noted below.

(a) Different mathematical formulations for stopping criterion may be tried and different level of stopping accuracy such as $\left[\frac{1}{k}\sum_{k=1}^{k}(\beta_{t+1,k}-\beta_{t,k})^2\right]^{1/2} < 10^{-6}, 10^{-7}$ can be adopted to make a significant difference in model estimation performance. This issue should be revisited in the future work.

(b) When researchers have to make a guess on initial parameters starting points in custom-built modeling, it would be time saving to make an initial guess of model's parameters fairly close to true values on the first trial. However almost all choices of initial guess of parameters set up by researchers are not the actual parameter estimates of the given model. In this context, it is a challenge for researchers to be more organized in choosing a vector of parameters, which may be a critical subject in custom-built modeling. This suggests that initial guessing of parameters can affect a model estimation performance to a high extent in terms of iteration numbers, convergence time and so on. A similar research study was reported by Liu and Mahmassani [11] that uses an empirical model estimation process for obtaining parameters of a probit model. They use the genetic algorithm for making a guess of initial parameter estimates. The authors of this paper have demonstrated that initial guessing of parameter can have significant effect on overall model estimation performance. All these issues addressed here are note worthy for future extension of research reported in this paper.

References

1. Hyuk-Jae R (2013) Mode choice behavior of various airport user groups for ground airport access. Open Transp J 7:43–55
2. Hyuk-Jae R (2013b) Empirical application of four fundamental algorithms for parameter estimates of multinomial logit discrete choice model. In: CSCE conference, Montreal
3. Kenneth T (2003) Discrete choice methods with simulation. Acadmic Press, Cambridge
4. Hyuk-Jae R, Ata MK (2013) Enhancing algorithmic base for discrete choice modelling. KSCE J Civ Eng 17(7):1798–1809
5. Moshe BA, Steven RL (1985) Discrete choice analysis: theory and application to travel demand. MIT Press, Cambridge
6. Green William H G (2003) Econometric analysis. Pearson Education, Inc, Upper saddle river
7. Daniel LM (1974) Conditional logit analysis of qualitative choice behaviour. Frontiers in econometrics. Academic press, New York
8. David AB (1979) On the computational competitiveness of full-information maximum likelihood and three-stage least-squares in the estimation of nonlinear, simultaneous-equations models. J Econom 9:315–342
9. David AB (1980) On the efficient computation of the nonlinear full-information maximum-likelihood estimator. J Econom 14:203–225
10. David SB (1988) A comparison of algorithms for maximum likelihood estimation of choice models. J Econom 38:145–167
11. Yu-Hsin L, Hani SM (2000) Global maximum likelihood estimation procedure for multinomial probit (MNP) model parameters. Transp Res Part B34:419–449

19

Adaptive neuro-fuzzy interface system for gap acceptance behavior of right-turning vehicles at partially controlled T-intersections

Jayant P. Sangole · Gopal R. Patil

Abstract Gap acceptance theory is broadly used for evaluating unsignalized intersections in developed countries. Intersections with no specific priority to any movement, known as uncontrolled intersections, are common in India. Limited priority is observed at a few intersections, where priorities are perceived by drivers based on geometry, traffic volume, and speed on the approaches of intersection. Analyzing such intersections is complex because the overall traffic behavior is the result of drivers, vehicles, and traffic flow characteristics. Fuzzy theory has been widely used to analyze similar situations. This paper describes the application of adaptive neuro-fuzzy interface system (ANFIS) to the modeling of gap acceptance behavior of right-turning vehicles at limited priority T-intersections (in India, vehicles are driven on the left side of a road). Field data are collected using video cameras at four T-intersections having limited priority. The data extracted include gap/lag, subject vehicle type, conflicting vehicle type, and driver's decision (accepted/rejected). ANFIS models are developed by using 80 % of the extracted data (total data observations for major road right-turning vehicles are 722 and 1,066 for minor road right-turning vehicles) and remaining are used for model validation. Four different combinations of input variables are considered for major and minor road right turnings separately. Correct prediction by ANFIS models ranges from 75.17 % to 82.16 % for major road right turning and 87.20 % to 88.62 % for minor road right turning. The models developed in this paper can be used in the dynamic estimation of gap acceptance in traffic simulation models.

Keywords Partially controlled intersections · Gap acceptance · Adaptive neuro-fuzzy interface system (ANFIS) · Membership function · Receiver operator characteristic (ROC) curves · Precision-recall (PR) curves

1 Introduction

Unsignalized intersections are the most common form of intersections in a road network. Unsignalized intersections include "priority unsignalized intersections" where priority is indicated by signs and "uncontrolled intersections" where no explicit priorities exist. Most of the studies on unsignalized intersections have been carried out assuming that the law of priority is fully respected. The traffic flow situations in India are considerably different from those in developed countries. There exist a mix of various types of vehicles with wide variations in their sizes and operating conditions. Priority rules are not respected and vehicles do not follow the lane movement. However, at a few intersections, where a minor road is joining a major road, the vehicles on the minor road will perceive higher priority for the major road vehicle, not because of signs but due to the existing traffic and geometric characteristics. This will result in observing partial priority by vehicles even in the absence of stop and yield signs. These cases of uncontrolled and partially controlled intersections are not well studied in the literature. Unlike highway capacity manual (HCM) 2010, there is no standard procedure available to evaluate the intersections in India. One of the main reasons for relatively less work on unsignalized intersection in India is the complex vehicle interactions resulting from

J. P. Sangole · G. R. Patil (✉)
Department of Civil Engineering, Indian Institute of Technology Bombay, Powai, Mumbai 400076, India
e-mail: gpatil@iitb.ac.in

J. P. Sangole
e-mail: sangolejayant@gmail.com

heterogeneous traffic, aggressive drivers' behavior, and inefficient enforcement of traffic rules. Fuzzy theory—widely used to model human behavior at situations involving uncertainty and imprecision—can be a good tool to evaluate and model driver's behavior at uncontrolled or partially controlled intersections.

Gap acceptance is an important objective characteristic of the behavior of drivers and it has influence on capacities, delays, and road safety at unsignalized intersections. The HCM uses gap acceptance concept for determining the capacity of unsignalized (sign controlled) intersections. Gap acceptance behavior is a choice by a driver to accept or reject a gap of particular size, and it is the result of a decision process of human brain. The choice usually depends on the driver's characteristics in addition to various external factors related to vehicles, traffic, and environment. The assessment of these factors is subject to uncertainty and imprecision. Fuzzy logic has been widely used to model human decisions in an environment of uncertainty and imprecision. Adaptive neuro-fuzzy interface system (ANFIS) combines the abilities of fuzzy set theory and neural network and provides an optimization scheme to find the parameters in the fuzzy system that best fit the data.

Gap acceptance at intersection is the process in which a vehicle in the secondary access accepts gaps available in the primary traffic stream. Thus, a vehicle at intersection intending to take right turn requires choosing between two alternative actions: to accept or reject an available gap. This decision is affected by driver characteristics and also by characteristics of the gap and the choice situation. The driver characteristics that affect the gap acceptance decision may include gender, age, driving skill, etc. [1–3]. Gap and choice situation factors may include gap size, speed of vehicles, type of subject and conflicting vehicle types, waiting time, type of sign control, etc. [1–6]. Many researchers dealt with gap acceptance at unsignalized intersections with the assumption that priority is fully respected, such as Hawkes [7], Ashworth [8, 9], Ashton [10], Miller [11], Polus et al. [12], Davis and Swenson [13], Xu and Tian [14].

Fuzzy logic and ANFIS have been successfully used in many engineering areas, for example, pedestrian behavior [15], capacity estimation [16], driving situation recognition [17], collision avoidance support system [18], traffic signal control at isolated intersection [19], rainfall prediction, signal co-ordination, and highway accident analysis [20]. Work related to application of fuzzy set theory for gap acceptance at priority intersection has been done by Rossi et al. [21], Rossi and Meneguzzer [22], and Rossi et al. [23]. Rossi and Meneguzzer [22] modeled the gap acceptance behavior at priority intersections using neuro-fuzzy system. In their work, the input variables considered are (i) if driver is facing gap or lag (binary variable), (ii) gap/lag size in seconds, and (iii) type of maneuver; the output variable is driver's decision to accept or reject the gap/lag. Further Rossi et al. [23] used data from driving simulator to develop gap acceptance fuzzy model. In another study, Rossi et al. [21] analyzed and compared random utility models and fuzzy logic models for representing gap acceptance behavior using data from driving simulator experiments. However, other parameters which affect the divers' decision such as vehicle type, geometry, and composition of vehicle type were not considered in these studies.

Most of the above studies are based on the assumption that the law of priority is fully respected. Limited work has been done for gap acceptance at unsignalized intersections with no priority or limited priority, which are prevalent in India. A few studies conducted for such conditions include Ashalatha and Chandra [24], Rao and Rengaraju [25, 26], etc. However, these studies do not focus on gap acceptance behavior at intersection. Kanagaraj et al. [27] developed probit-based gap acceptance models merging vehicles. Four different types of merging has been considered, namely, normal merging, forced merging, group merging, and vehicle cover merging for left-turning vehicles at uncontrolled T-intersection. The gap acceptance behavior of right-turning vehicles which involve crossing of major road through vehicles is not considered.

The literature discussed above shows that more studies are required for proper understanding of gap acceptance behavior at uncontrolled intersections. The traffic flow characteristics at intersection are the result of complex interactions between drivers, vehicles, transport facilities, and the environment, which are not easy to model. Considering the capabilities of ANFIS and some of its applications to traffic flow modeling, we adopted the ANFIS proposed by Jang [28] to model the gap acceptance behavior of right-turning vehicles at partially controlled T-intersections in this study. Vehicles are driven on the left side in India, thus right turns are critical. Here, we are considering gap acceptance of two right turning movements at T-intersection, one from major to minor approach and the other from minor to major approach. When a vehicle is taking right turn, it conflicts with the through vehicles on the major approach. Note that there is a possibility of conflicts between right-turning vehicles of major and minor approaches. However, in this study, we did not observe many such instances. Moreover, we did not select such data for analyzing gap acceptance behavior. Gap is defined as the time interval between passing of the rear bumper of a leading vehicle and front bumper of the immediate following vehicle at a point. Lag is defined as the time elapsed after a right turn intended vehicle reaches the stop line until a major approach conflicting vehicle reaches the conflict point. Details of gap/lag measurement at intersections can be referred to in [29].

2 Data collection

Four T-intersection sites were selected for this study: three in Aurangabad city (intersection 1, 2, 4) and one in Thane city (intersection 3). Both cities are in Maharashtra state; Thane city is in Mumbai region, where two-wheelers composition is less compared to other cities in India. Intersections are selected such that at each site a nearby building is available to place a camera. Other points considered were plain terrain, adequate sight distance at intersection, no parking, no bus bays, and sufficient distance from adjacent intersections. Geometry of intersection is as shown in Fig. 1. Geometry of all intersections is similar except intersection 2, where minor road is two-lane divided. However, since the major roads to be crossed by the right turning movements are 4-lane divided at all intersections, the gap acceptance behavior is expected to be similar. Data were collected during morning hours (approximately between 10:00 to 11:00 am) of typical weekdays. Video recording was done for about 60 min at each intersection covering all three approaches. Data was extracted in laboratory by playing the video on computer screen. Data extracted include gap/lag in seconds, subject vehicle type (vehicle intending to take right turn) and conflicting vehicle type. The vehicles are divided into 4 categories: two-wheeler, auto-rickshaw, car, and bus/truck. Extracted data include observations of 384 and 530 vehicles for major and minor road right turning movements, respectively. The total lags/gaps (both accepted and rejected) are 722 for major road right turning movements (WS movement in Fig. 1). For minor road right turning (SE movement), the total lags/gaps (both accepted and rejected) are 1,066.

3 Preliminary data analysis

Table 1 shows the traffic composition at all intersections as well as the mode-wise share (in %). It can be observed that

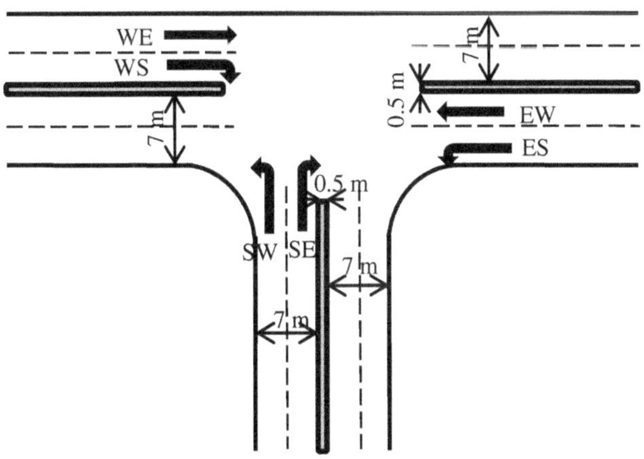

Fig. 1 Geometry of intersections

the proportion of two-wheelers at all intersections is much higher than other modes. This proportion is significantly high at the three intersections in Aurangabad city (74.47 %, 69.93 %, and 78.76 %, respectively, at intersections 1, 2, and 4). At the intersection in Thane city (intersection 3), the proportions of auto-rickshaws and cars are nearly close to two-wheelers proportion. Thane city is within the Mumbai Metropolitan Region (MMR) where public transport is heavily used, thus has relatively less proportion of two-wheelers compared to other three intersections.

Figure 2 through Fig. 4 shows the cumulative percentage of gap accepted and rejected by right turning two-wheelers, auto-rickshaws, and cars (called as *subject vehicle type*) with other *conflicting vehicle types* (the major road through vehicle type which may conflict with right-turning vehicles). Buses and trucks constitute a very small fraction of the total traffic as subject vehicle type and are not included in the vehicle type to vehicle type interaction analysis. Note that in the cumulative accepted gap graphs, the curves do not reach 100 % on y-axis. This is because the gaps more than 8 s are not plotted on x-axis. It is clear from these graphs that the conflicting vehicle types affect the gap acceptance behavior. As seen in Fig. 2, for a gap of 3 s, about 22 % of two-wheelers are accepting the gaps if the conflicting vehicles are two-wheelers, but the percentage drops to about 10 % if the conflicting vehicles are cars. An interesting revelation from this analysis is that less percentage of cars or three-wheelers accepts a given gap if the conflicting vehicle is a two-wheeler compared to when conflicting vehicles are cars or three-wheelers. For example, in Fig. 3, about 30 % of three-wheelers accept a gap of 3 s when the conflicting vehicles are cars; but the acceptance rate is only about 10 % when the conflicting vehicles are two-wheelers. Similarly, as seen in Fig. 4, about 35 % of cars accept gaps of 3 s when the conflicting vehicles are three-wheelers, but the percentage reduces to about 18 % acceptance when two-wheelers are conflicting. This behavior of three-wheelers and cars when the conflicting vehicle is a two-wheeler may be because of the general perception that two-wheelers require less time to clear the intersection, they are aggressive, and their actions are unpredictable. Another reason for the pessimistic behavior from cars and three-wheelers toward two-wheelers is that in case of collision, often, bigger vehicles are held responsible.

4 Development of ANFIS models

The gap acceptance of the right turning movement from major to minor approach (WS) and the right turning movement from minor to major approach (SE) movements are the focus in this study (see Fig. 1). The conflicting movement for both WS and SE is the through traffic from

Table 1 Traffic composition and mode-wise share at all intersections

Intersection no.	Mode movement	Two-wheeler	Auto-rickshaw	Car	Bus/truck	Sub-total	Total
Intersection 1	EW	886	126	109	72	1,193	3,549
	ES	127	28	43	14	212	
	WE	780	99	98	63	1,040	
	WS	242	25	45	5	317	
	SW	375	32	49	13	469	
	SE	233	18	57	10	318	
	Total	2,643	328	401	177	3,549	
	Mode Share (%)	74.47	9.24	11.30	4.99		
Intersection 2	EW	489	94	78	58	719	2,138
	ES	8	2	23	0	33	
	WE	457	98	85	49	689	
	WS	232	13	28	2	275	
	SW	214	62	28	1	305	
	SE	95	2	20	0	117	
	Total	1,495	271	262	110	2,138	
	Mode share (%)	69.93	12.68	12.25	5.14		
Intersection 3	EW	173	156	203	55	587	3,258
	ES	316	185	197	29	727	
	WE	169	128	147	52	496	
	WS	83	96	98	3	280	
	SW	91	86	155	5	337	
	SE	365	167	238	61	831	
	Total	1,197	818	1,038	205	3,258	
	Mode share (%)	36.74	25.11	31.86	6.29		
Intersection 4	EW	519	130	49	5	703	3,075
	ES	578	44	46	2	670	
	WE	395	98	49	7	549	
	WS	179	31	36	0	246	
	SW	212	32	37	1	282	
	SE	539	43	41	2	625	
	Total	2,422	378	258	17	3,075	
	Mode share (%)	78.76	12.29	8.39	0.55		

east to west on the major road (EW). Based on the preliminary analysis, we identified three input variables: *size of gap* (s), *subject vehicle type*, and *conflict vehicle type* for developing ANFIS. The output variable is driver's decision—a binary variable indicating acceptance or rejection of the gap. The Fuzzy Logic toolbox available in MATLAB is used for developing the model. The variable *size of gap* is a fuzzy variable, and is divided into three linguistic variables: *small gap, medium gap,* and *large gap*. In fuzzy logic, a fuzzy variable is the variable which takes a quantity that can take on linguistic rather than precise numerical values. Initially, a membership function of input variables is supplied exogenously. This membership function will get tuned or adjusted during backward learning process of ANFIS to minimize the training error. Tuned membership function for gap, which defines the range of linguistic terms, is as shown in Fig. 5. The remaining two input variables, *subject vehicle type* and *conflicting vehicle type*, are considered as crisp variables. Variables that are measured precisely are called crisp variables. These crisp variables require the same treatment as a fuzzy variable to introduce them in neuro-fuzzy system. For these variables, two-wheeler is coded as 1, auto-rickshaw as 2, and car as 3. Observations for heavy vehicles (bus/truck) as subject vehicle type are very less and are not considered. However, a few observations are available for heavy vehicles as conflicting vehicle type; this category is considered for conflicting vehicle type and coded as 4.

ANFIS usually consists of a six-layer feedforward neural network structure. Figure 6 shows the ANFIS

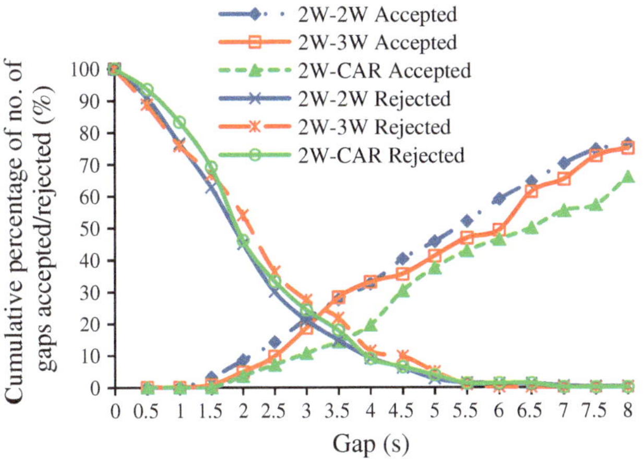

Fig. 2 Cumulative percentage of no. of gaps accepted and rejected for two-wheelers with other conflicting vehicle types

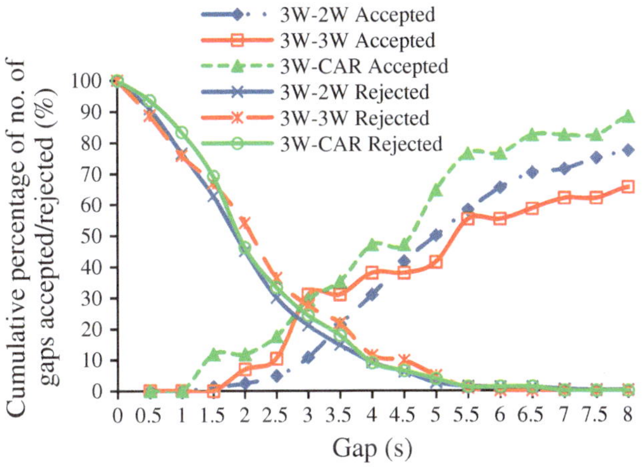

Fig. 3 Cumulative percentage of no. of gaps accepted and rejected for auto-rickshaw with other conflicting vehicle types

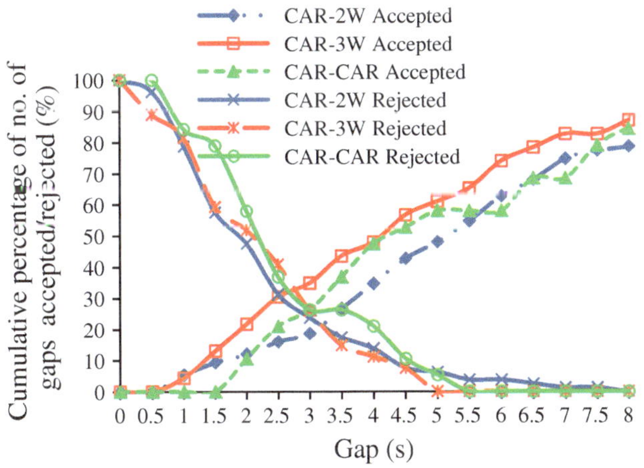

Fig. 4 Cumulative percentage of no. of gaps accepted and rejected for car with other conflicting vehicle types

architecture for a model with two inputs: gap size (GAP or x_1) and subject vehicle type (SVT or x_2), and one output: possibility of accepting gap (y) (for detailed structure of ANFIS see Jang et al. [20]). Brief explanation of each layer is as discussed below:

Layer 1 is called the input layer. In this layer, neurons pass the input signals (also known as crisp) to next layer. *Layer 2* is known as the fuzzification layer. The fuzzification is a process in which the variable is divided into linguistic variables defined by membership function. *Layer 3* is the rule layer. In this layer, each neuron corresponds to a Sugeno-type fuzzy rule [30]. Rules are defined as follows:

If GAP is *Small* and SVT is CAR then possibility of accepting gap is 0

If GAP is *High* and SVT is CAR then possibility of accepting gap is 1

...

and so on.

The *layer 3* neurons receive inputs from the *layer 2* neurons and calculate the output of the layer known as the firing strength of the rule. The output of neuron i in *Layer 3* is obtained as

$$y_i^{(3)} = \prod_{j=1}^{k} x_{ji}^{(3)}, \tag{1}$$

where $x_{ji}^{(3)}$ is the input from neuron j located in *layer 2* to neuron i in *layer 3; k* is the number of fuzzy set neurons; and $y_i^{(3)}$ is the output of rule neuron i in *layer 3* (i.e., firing strength). In *layer 4,* called the normalization layer, each neuron receives inputs from all neurons in the rule layer, and calculates the normalized firing strength—the ratio of the firing strength of a given rule to the sum of firing strengths of all rules—of a given rule. *Layer 5* is the defuzzification layer. Defuzzification is the process in which the weighted consequent value is determined for a given rule. *Layer 6* is the output layer having single neuron. In this layer, overall output y is determined by summing all the outputs from *layer 5*. As in neural network, the learning

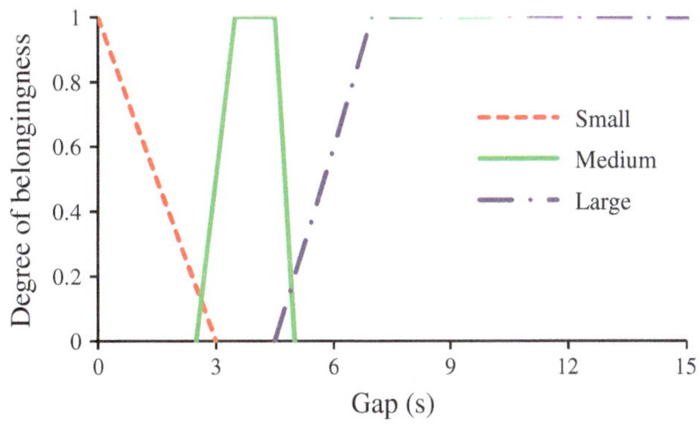

Fig. 5 Membership function of gap/lag for major and minor road vehicle model

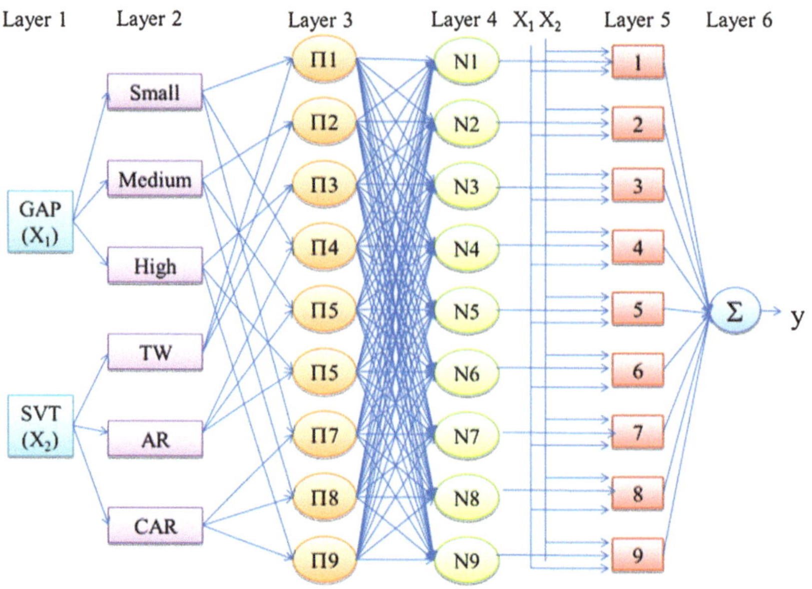

Fig. 6 Structure of ANFIS

Table 2 Models for major road right-turning vehicles

Model	Input variables considered	No. of rules	Training prediction (%)
Model A1	Gap (s)	3	81.80
Model A2	Gap (s) and subject vehicle type	9	73.83
Model A3	Gap (s) and conflicting vehicle type	12	73.13
Model A4	Gap (s), subject vehicle type, and conflicting vehicle type	36	73.65

algorithm of ANFIS includes a forward pass and a backward pass. In the forward pass, a set of input parameters is presented to ANFIS and outputs are calculated layer by layer. If we have a training set of P input–output patterns, the output can be represented in matrix notation as below [28]:

$$\boldsymbol{y}_\mathrm{d} = \boldsymbol{Ak}, \tag{2}$$

where $\boldsymbol{y}_\mathrm{d}$ is a $P \times 1$ output vector used for training; $\boldsymbol{A}$ is a $P \times n(1 + m)$ matrix and $\boldsymbol{k}$ is a $n(1 + m) \times 1$ vector of unknown consequent parameters; m is the number of input variables; and n is the number of neurons in the rule layer. Generally the elements in $\boldsymbol{y}_\mathrm{d}$ are much larger than the number of unknowns, that is P is greater than $n(1 + m)$. Thus, exact solutions to Eq. (2) may not exist and a least-square estimate of $\boldsymbol{k}$, $\boldsymbol{k}^*$, is obtained that minimizes the squared error $(\boldsymbol{Ak}\text{-}\boldsymbol{y}_\mathrm{d})^2$. It is done by using the pseudo-inverse technique [28]:

$$\boldsymbol{k}^* = (\boldsymbol{A}^\mathrm{T}\boldsymbol{A})^{-1}\boldsymbol{A}^\mathrm{T}\boldsymbol{y}_\mathrm{d}, \tag{3}$$

where $\boldsymbol{A}^\mathrm{T}$ is the transpose of $\boldsymbol{A}$ and $(\boldsymbol{A}^\mathrm{T}\boldsymbol{A})^{-1}\boldsymbol{A}^\mathrm{T}$ is the pseudo inverse of $\boldsymbol{A}$, if $(\boldsymbol{A}^\mathrm{T}\boldsymbol{A})$ is non-singular. Once the consequent parameters are obtained, an actual network output vector, $\boldsymbol{y}$, can be computed and the error vector, $\boldsymbol{e}$, is determined by

$$\boldsymbol{e} = \boldsymbol{y}_\mathrm{d} - \boldsymbol{y}. \tag{4}$$

The back-propagation algorithm is used for backward pass and antecedent parameters are updated. All extracted data are converted to a desired format required for ANFIS; about 80 % of the data is used for training purpose and the rest are kept for validation. For a given set of inputs, the ANFIS gives output in the range of 0–1. The training prediction is calculated to know how well the model is trained with given input training data. Different combinations of input variables are used to develop four different models for major and minor road right-turning vehicles. Separate models are developed for major road right turning and minor road right-turning vehicles (see Tables 2 and 3). The number of rules created and training predictions are presented in the Tables. It can be observed that the prediction accuracy by all models is greater than 70 %.

5 Validation and comparison of models

As mentioned above, about 20 % of the data are randomly selected and kept for the validation of the various models

Table 3 Models for minor road right-turning vehicles

Model	Input variables considered	No. of rules	Training prediction (%)
Model B1	Gap (s)	3	77.97
Model B2	Gap (s) and subject vehicle type	9	77.85
Model B3	Gap (s) and conflicting vehicle type	12	79.04
Model B4	Gap (s), subject vehicle type, and conflicting vehicle type	36	78.57

Table 4 Data for model validation

Gap (s)	Major road		Minor road	
	Accepted	Rejected	Accepted	Rejected
0–5	20	73	16	119
5–10	30	4	49	3
10–21	13	0	24	0

developed (see Table 4). Provost et al. [31] suggested the use of receiver operator characteristic (ROC) curves for binary decision problem, because simply using prediction by model can be misleading. ROC curve shows how correct predictions of model vary with incorrect predictions of model. Precision-recall (PR) curve is an alternative to ROC curves, and can be used if data are highly skewed. In ROC curves, data should be present in the upper left-hand corner, and in PR curves, it should be in the upper right-hand corner [31]. The output of ANFIS model is the possibility of accepting the lag/gap and it varies from 0 to 1. If the possibility of accepting lag/gap is greater than 0.5, then the particular lag/gap is taken as accepted.

From the validation results, we need to group the responses into the following categories: (i) true positive (TP) (response correctly labeled as positives), (ii) false positives (FP) (response incorrectly labeled as positive), (iii) true negatives (TN) (negatives correctly labeled as negative), and (iv) false negatives (FN) (response incorrectly labeled as negative). For gap acceptance problem, gap accepted is considered as condition positive and gap rejected is considered as condition negative. Calculations are done by considering three cut-off points for gap: (i) less than or equal to 5 s, (ii) greater than 5 s and less than or equal to 10 s, and (iii) greater than 10 s and less than or equal to 21 s (see Table 4). The false positive rate (FPR) is the proportion of negative responses that are misclassified as positive, whereas the true positive rate (TPR) is the proportion of positive responses that are correctly labeled [31]. Recall used in the PR curves is the same as TPR, and the precision is the fraction of responses classified as positive that are truly positive, i.e., positive predictive value.

The ROC and PR curves for major road right turning model are shown in Fig. 7 and for minor road right turning in Fig. 8. The ROC curves clearly show that they are

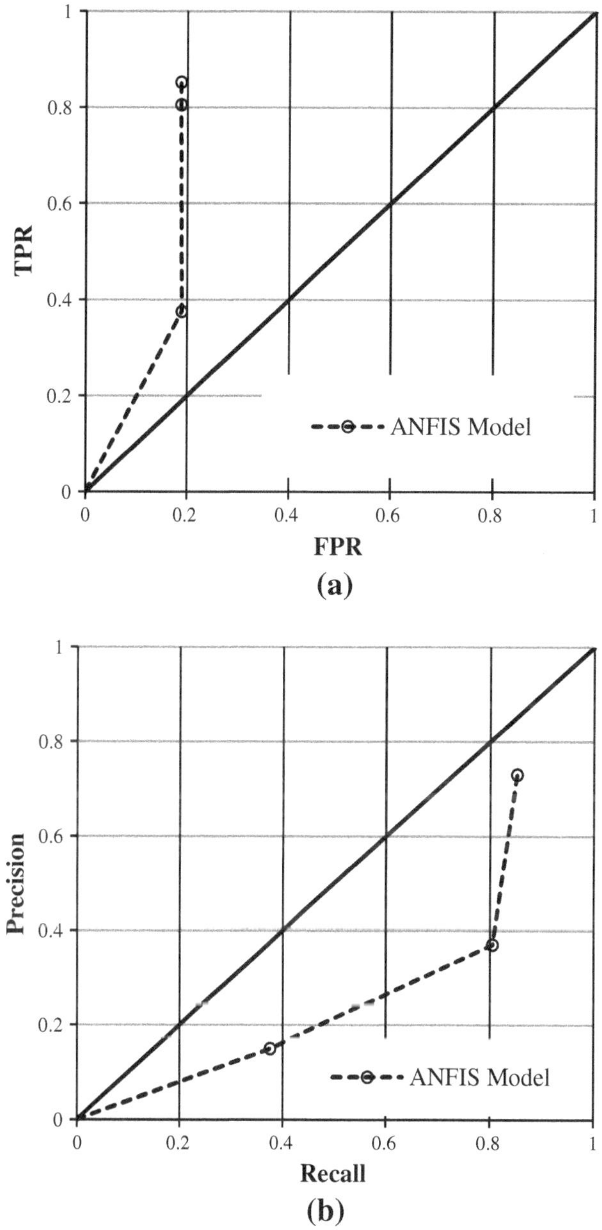

Fig. 7 ROC curve and PR curve for major road right-turning vehicles. **a** ROC curve. **b** PR Curve

dominating upper-left corner side; it is 1 for minor road right-turning vehicles. Also PR curves show that all data points tend toward right upper corner. Comparison of predictions of ANFIS models with field observations for

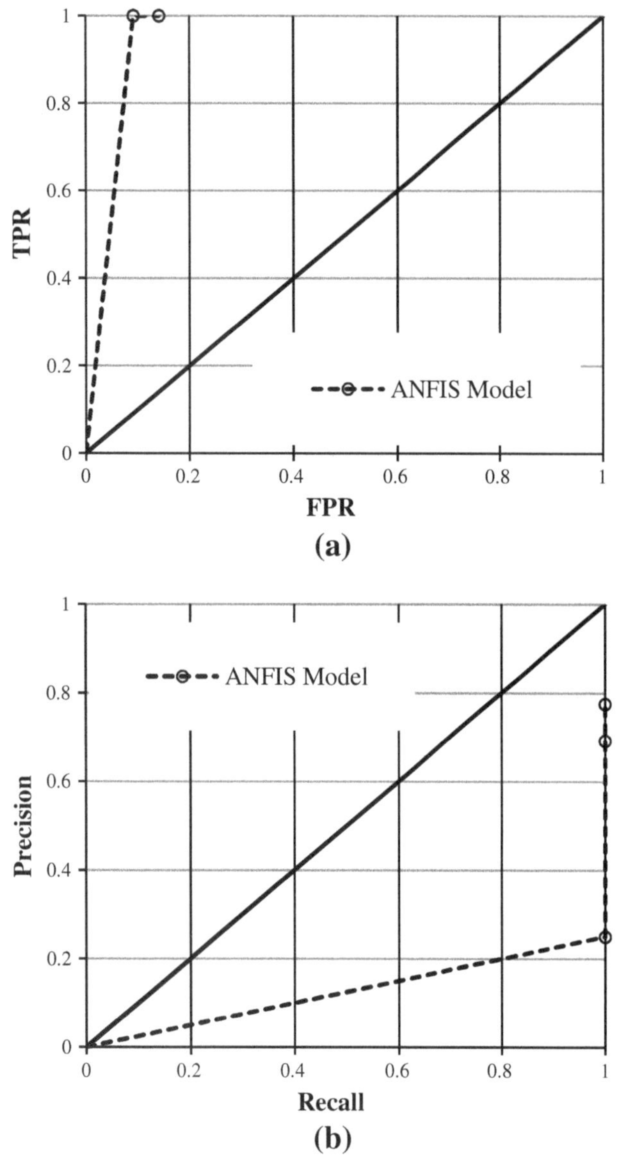

Fig. 8 ROC curve and PR curve for minor road right-turning vehicles. **a** ROC curve. **b** PR Curve

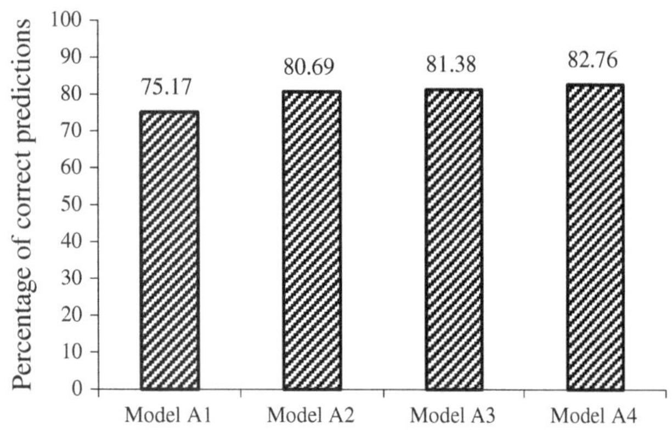

Fig. 9 Prediction of major road models

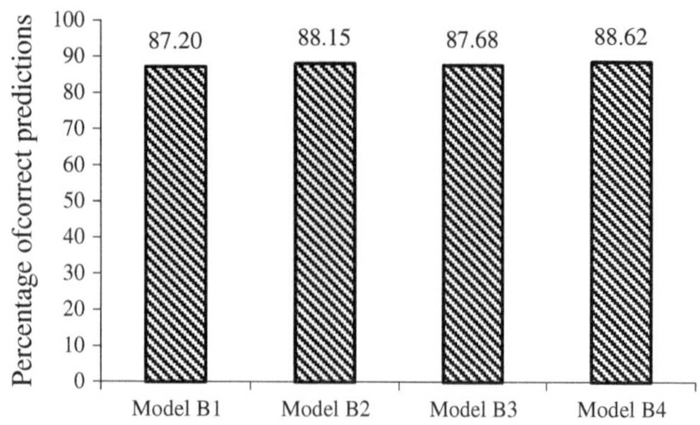

Fig. 10 Prediction of minor road models

major and minor right-turning vehicles are presented in Fig. 9 and Fig. 10, respectively. The percentages of correct predictions by ANFIS range from 75.17 % to 82.16 % for major road right turning and 87.20 % to 88.62 % for minor road right turning. For major road, ANFIS model A4 gives the highest prediction rate. For minor road, ANFIS model B4 has maximum predicated rate.

6 Summary and conclusions

Unlike in developed countries, the unsignalized intersections in India are not properly controlled; that is, the priorities of different movements are not fully respected. At a few intersections, limited priorities are observed where the right-turning vehicles look for suitable gaps in through vehicles. The primary focus of this paper is to model the gap acceptance behavior of vehicles at partially controlled intersections in India with the application of ANFIS. Gap acceptance data are collected at four T-intersections with the help of video camera. From preliminary analysis of data, it is found that less percentage of cars or three-wheelers accepts a given gap if the conflicting vehicle is a two-wheeler compared to when conflicting vehicles are cars or three-wheelers.

An ANFIS is developed for the extracted data considering various input variables (gap size, subject vehicle type, and conflicting vehicle type). Four different combinations of input variables have been considered for major and minor road right turnings separately. Training prediction accuracy by all models is greater than 70 %. ANFIS models also predicted well for the 20 % data kept for validation. Predictions of minor road models are slightly better than major road models.

ANFIS modeling can easily be generalized to various gap acceptance situations such as four-legged intersections including variables related to geometry, by changing if…then rules or form of membership functions. The model can be a potential supplement to microscopic traffic simulators. The results indicate that the ANFIS can be used to model the gap acceptance behavior at partially controlled intersections. In this study, we have considered only right turning movement at T-intersection; the study can be extended to four-legged intersections and more movements can also be studied. Effect of some other parameters such as approach speed, geometric characteristics, driver's characteristics, etc. will further increase the understanding of the traffic behavior at uncontrolled intersections.

Acknowledgments This study is partially funded by Department of Science and Technology (DST), Govt. of India, through project SR/FTP/ETA-61/2010.

References

1. Sheikh Abu MA-MI (1997) Developing behavioral models for driver gap acceptance at priority intersections. Doctoral Thesis, King Fahd University of Petroleum and Minerals, Dhahran, Saudi Arabia
2. Kaysi IA, Abbany AS (2007) Modeling aggressive driver behaviour at unsignalized intersections. Accid Anal Prev 39:671–678
3. Yan X, Radwan E, Guo D (2007) Effects of major-road vehicle speed and driver age and gender on left-turn gap acceptance. Accid Anal Prev 39:843–852
4. Hamed MM, Easa SM, Batayneh RR (1997) Disaggregate gap-acceptance model for unsignalized T-intersections. J Transp Eng 123(1):36–42
5. Pant P, Balakrishnan P (1994) Neural network for gap acceptance at stop-controlled intersections. J Transp Eng 120(3):432–446.
6. Moshe AP, Abishai P, Moshe L (2002) A decision model for gap acceptance and capacity at intersections. Transp Res Part B: Methodol 36(7):649–663.
7. Hawkes A (1968) Gap-acceptance in road traffic. J Appl Probab 5:84–92
8. Ashworth R (1968) A note on the selection of gap acceptance criteria for traffic simulation studies. Transp Res 2(2):171–175
9. Ashworth R (1970) The analysis and interpretation of gap acceptance data. Transp sci 4(3):270–280
10. Ashton WD (1971) Gap-acceptance problems at a traffic intersection. Appl Stat 20(2):130–138
11. Miller A (1974) A note on the analysis of gap-acceptance in traffic. Appl Stat 23(1):66–73
12. Polus A, Lazar SS, Livneh M (2003) Critical gap as a function of waiting time in determining roundabout capacity. J Transp Eng 129(5):504–509
13. Davis GA, Swenson T (2004) Field study of gap acceptance by left-turning drivers. Transp Res Rec: J Transp Res Board 1899(1): 71–75
14. Xu F, Tian ZZ (2008) Driver behavior and gap-acceptance characteristics at roundabouts in California. Transportation Research Record: Journal of the Transportation Research Board 2071(1):117–124
15. Ottomanelli M, Caggiani L, Iannucci G, Sassanelli D (2010) An adaptive neuro-fuzzy inference system for simulation of pedestrians behaviour at unsignalized roadway crossings. In: soft computing in industrial applications. Springer, Netherlands, pp 255–262
16. Adeli H, Jiang X (2003) Neuro-fuzzy logic model for freeway work zone capacity estimation. J Transp Eng 129(5):484–493
17. Nigro JM, Loriette-Rougegrez S, Rombaut M (2002) Driving situation recognition with uncertainty management and rule-based systems. Engineering Applications of Artificial Intelligence 15 (3–4):217-228.
18. Valdés-Vela M, Toledo-Moreo R, Terroso-Sáenz F, Zamora-Izquierdo MA (2013) An application of a fuzzy classifier extracted from data for collision avoidance support in road vehicles. Eng Appl Artif Intell 26(1):173–183.
19. Keyarsalan M, Ali Montazer G (2011) Designing an intelligent ontological system for traffic light control in isolated intersections. Eng Appl Artif Intell 24(8):1328–1339.
20. Jang JSR, Sun CT, Mizutani E (1997) Neuro-fuzzy and soft computing-a computational approach to learning and machine intelligence. Prentice Hall, NJ
21. Rossi R, Massimiliano G, Gregorio G, Claudio M (2012) Comparative analysis of random utility models and fuzzy logic models for representing gap-acceptance behavior using data from driving simulator experiments. Procedia–Soc Behav Sci (Elsevier) 54:834–844.
22. Rossi R, Meneguzzer C (2002) The effect of crisp variables on fuzzy models of gap-acceptance behaviour. In: proceedings of the 13th Mini-EURO Conference: handling uncertainty in the analysis of traffic and transportation systems, pp 240–246
23. Rossi R, Gastaldi M, Gecchele G (2011) Development of gap acceptance fuzzy models using data from driving simulator experiments. In: vulnerability, uncertainty, and riskm, pp 138–147.
24. Ashalatha R, Chandra S (2011) Critical gap through clearing behavior of drivers at unsignalised intersections. KSCE J Civ Eng 15(8):1427–1434.
25. Rao V, Rengaraju V (1997) Probabilistic model for conflicts at urban uncontrolled intersection. J Transp Eng 123(1):81–84.
26. Rao V, Rengaraju V (1998) Modeling conflicts of heterogeneous traffic at urban uncontrolled intersections. J Transp Eng 124(1):23–34.
27. Kanagaraj V, Srinivasan K, Sivanandan R (2010) Modeling Vehicular Merging Behavior Under Heterogeneous Traffic Conditions. Transp Res Rec: J Transp Res Board 2188(1):140–147.
28. Jang JS (1993) ANFIS: adaptive-network-based fuzzy inference system. IEEE Trans Sys Man Cybern 23(3):665–685
29. Sangole JP, Patil GR, Patare PS (2011) Modelling gap acceptance behaviour of two-wheelers at uncontrolled intersection using neuro-fuzzy. Procedia Soc Behav Sci (Elsevier) 20:927–941
30. Zadeh LA (1965) Fuzzy sets. Inf Control 8:338–353.
31. Provost FJ, Fawcett T, Kohavi R (1998) The case against accuracy estimation for comparing induction algorithms. In: proceedings of fifteenth International Conference on Machine Learning, Madison, MI, pp 445–453

20

Fe-doped epitaxial YBCO films prepared by chemical solution deposition

Hong Zhang · Yong Zhao · Wentao Wang · Min Pan · Ming Lei

Abstract $YBa_2Cu_3O_{7-\delta}$ (YBCO)-coated conductors have wide-ranging potential in large-scale applications such as superconducting maglev trains and superconducting electric cables, but low current carrying capability restrains the practical application of YBCO-coated conductors at high temperatures and high magnetic fields. It is crucial to develop YBCO-coated conductors with high critical current density. In this paper, epitaxial, dense, smooth, and crack-free Fe-doped YBCO films were prepared on a $LaAlO_3$ single crystal substrate via a fluorine-free polymer-assisted metal organic deposition method. The effects of the dilute Fe doping on microstructure and superconducting character of YBCO films were investigated. The critical temperature for superconducting of the Fe-doped YBCO films decreases slightly. However, the in-field critical current density of YBCO films improves with dilute Fe doping of amounts less than $x = 0.005$, compared to the pure YBCO film. Therefore, the current carrying capability of YBCO film can improve by doping with appropriate amounts of Fe. This means that dilute Fe doping in YBCO films may be a feasible way to prepare high-performance coated conductors.

Keywords Epitaxial films · Fe doping · Superconductor · Coated conductor

H. Zhang (✉) · Y. Zhao (✉) · W. Wang · M. Pan · M. Lei
Superconductivity and New Energy R&D Center, Southwest Jiaotong University, Chengdu 610031, China
e-mail: zhanghong@home.swjtu.edu.cn

Y. Zhao
e-mail: yzhao@home.swjtu.edu.cn

1 Introduction

Recently, worldwide attention has been paid to the preparation of $YBa_2Cu_3O_{7-\delta}$ (YBCO)-based coated conductors, i.e., the second generation tapes because of their wide-ranging potential in large-scale application such as superconducting maglev trains and superconducting electric cables [1–7]. The major approaches for fabricating YBCO-coated conductors include in situ processes like pulse laser deposition (PLD), magnetic sputtering, etc., as well as ex situ wet-chemistry approaches. Chemical solution deposition (CSD) and related techniques provide versatile and cost-effective processing routes with controlled stoichiometry and microstructure [8–12]. Therefore, CSD is a preferred fabrication method. In the early work preparing solution-derived REBCO thin films, the critical current densities were quite limited. It was widely believed that organic-based precursors might lead to the formation of stable $BaCO_3$, which accumulates and precipitates at the grain boundaries (GBs) of REBCO and thus significantly reduces the transport critical current density J_c [13]. In an effort to solve this problem, a promising approach using trifluoroacetates metal organic deposition was developed to deposit YBCO films [14, 15]. However, the process leads to the formation of HF which is hazardous, and the microstructure of YBCO films were relatively porous [16], which limits the current transport properties. Alternatively, if the formation of $BaCO_3$ can be well controlled, a kind of novel non-fluorine chemical solution deposition method is a promising and cost-effective approach to the preparation and application of high performance YBCO films. In this paper, YBCO films were prepared using a non-fluorine chemical solution deposition method which was devised in our laboratory [17–19].

The large-scale application of YBCO-coated conductors in superconducting magnets, generators, motors, etc., is prevented because of relatively low current carrying capability with increasing magnetic field and temperature. Therefore, enhancing the J_c is crucial for the practical applications of the YBCO-coated conductors at high temperatures and high magnetic fields. Several methods for increasing the density and effectiveness of vortex pinning sites to enhance J_c were developed by generating artificial defects [20–23]. Studies have confirmed that incorporation of impurities such as Co, Fe, Ga, Zn, Ni, etc., to the Cu sites of YBCO can significantly increase the critical current density versus magnetic field (J_c–H) characteristics of high temperature superconductor bulks [24–26]. Dilute impurity doping to the CuO chain not only enhances flux pinning by lattice deformation but also results in a decrease in critical temperature (T_c) [24]. There are few reports about the doping effects of these dilute impurities in YBCO films [27, 28], especially those prepared with fluorine-free metal organic deposition (MOD).

The authors have previously published results of cobalt and zinc doping to the copper sites of YBCO [17]. It is a perspective way to essentially improve the current carrying capability of YBCO film. There is no work of Fe doping of YBCO in their early work. Fe is cheap and easy to obtain, so it is a perspective material for superconductor production. In this study, different dilute quantities of Fe doping of YBCO were introduced into the MOD YBCO films. Flux pinning properties of dilute Fe-doped YBCO films were investigated. The films were deposited by a fluorine-free polymer-assisted metal organic deposition (PA-MOD) method [17]. Epitaxial, dense, smooth, and crack-free YBCO films were prepared on a $LaAlO_3$ (LAO) single crystal substrate. In-field J_c of YBCO films was improved by this dilute impurity doping method.

2 Experimental

Fe-doped $YBa_2Cu_{3-x}Fe_xO_{7-\delta}$ films were deposited on a $LaAlO_3$ (LAO) single crystal substrate using a fluorine-free method [17]. The precursor solutions were synthesized by dissolving acetates of iron, in addition to acetates of yttrium, barium, and copper, in propionic acid with $x = 0$, 0.0005, 0.001, 0.002, and 0.005. Then polyvinyl butyral was added into the solution, which was subjected to continuous stirring to adjust the viscosity in order to obtain the final coating solution. The final cation concentration of the solution is 0.6 M. Then the solution was coated to LAO using a spin coater with a rotation speed of 6,000 r/min and dried at 150–200 °C for 5–10 min. The coated samples were fabricated in humid Ar/O_2 mixture gas at 160–500 °C for 11 h and then fired at 770–800 °C in dry Ar/O_2 mixture gas for 1 h. Finally, the samples were annealed in dry O_2 gas at 400–450 °C for 1 h. Finally, the obtained $YBa_2Cu_{3-x}Fe_xO_{7-\delta}$-coated films with x of 0–0.005 were about 500 nm in thickness.

A *Philips* X'Pert MRD diffractometer with Cu-Kα radiation was used to record the θ-2θ X-ray diffraction (XRD) patterns. The microstructure analyses of the YBCO layer were performed using an environmental scanning electron microscope (ESEM). Superconducting transition and magnetic hysteretic loop were measured by Quantum-Design SQUID XL. The J_c value of the YBCO film was determined using Bean critical state model formula.

3 Results and discussion

Figure 1 shows the typical θ-2θ X-ray diffraction patterns of $YBa_2Cu_{3-x}Fe_xO_{7-\delta}$ ($x = 0$, 0.0005, 0.001, 0.002, 0.005) films. As can be seen, only (0 0 l) YBCO reflection peaks were detected, excluding the peaks of LAO single crystal substrate. This indicates a strong c-axis orientation for all films.

The microstructure analyses of the YBCO layer were performed, and the results are shown in Fig. 2. Dense, smooth, and crack-free surface morphologies can be observed in all the films. A few particles floating on the film surface may be attributed to texture degradation with the growth of film during the firing procedure. We can see the particles in the Fe-doped YBCO films less than the pure YBCO films. There are some very small pores in the

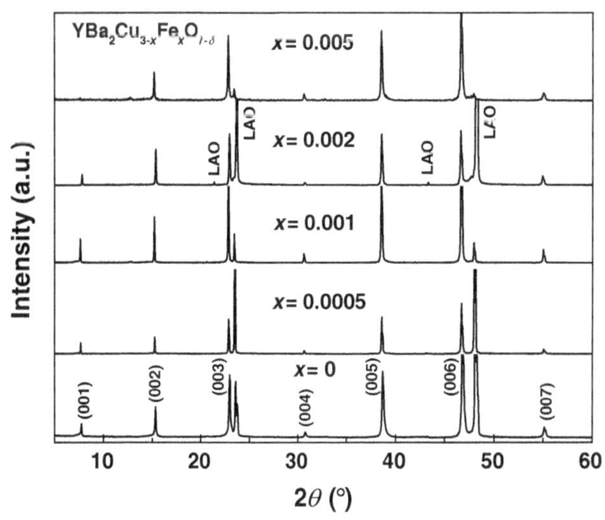

Fig. 1 Typical X-ray diffraction θ-2θ patterns of $YBa_2Cu_{3-x}Fe_xO_{7-\delta}$ ($x = 0$, 0.0005, 0.001, 0.002, 0.005) films

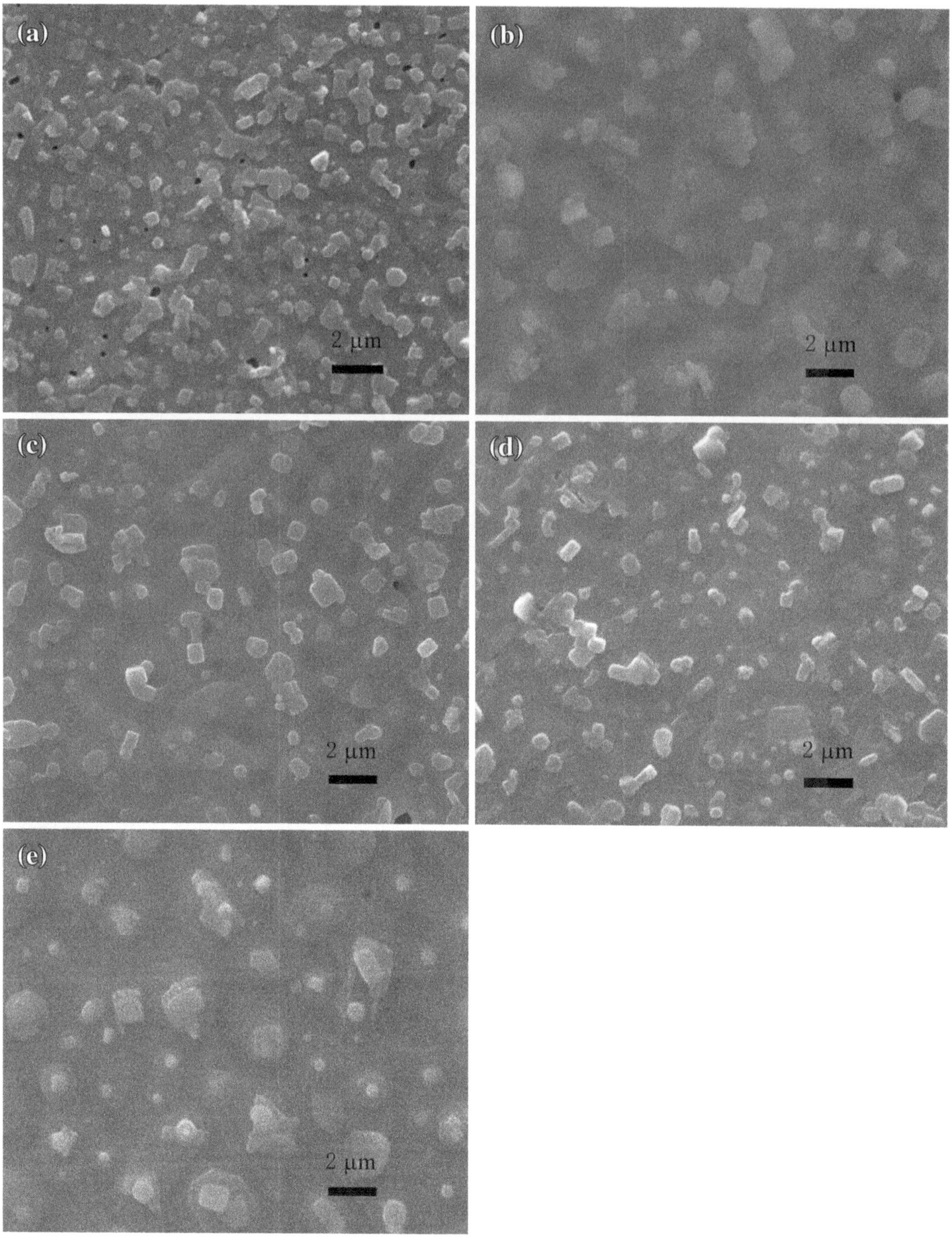

Fig. 2 The surface morphology of $YBa_2Cu_{3-x}Fe_xO_{7-\delta}$ ($x = 0$, 0.0005, 0.001, 0.002, 0.005) films. **a** $x = 0$, **b** $x = 0.0005$, **c** $x = 0.001$, **d** $x = 0.002$, **e** $x = 0.005$

pure YBCO film, but these are not seen in the Fe-doped films. It can be seen that a very small amount of Fe doping improves the surface connectivity of the YBCO films.

Figure 3 shows the superconductivity critical temperature of $YBa_2Cu_{3-x}Fe_xO_{7-\delta}$ ($x = 0$, 0.0005, 0.001, 0.002, and 0.005) films. The inset shows the temperature dependence of magnetic moment for $YBa_2Cu_{3-x}Fe_xO_{7-\delta}$

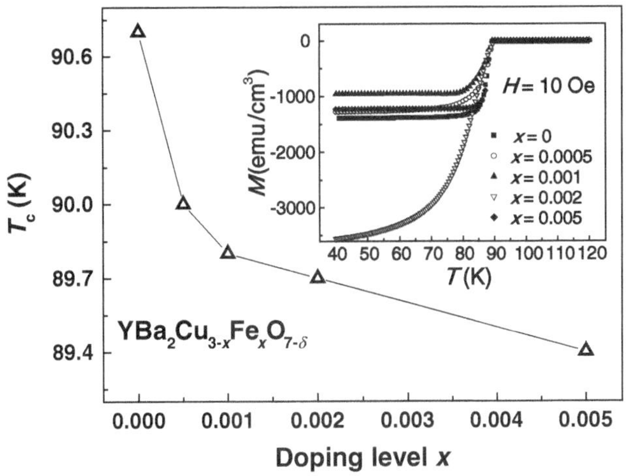

Fig. 3 The superconductivity critical temperature of $YBa_2Cu_{3-x}Fe_xO_{7-\delta}$ ($x = 0$, 0.0005, 0.001, 0.002, and 0.005) films. *Inset* is the temperature dependence of magnetic moment for $YBa_2Cu_{3-x}Fe_xO_{7-\delta}$ ($x = 0$, 0.0005, 0.001, 0.002, and 0.005) films

films. The pure YBCO sample exhibits a sharp superconducting transition; while for the doped samples, the superconducting transition width broadens, showing that such Fe doping leads to a decrease of superconductivity. The decrease may be due to the direct iron substitution for copper in the Cu–O chain. The substitution affects the oxygen vacancies which play an important role in determining the transport properties of YBCO. It is found that the superconductivity critical temperature of $YBa_2Cu_{3-x}Fe_xO_{7-\delta}$ decreases with Fe doping. When the value of $x = 0.005$ for $YBa_2Cu_{3-x}Fe_xO_{7-\delta}$, the T_c value is 1.2 K smaller than that of pure YBCO film. It exhibits a very small decrease of T_c because of the small amount of Fe doping.

Figure 4 shows the magnetic field dependence of J_c for $YBa_2Cu_{3-x}Fe_xO_{7-\delta}$ ($x = 0, 0.0005, 0.001, 0.002, 0.005$) films at different temperatures with the magnetic field parallels to the c-axis. The J_c values of Fe-doped YBCO films are not significantly higher than that of pure YBCO films at 30 and 50 K. However, at 77 K, Fe-doped YBCO films possess higher J_c values than those of pure YBCO films, except for the sample with $x = 0.005$. As 77 K is a feasible temperature for application of YBCO superconductors by using liquid nitrogen, this improved J_c value is significant. It means that the current carrying capability of YBCO film can be improved by doping with appropriate amounts of Fe.

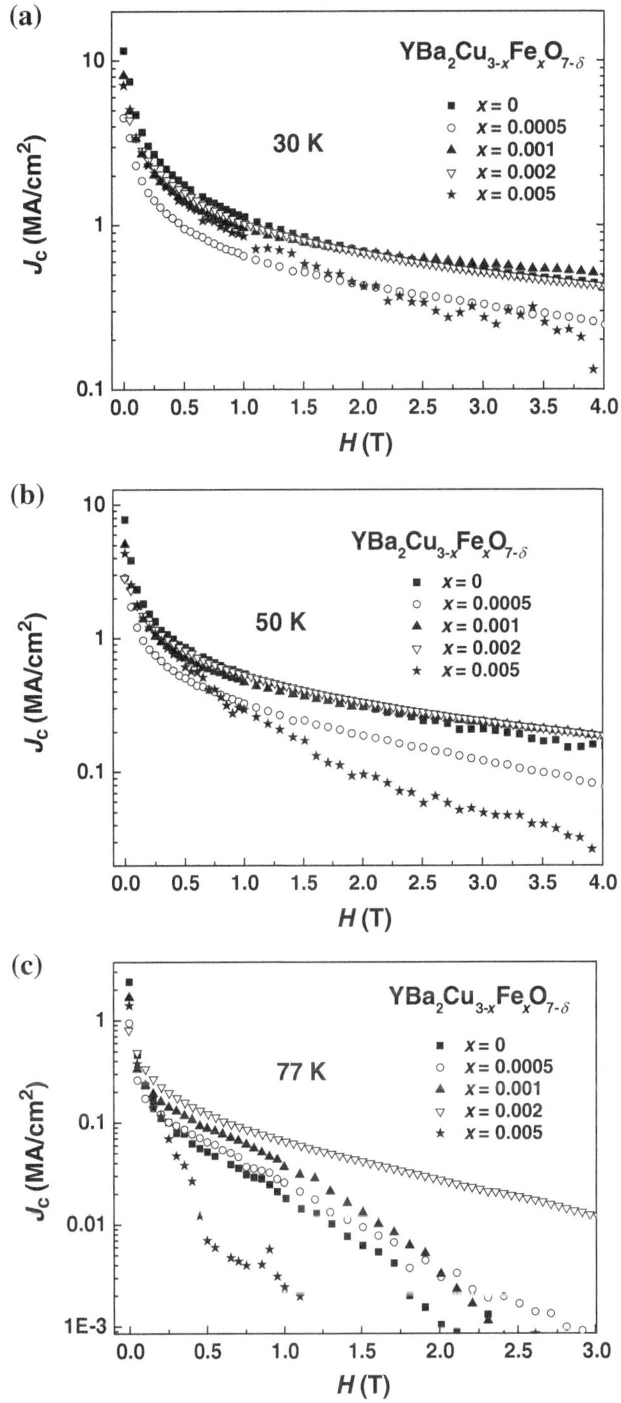

Fig. 4 The magnetic field dependence of critical current densities for $YBa_2Cu_{3-x}Fe_xO_{7-\delta}$ ($x = 0$, 0.0005, 0.001, 0.002, 0.005) films. **a** $T = 30$ K, **b** $T = 50$ K, **c** $T = 77$ K

4 Conclusion

In this paper, from the investigation of the Fe doping effects on YBCO films, it can be seen that the critical temperature for superconducting decreases slightly, and J_c improves at 77 K for doping quantities of x less than 0.005.

Acknowledgments This research was supported by the Specialized Research Fund for the Doctoral Program of Higher Education (200806131034, 200806130023), Natural Science Foundation of China under Contract Nos. 50672078 and 50872116, the National Science Fund for Distinguished Young Scholars under Contract No. 50588201, and 51102199, the National High-Tech Program of China (863 Program) under Contract No. 2007AA03Z203, the PCSIRT of the Ministry of Education of China (IRT0751), Research and Development Foundation of Southwest Jiao tong University under Grant Contract No. 2004A02, Fundamental Research Funds for the Central Universities (SWJTU12CX019), the National Natural Science Foundation (51202202), and Fundamental Research Funds for the Central Universities of China (SWJTU2682013CX005).

References

1. Yao X, Oka A, Izumi T, Shiohara Y (2000) Crystal growth and superconductivity of Fe-doped YBCO single crystals. Physica C 339(2):99–105
2. Espinosa-Arronte B, Djupmyr M, Andersson M (2005) Scaling of the B-dependent resistivity for different orientations in Fe doped $YBa_2Cu_3O_{7-\delta}$. Physica C 423(1–2):69–73
3. Liu LH, Dong C, Zhang JC, Chen H, Chen L (2002) A simple volumetric method for oxygen content determination in high-Tc doped YBCO compositions. Physica C 383(1–2):17–22
4. Qiao Y, Li Y, Sathiraju S, Reeves J, Lenseth K, Selvamanickam V (2002) An overview of the coated conductor progress at IGC-SuperPower. Physica C 382(1):48–51
5. Obradors X, Puig T, Pomar A, Sandiumenge F, Mestres N, Coll M, Cavallaro A, Rom N, Gazquez J, Gonzalez JC, Castano O, Gutierrez J, Palau A, Zalamova K, Morlens S, Hassini A, Gibert M, Ricart S, Moreto JM, Pinol S, Isfort D, Bock J (2006) Progress towards all-chemical superconducting $YBa_2Cu_3O_{7-y}$ coated conductors. Supercond Sci Technol 19(3):s13–s15
6. Izumi T, Yamada Y, Shiohara Y (2003) Activity of R&D for HTS coated conductors in Japan. Physica C 392–396(1):9–16
7. Motowidlo LR, Selvamanickam V, Galinski G, Haldar P, Sokolowski RS (2000) Recent progress in high-temperature superconductors at Intermagnetics General Corporation. Physica C 335(1–4):44–50
8. Parans Paranthaman M, Izumi T (2004) High-performance YBCO-coated superconductor wires. MRS Bull 29(9):533–536
9. Parans Paranthaman M (2006) Superconductor wires, McGraw-Hill 2006 year book of science and technology. McGraw-Hill, New York, pp 319–322
10. Goyal A, Norton DP, Budai JD, Paranthaman M, Specht ED, Kroeger DM, Christen DK, He Q, Saffian B, List FA, Lee DF, Martin PM, Klabunde CE, Hatfield E, Sikka VK (1996) Fabrication of long range, biaxially textured, high temperature superconducting tape on rolled Ni substrates. Appl Phys Lett 69(12): 1795–1797
11. Goyal A, Paranthaman M, Schoop U (2004) The RABiTS approach: using rolling-assisted biaxially textured substrates for high-performance YBCO superconductors. MRS Bull 29(9): 552–561
12. Rupich MW, Verebelyi DT, Zhang W, Kodenkandath T, Li X (2004) Metalorganic deposition of YBCO films for second-generation high-temperature superconductor wires. MRS Bull 29(9):572–578
13. Parmigiani F, Chiarello G, Ripamonti N, Goretzki H, Roll U (1987) Observation of carboxylic groups in the lattice of sintered $Ba_2YCu_3O_{7-y}$ high-T_c superconductors. Phys Rev B 36(13): 7148–7150
14. Honjo T, Nakamura Y, Teranishi R, Fuji H, Shibata J, Izumi T, Shiohara Y (2003) Growth mechanism of YBCO films in metal organic deposition method using trifluoroacetates. IEEE Trans Appl Supercond 13(2):2516–2519
15. Minei T, Tanaka T, Ogata M, Mori N, Yamada K, Mukaida M (2006) Growth process and microstructure of Y123 film fabricated by advanced TFA-MOD process. Physica C 445–448(1): 570–573
16. Castano O, Cavallaro A, Palau A, Gonzalez JC, Rosell M, Puig T, Pinol S, Mestres N, Sandiumenge F, Pomar A, Obradors X (2003) Influence of porosity on the critical currents of trifluoroacetate-MOD $YBa_2Cu_3O_7$ films. IEEE Trans Appl Supercond 13(2): 2504–2507
17. Wang WT, Pu MH, Yang Y, Zhang H, Cheng CH, Zhao Y (2010) Enhanced flux-pinning in fluorine-free MOD YBCO films by chemical doping. Physica C 470(20):1261–1265
18. Wang WT, Li G, Pu MH, Sun RP, Zhou HM, Zhang Y, Zhang H, Yang Y, Cheng CH, Zhao Y (2008) Chemical solution deposition of YBCO thin film by different polymer additives. Physica C 468(15–20):1563–1566
19. Zhao Y, Pu MH, Li G, Du XH, Zhou HM, Zhang YB, Yang XS, Wang Y, Sun RP, Cheng CH (2007) Development of a new series of buffer layers for REBCO coated conductors. Physica C 463–465(1):574–579
20. Feng Y, Pradhan AK, Zhao Y, Wu Y, Koshizuka N, Zhou L (2001) Improved flux pinning in $Y_xHo_{1-x}Ba_2Cu_3O_y$ fabricated by powder melting process. Supercond Sci Technol 14(4):224–228
21. Babu NH, Shi YH, Iida K, Cardwell DA (2005) A practical route for the fabrication of large single-crystal (RE)–Ba–Cu–O superconductors. Nat Mater 4(1):476–480
22. Werner M, Sauerzopf FM, Weber HW, Wisniewski A (2000) Fishtail effect in the magnetization of superconducting $RBa_2Cu_3O_{7-\delta}$ (R = Y, Nd, Yb) and $Y_2Ba_4Cu_8O_{16}$ single crystals. Phys Rev B 61(21):14795–14803
23. Sauerzopf FM (1998) Anisotropic flux pinning in $YBa_2Cu_3O_{7-\delta}$ single crystals: the influence of defect size and density as determined from neutron irradiation. Phys Rev B 57(17):10959–10971
24. Ishii Y, Shimoyama JI, Tazaki Y, Nakashima T, Horii S, Kishio K (2006) Enhanced flux pinning properties of $YBa_2Cu_3O_y$ by dilute impurity doping for CuO chain. Appl Phys Lett 89(11): 202514–202517
25. Zhou YX, Scruggs S, Salama K (2006) Effects of ionic doping superconducting properties of melt textured $YBa_2(Cu_{1-x}M_x)_3O_{7-\delta}$ (M = Co, Ni, Zn or Ga) large grains. Supercond Sci Technol 19(7):S556–S558
26. Pan SH, Hudson EW, Lang KM, Eisaki H, Uchida S, Davis JC (2000) Imaging the effects of individual zinc impurity atoms on superconductivity in $Bi_2Sr_2CaCu_2O_{8+\delta}$. Nature 403(2):746–750
27. Ohazama T, Mukaida M, Horii S, Saito A, Kita R, Ichinose A, Yoshida Y, Matsumoto K, Ohshima S (2005) Low surface resistance Zn-doped $ErBa_2Cu_3O_{7-\delta}$ thin films. Physica C 426–431(2):1611–1615
28. Horii S, Ichino Y, Yoshida Y, Matsumoto K, Mukaida M, Ichinose A, Shimoyama J, Kishio K (2006) Dilute Co-doping effects on critical current properties in PLD-$ErBa_2Cu_3O_y$ films. Physica C 445–448(1):837–840

21

Structural model for the first wall W-based material in ITER project

Dehua Xu · Xinkui He · Shuiquan Deng · Yong Zhao

Abstract The preparation, characterization, and test of the first wall materials designed to be used in the fusion reactor have remained challenging problems in the material science. This work uses the first-principles method as implemented in the CASTEP package to study the influences of the doped titanium carbide on the structural stability of the W–TiC material. The calculated total energy and enthalpy have been used as criteria to judge the structural models built with consideration of symmetry. Our simulation indicates that the doped TiC tends to form its own domain up to the investigated nano-scale, which implies a possible phase separation. This result reveals the intrinsic reason for the composite nature of the W–TiC material and provides an explanation for the experimentally observed phase separation at the nano-scale. Our approach also sheds a light on explaining the enhancing effects of doped components on the durability, reliability, corrosion resistance, etc., in many special steels.

Keywords Tungsten-based material · First wall material · First principles

D. Xu · X. He · S. Deng (✉) · Y. Zhao
Key Laboratory of Magnetic Levitation Technologies and Maglev Trains, Ministry of Education of China, and Superconductivity and New Energy R&D Center, Southwest Jiaotong University, Chengdu 610064, China
e-mail: s.deng@swjtu.edu.cn

D. Xu
School of Physical Science and Technology, Southwest Jiaotong University, Chengdu 610064, China

1 Introduction

In facing the imminent depletion of fossil fuels and the mounting demand of energy resources, various new sustainable energy resources have been proposed. Among which, the fusion energy has been considered the only way to solve the constantly growing demands of energy and the increasingly stringent requirements of environment. The ITER (International Thermonuclear Experimental Reactor) project is one important step for humankind to harness the fusion energy [1]. The first wall material is crucially important in the plasma facing components, which work in an extreme environment exposing to high energy particles and heat flux from the plasma [2]. Roughly speaking, the first wall materials can be classified into low Z materials and high Z materials. Tungsten belongs to the high Z category, which has the advantage of high strength at high temperature, and good resistance against erosion and irradiation due to high energy particles such as neutrons, He, etc. However, pure tungsten has a low fracture toughness which can be improved by changing its chemical composition through alloying or sintering with low Z materials to form composite materials. In the last few years, many efforts have been made in studying the tungsten-based materials such as W–(0.3, 0.9, 1.7) TiC (wt%), W–(0.3, 1.0, 2.0) Y, and W–(0.3, 1.0, 2.0) Y_2O_3, etc. [3, 4].

The W–TiC system was first discovered and systematically studied by the research group of Kurishita. They have not only studied the preparation conditions, in particular, the effects of Ar, N_2, H_2, etc., atmosphere on the final mechanical properties [5–8], but also have shown that the addition of TiC can result in significant refinement of grains of the alloys, which can thus enhance the ductility and resistance to the neutron irradiation. In addition, the high temperature mechanic properties and the crack and fracture mechanical properties can also be improved. [9–13]. The possible reason as suggested by Kurishita et al. [14] is that the addition of TiC and the recrystallization of W–TiC alloy result in much efficient packing of the grains and strengthening grain boundaries. The other research activities [15–18] have also confirmed that the final properties of this system depend heavily on the preparation routes and details, a phenomenon indicating the role of multi-scale microstructures. Although many experimental works have been done on this system, the microscopic mechanism which connects the multi-scale structures and the properties has not been established. So far as we know, there has been hitherto no theoretical work done on W–TiC system. This work focuses on the inhomogeneity at the nano-scale and attempts to understand the experimental findings from the electronic structures.

2 Computational details

In this work, we have used the state-of-the-art quantum code CASTEP for all computations. In our calculations, the density functional theory (DFT) is approximated using local density approximation (LDA) and its generalized gradient corrections (GGA) as implemented in the PBE formulation [19]. The built-in ultra-soft pseudo-potentials have been used throughout this work to treat the core electrons, which leads to a relatively low energy cutoff of 310.0 eV for plane wave expansions of wave functions, potentials, etc. A $4 \times 4 \times 8$ k-mesh in the first Brillouin zone is used for all integrations in the momentum space. Throughout this work, the DFT total energy, enthalpy if external pressure is present, has been taken as the object function for geometry optimization with respect to each test structure model. The high-dimensional multivariable minimization problem is solved using the Broyden–Fletcher–Goldfarb–Shanno (BFGS) quasi-Newton scheme [20].

3 Initial structure models

The experimental researches on tungsten-based materials have studied a wide range of composition for TiC and other chemicals. In this work, we investigate a specific composition, W+ ~ 3 % TiC in weight percentage, which corresponds to a mole composition of $W_7(TiC)$. A supercell technique is used to study the possible microstructures at the nano-length scale. As shown in Fig. 1b, a basic supercell of $2 \times 2 \times 1$ with respect to that of the α-W (Fig. 1a) contains 8 W atoms.

According to the symmetry of the supercell as shown in Fig. 1b, we have built four different models as shown in Fig. 2. In Fig. 2 (a) the Ti atom is put on the edge of the supercell, while the C atom is put inside the supercell, but at an asymmetric position to the two Ti atoms (Ti–C: 2.162 Å, 2.390 Å); (b) Ti atom is put at the center position of the α-W with Ti and C atoms (Ti–C: 2.162 Å) lying in the (001) plane but not in the (010) plane in contrast to that in model (c); (c) Both Ti and C atoms are put on the (001) surface of the supercell with Ti–C distance at 2.162 Å and Ti, C atoms lying in (010) plane; (d) Same as for (c) except that the C atom lies equally distant (Ti–C: 2.162 Å) to the two Ti atoms.

4 Results and discussion

The optimization on model (d) results in a configuration with the minimum energy as compared to the other three models. The optimization starting from the other three models lead to higher energy and larger stresses. Considering the periodicity of the models, model (a) and (d) are actually very similar, while models (b) and (c) belong to one class. Our calculations indeed obtain almost the same geometrical details for models (b) and (c). In both cases, the calculated Ti–C distance is 1.974 Å, which is a little shorter than that, 2.162 Å, in the TiC structure. The next Ti–C distance in both models is 3.709 Å, which clearly shows that the Ti–C units in both models remain isolated with each other. The optimized W–C distance in (b) and (c) is 2.004 Å, shorter than that in the WC structure (2.197 Å). The separation of TiC units with each other may cost more energy for the W host structure to accommodate such units. Our calculations indicate that models (b) and (c) are both energetically unfavorable by an energy of about 0.23 eV higher than that of model (d). As we mentioned above, models (a) and (d) are structurally very similar. They differ with each other only in the locations of the C atoms. In model (d) the Ti–C units lie in the (010) plane while those in model (a) cross the (010) plane. In these two structures, the TiC forms zigzag chains which run along the c direction. This character is similar to that observed in the crystal of TiC. For model (a), the geometric optimization shifts the two Ti atoms away from the (010) W plane, while the original longer Ti–C distance, 2.392 Å, in the zigzag chain becomes 1.912 Å, with the other Ti–C distance almost unchanged. For model (d), the basic geometrical character remains unchanged except that the Ti–C distance (2.221 Å) becomes longer than that in the initial model. The optimized W–C distance in model (d) is 1.997 Å which is shorter than the normal W–C bond length. This fact indicates a stronger W–C bonding, which thus weakens the Ti–C bonding and results in a longer Ti–C distance (2.221 Å). The energy differences between the models in each category are negligibly small, for example,

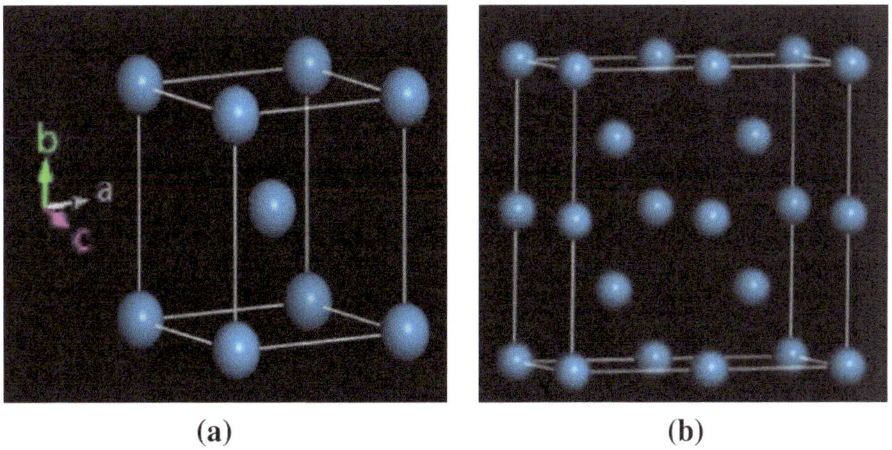

Fig. 1 **a** The unit cell of α-W; **b** the 2 × 2 × 1 supercell of α-W, the coordinate system is for both figures

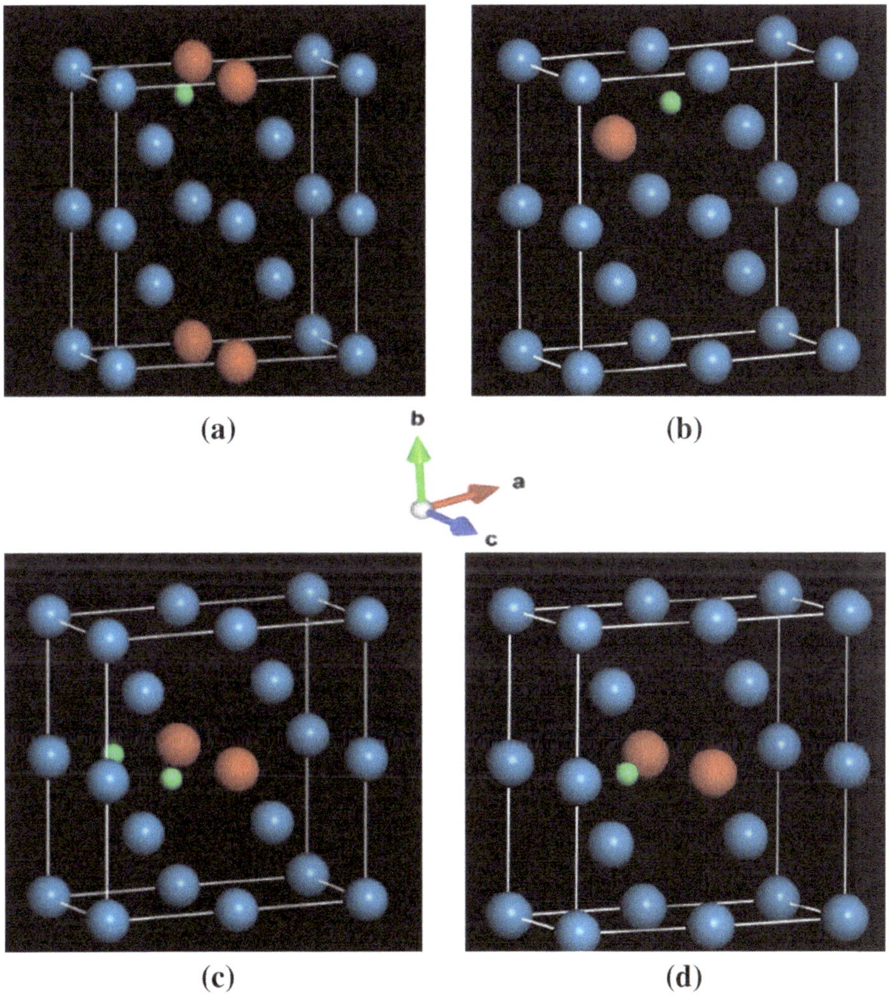

Fig. 2 In **a–d**, the *blue balls* represent W, *red balls* represent Ti, and *green balls* represent C. The crystallographic coordinate system shown in the center is for all models from **a** to **d**

the energy difference between (b) and (c) is only 0.0418 eV, which is at the order of phonon energy. In all the cases, the Ti–C bonding is weakened due to the formation of W–C bonding, which indicates that the W–TiC composite materials are not pure physical mixing, chemical interactions occur in all the cases. Nevertheless, this W–C bonding is not large enough to compensate the energy lost due to the weakening of Ti–C bonding. As we have described above that the configurations (a, d) which keep the character of TiC structure have indeed energy advantage. This energy gain favors the phase separation at least at the unit-cell length that we investigated. We thus have given a theoretical explanation to the observed experimental fact. [10, 21].

Recently, we have used more accurate VASP code [22, 23] to calculate the above models as well as much larger models. When the chosen models are large enough to be comparable with the experimental W–TiC ratio, the Ti–C distance converges to a value of $\sim$2.16 Å. This is consistent with the result obtained by CASTEP code. The full analyses on results of VASP calculations are undergoing and have not been finished.

It has already been well known that some tiny amounts of elements such as C, Mn, Si, P, S, Cr, Mo, V, etc., play essential roles in enhancing the hardness, durability, fracture toughness, and good thermal performance of some special steels under extreme weather conditions [24]. A typical example is the widely used rail-track steel which contains $\sim$0.4 %–0.8 % C, $\sim$0.65 %–1.7 % Mn, $\leq$0.6 % Si, and other trace elements depending on the type of the steels. However, as in the W–TiC case, the microscopic mechanisms which depend on the multi-scale structures of the interested composite materials remain unsolved. The recent experimental and theoretical studies [25–27] have shown the toughness of the problem. Our first-principles approach as used in this work provides a tool to attack the problem at the atomic and subatomic length scales. Because of its universality, we do expect its important applications in studying complex extreme fusion materials as well as the railway track materials.

5 Conclusion

Through using the first-principle method as implemented in the CASTEP code, we have optimized four models which simulate the relative configurations of TiC in the W lattice. By using the total energy and enthalpy as the standard, we have analyzed the obtained results. We conclude that although there exists W–C interactions which weaken the Ti–C bonding, the Ti–C units tend to stay together to keep the structural character in the TiC structure. This provides an explanation for the experimental phenomenon of phase separation between W and TiC at nano-scale. However, we must point out that our conclusion is based on the relative smaller models which limit the choice of the supercell. Besides, our models involve only the chain configuration and the two atom Ti–C unit, therefore, it is worth further studies on more and larger realistic models to test the reliability of the conclusions that we drawn here.

Acknowledgments This work is finantially supported by the Science Foundation for International Cooperation of Sichuan Province (2014HH0016), the Fundamental Research Funds for the Central Universities (SWJTU2014: A0920502051113-10000), National Magnetic Confinement Fusion Science Program (2011GB112001).

References

1. Barabash V (2007) Materials challenges for ITER-current status and future activities. J Nucl Mater 367:21–32
2. Federici G, Wuerz H, Janeschitz G et al (2002) Erosion of plasma-facing components in ITER. Fusion Eng Des 61:81–94
3. Baluc N, Schäublin R (2011) From materials development to their test in IFMIF: an overview. Nucl Fusion 51(11):113–116
4. Rieth M, Dudarev SL (2013) Recent progress in research on tungsten materials for nuclear fusion applications in Europe. J Nucl Mater 432(1–3):482–500
5. Ishijima Y, Kurishita H, Yubuta K et al (2004) Current status of ductile tungsten alloy development by mechanical alloying. J Nucl Mater 329–333, Part A(0):775–779
6. Kajioka M, Sakamoto T, Nakai K et al (2011) Effects of plastic working and MA atmosphere on microstructures of recrystallized W-1.1%TiC. J Nucl Mater 417(1–3):512–515
7. Kawai M, Kurishita H, Kokawa H et al (2012) Development of advanced materials for spallation neutron sources and radiation damage simulation based on multi-scale models. J Nucl Mater 431(1–3):16–25
8. Matsuo S, Kurishita H, Arakawa H et al (2008) Deformability enhancement in ultra-fine grained, Ar-contained W compacts by TiC additions up to 1.1%. Mater Sci Eng A 492(1–2):475–480
9. Kitsunai Y, Kurishita H, Kayano H et al (1999) Microstructure and impact properties of ultra-fine grained tungsten alloys dispersed with TiC. J Nucl Mater 271–272:423–428
10. Kurishita H, Amano Y, Kobayashi S et al (2007) Development of ultra-fine grained W–TiC and their mechanical properties for fusion applications. J Nucl Mater 367–370, Part B(0):1453–1457
11. Kurishita H, Kobayashi S, Nakai K et al (2008) Development of ultra-fine grained W-(0.25–0.8)wt%TiC and its superior resistance to neutron and 3 MeV He–ion irradiations. J Nucl Mater 377(1):34–40
12. Kurishita H, Matsuo S, Arakawa H et al (2008) Superplastic deformation in W-0.5wt.% TiC with approximately 0.1 μm grain size. Mater Sci Eng A 477(1–2):162–167
13. Kurishita H, Matsuo S, Arakawa H et al (2009) High temperature tensile properties and their application to toughness enhancement in ultra-fine grained W-(0–1.5)wt% TiC. J Nucl Mater 386–388:579–582
14. Kurishita H, Matsuo S, Arakawa H et al (2010) Development of re-crystallized W-1.1%TiC with enhanced room-temperature ductility and radiation performance. J Nucl Mater 398(1–3):87–92
15. Genç A, Coşkun S, veçoğlu ML (2010) Microstructural characterizations of Ni activated sintered W-2wt% TiC composites

produced via mechanical alloying. J Alloy Compd 497(1–2):80–89
16. Genç A, Coşkun S (2010) M.L. veçoğlu. Decarburization of TiC in Ni activated sintered W-xTiC (x = 0, 5, 10, 15 wt%) composites and the effects of heat treatment on the microstructural and physical properties. Int J Refract Metal Hard Mater 28(3):451–458
17. Matsuo S, Kurishita H, Arakawa H et al (2008) Deformability enhancement in ultra-fine grained, Ar-contained W compacts by TiC additions up to 1.1%. Mater Sci Eng A 492(1–2):475–480
18. Chen Y, Wu YC, Yu FW et al (2008) Microstructure and mechanical properties of tungsten composites co-strengthened by dispersed TiC and La_2O_3 particles. Int J Refract Metal Hard Mater 26(6):525–529
19. Perdew JP, Burke K, Ernzerhof M (1996) Generalized gradient approximation made simple. Phys Rev Lett 77(3):865–868
20. Shanno DF (1978) Conjugate gradient methods with inexact searches. Math Oper Res 3(2):244–256
21. Veleva L (2011) Contribution to the production and characterization of W–Y, W–$Y_2O_3$and W–TiC materials for fusion reactors. [D] Lausanne Ecole Polytechnique Federale de Lausanne
22. Kresse G, Joubert D (1999) From ultrasoft pseudopotentials to the projector augmented-wave method. Phys Rev B 59:1758
23. Kresse G (2007) Vienna ab-initio simulation package (VASP), http://cms.mpi.univie.ac.at/vasp/vasp/vasp.html
24. ArcelorMittal Products Information (2013) http://www.arcelormi 'ttal.com/rails specialsections/en/
25. Missori S, Sili A (2000) Structural characterization of C–Mn steel laser beam welded joints with powder filler metal. Weld Res Suppl 79(11):317–323
26. Song R, Ponge D, Rabbe D (2005) Influence of Mn content on the microstructure and mechanical properties of ultrafine grained C–Mn Steels. ISIJ Int 45(11):1721–1726
27. Eden HC, Garnham JE, Davis CL (2007) Influential microstructural changes on rolling contact fatigue crack initiation in pearlitic rail steels. Mater Sci Technol 21(6):623–629

22

Identifying the relationship between suspension parameters of underframe equipment and carbody modal frequency

Guangbing Luo · Jing Zeng · Qunsheng Wang

Abstract As a critical component of the railway vehicle, underframe equipment has a great influence on the ride comfort of railway vehicles due to their big mass and active vibration. Therefore, study on the relationship between suspension parameters of underframe equipment and the modal frequency of carbody is extremely crucial for controlling the ride quality of railway vehicles. In this paper, a finite element model of the carbody was developed to investigate the effects of the suspension location, the mass of the suspension equipment, and the suspension frequency on the mode of the carbody. Then, the matching relationship between the suspension parameters and the modal frequency of the carbody was studied through the transfer function. In addition, roller rig tests were performed to verify the numerical simulation model of the carbody. The results show that the suspension parameters of the underframe equipment have a great influence on the mode of the carbody, especially for the frequency of the first bending mode. To improve the frequency of carbody high-frequency bending and reduce energy transfer, equipment with a large mass should be suspended toward the middle of the carbody. The weight of the equipment strongly affects the first bending frequency and energy transfer of the carbody. The frequency of heavy suspended equipment should be sufficiently low to increase the transmissibility of high frequencies and improve the vibration characteristics of the carbody. Although the bending frequency of the carbody can be improved effectively by increasing the suspension stiffness of the suspension equipment, in order to reduce carbody vibration effectively, the suspension frequency of the equipment should be slightly lower than the carbody bending frequency.

Keywords Underframe equipment · Suspension parameters · Modal frequency · Carbody · Bending frequency · Suspension types

1 Introduction

In modern railway operations, lightweight car bodies are vital to improving average train speeds. However, lowering the carbody weight reduces the stiffness of the structure, and increases accelerations within the vehicle system. This may have a negative impact on ride comfort [1, 2]. To handle these problems, a number of solutions have been proposed. For example, an essentially nonlinear vibration absorber was shown to mitigate the large acceleration transmitted to the passenger compartment of a vehicle and improve ride comfort [3], and a dynamic vibration absorber was installed in the underframe of a high-speed passenger car to suppress vibration [4]. To reduce bending vibration, some authors have suggested the use of piezoelectric elements shunted by an external electric circuit or active damping via a piezo stack of actuators [5, 6]. A constraint layer on the carbody was researched in [7], and a heavy additional mass suspended under the center of the carbody has also been considered [8].

To optimize vibrational comfort in the design stage, the dynamics of the whole vehicle must be taken into account [9, 10]. The natural frequency of rolling stock is not only an important parameter in the dynamic loading of a carbody, but also forms the basis of the vehicle system dynamics. Elastic vibration in the carbody will be excited when the external excitation frequency equals the natural mode of the structure.

G. Luo (✉) · J. Zeng · Q. Wang
State Key Laboratory of Traction Power, Southwest Jiaotong University, Chengdu, China
e-mail: lguangbing2009@163.com

J. Zeng
e-mail: zeng@swjtu.edu.cn

The relationship between the suspension parameters of the equipment and the inherent carbody modal frequency have attracted widespread attention, as they are essential in ensuring system stability and passengers' ride comfort. Carbody modeling is now the starting point of such research. By considering the carbody accessory structure, some progress has already been achieved. For instance, the influence of three types of equipment suspended from the carbody and the avoidance of structural resonance by adjusting the stiffness distribution has been reported [11]. Kang and Zeng [12] proposed a rigid coupling of the nonlinear system dynamics model. They discussed the vibration character of suspension equipment and studied the matching relation between the carbody modal frequency and suspension parameters, determining the optimal suspension location from the dynamic response of the underframe equipment. Taking CRH3 electric multiple units (EMU) as an example, Zhao et al. [13] analyzed the influence of underframe equipment on the carbody mode, and suggested that the elastic suspension of equipment could effectively improve the bending and rhombic modes. The modeling of a carbody structure and its suspension units was elaborately described in [14], which presented a reasonable modeling method by changing the mass distribution and boundary conditions. The carbody, wheelset, and frame have been considered as elastomers to analyze the modal frequency of a 270 km/h high-speed EMU [15], and to compare the stiffness matching conditions of the whole vehicle. The modal frequency of passenger cars has been analyzed using a finite element method, with the effect of the mass and stiffness of additional parts also considered [16]. A linear prediction method has been applied to the modal properties of the carbody, and this approach was validated by data from both stationary and running tests [17]. However, most research only considers a modal analysis of the initial frequency. The relation between the modal frequency and vibration has not been studied. High-speed EMUs in China often use both elastic and rigid suspension. Research has shown that elastic suspension effectively reduces the elastic vibration, changing the modal frequency of the carbody. The equipment is equivalent to a dynamic vibration absorber when elastic suspension is used. The two peak values will appear in the curve of the first bending frequency of the carbody. The key problem in this paper is to determine the best combination of carbody mode suspension parameters and vibration. A series of simulations enables us to identify the matching principles between the equipment suspension parameters and the modal frequency of the carbody.

2 Finite element analysis of carbody

2.1 Finite element model of carbody

Taking a high-speed test train as an example, consider the matching relationship between the carbody mode and the equipment suspension parameters. As the main bearing structure, the carbody consists of the underframe, corrugated floor, side sill, and so on. According to the structural characteristics of the carbody plate and girders, the real structure can be simulated using the ANSYS software. The corrugated sheet and side sills are simulated by Shell63 and beam elements, respectively. Using these elements, the structure of an aluminum alloy carbody can be accurately simulated. For the initial conditions, a mass unit is used to simulate the counterbalance. The equipment suspended from the underframe of the carbody is simulated by mass21 at the actual mass center, and the real connecting positions are rigidly coupled. The sides of each element measure 40–80 mm. The model includes around 122,000 nodes and 197,000 units. The finite element model is shown in Fig. 1.

2.2 Modal analysis

Modal analysis is used to solve the motion equations of the linear system without any damping, and to determine the natural frequency of the system. Because of the complex structure of the carbody, the natural frequency cannot be accurately obtained by analytical methods. However, it can be solved by finite element methods. The motion equation of the carbody is

$$M\ddot{x} + C\dot{x} + Kx = f, \tag{1}$$

where M denotes the mass matrix of the system, C represents the damping matrix, K represents the stiffness matrix, f represents the loading vector, and x is the deformation shape vector of the carbody structure.

Ignoring the influence of structural damping on the modal frequency, the free vibration solution of the system is

$$x = \delta \sin(\omega t + \varphi_0), \tag{2}$$

where ω and φ_0 are the squared circular eigenfrequency and initial phase, respectively, and δ is a magnitude vector. We substitute Eq. (2) into Eq. (1), and seek non-zero solutions. For such solutions, we require a zero coefficient determinant, that is

$$\left|K - \omega^2 M\right| = \mathbf{0}. \tag{3}$$

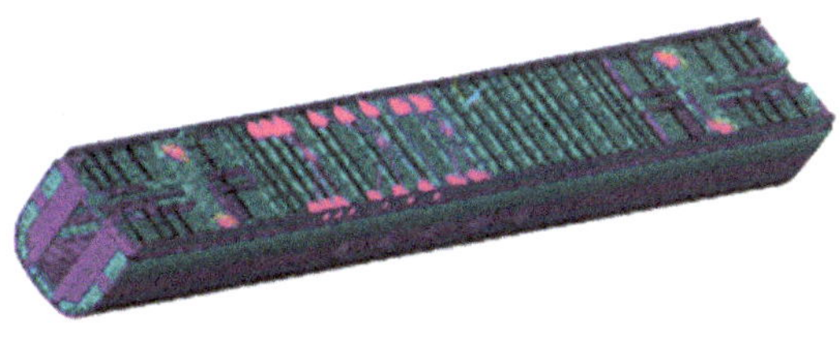

Fig. 1 Finite element model

The eigenvalues and eigenvectors of Eq. (3) give the intrinsic modes of the system.

3 Validation of modal simulation

3.1 Modal test method for a carbody

We now investigate a modal test for an aluminum alloy carbody. Seats, racks, and suspension equipment are not considered in the test, and the unladen body mass is about 9,000 kg. We also test the preparation conditions, using a traction transformer of 6,000 kg and a traction converter of 3,000 kg. Mass blocks lying on the floor are used to simulate the seats, giving a total carbody mass of about 34,000 kg. Details are shown in Figs. 2 and 3.

We consider seven cross-sections to reflect the vibration form of the carbody, as shown in Fig. 3, and place four acceleration sensors at the top and bottom of each section to measure the vertical and lateral acceleration. Thus, the vibration in the carbody can be obtained by analyzing the acceleration. An excitation force of about 120 N is applied sinusoidally with a scanning frequency range of 1–100 Hz.

3.2 Comparison between simulation and test

The limitations of the test method make it difficult to measure each modal frequency of the carbody. Only low-order modes can be accurately determined by the conventional test method, whereas high-order modes of local vibrations can hardly be measured. However, by comparing the test results with the simulation calculations, we can find all the modes. This comparison validates the rationality and accuracy of the modeling. From Table 1, we can see that the error in the low-order mode is relatively small, whereas higher mode suffer bigger errors. This is because local deformations appear in high-order mode, and the local stiffness of the carbody is not well simulated. However, under the preparation condition, the error in low-order mode is about 10 %, which is acceptable for engineering requirements. Based on the system model, we compare the test results and the simulations in terms of the sweeping frequency. Figure 4 shows that the simulation results can effectively reflect the actual motion.

Fig. 2 Test carbody

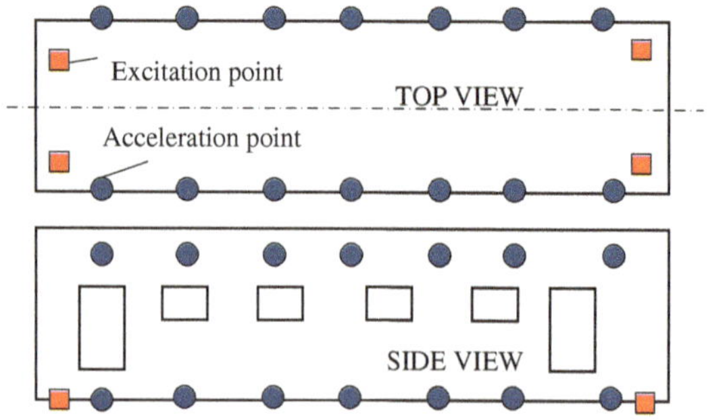

Fig. 3 Accelerometer positions on carbody

Thus, the modeling method has been shown to be efficient. This forms the foundation of the following research into the relationship between suspension parameters and carbody modes.

4 Relationship between suspension parameters and carbody modes

High-speed EMUs generally adopt a distributed power system. The equipment is suspended on the crossbeam of the underframe, and weighs from tens of kilograms to several tons. The suspension method and distribution of the equipment will affect the modal frequency of the carbody. To research the relationship between the suspension of equipment and carbody modes, we consider just one device. Some basic laws are then acquired by analyzing the influence of different suspension parameters on the carbody modes.

It has been shown that the elastic suspension of underframe equipment can effectively reduce carbody vibration and improve ride comfort [18–20]. The vibration of the carbody is closely related to the modal frequency; namely, elastic vibration of the carbody will be aroused only when the external excitation frequency matches the frequency of the carbody. Therefore, the relationship between the main mode of the carbody and the two suspension styles (elastic and rigid) is discussed in this section. Table 2 presents simulation results for both rigid and elastic suspension. As we can see, vertical and lateral first-order bending mode appear under the condition of elastic suspension. Using modal analysis, it is easy to determine that the low-frequency mode is mainly coupled with bending movement in the carbody, caused by the natural frequency of the equipment, whereas high-frequency mode is mainly caused by the interaction of the carbody's natural frequency and the equipment. As a whole, elastic suspension has a

Table 1 Main modal frequency comparison of experimental and simulation results

Modal name	Aluminum alloy car body			Preparation condition car body		
	Exper-(Hz)	Simul-(Hz)	Error (%)	Exper-(Hz)	Simul-(Hz)	Error (%)
1st bending	19.45	19.66	1.08	12.35	12.59	1.94
Rhombic	20.43	20.10	1.62	16.47	14.68	10.87
Torsion	29.33	34.90	18.99	22.64	25.68	13.42

Exper Experiment, *Simul* Simulation

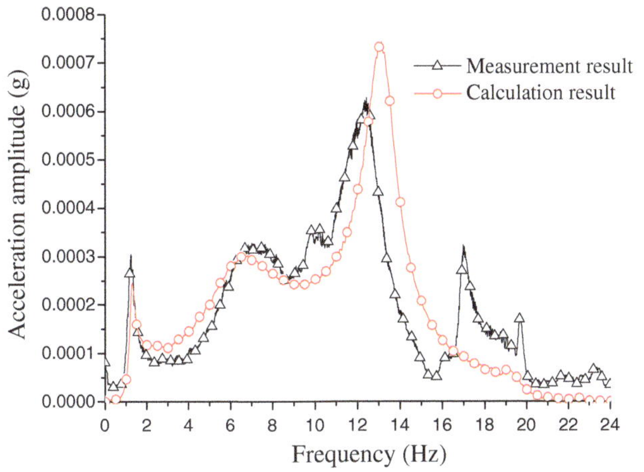

Fig. 4 Vibration characteristics of carbody

Table 2 Influence of equipment suspension style on carbody modes

Modal name	Modal frequency (Hz)	
	Elastic	Rigid
LF vertical bending	10.15	–
LF lateral bending	10.42	–
Rhombic	14.59	14.68
HF vertical bending	16.82	12.59
HF lateral bending	17.61	13.87
1st breath	19.99	21.11
Torsion	24.98	25.68
2nd vertical bending	30.29	30.27

LF Low frequency, *HF* High frequency

significant effect on the first-order bending mode. Compared with the carbody bending mode when the equipment is under rigid suspension, one of the two new frequencies rises while the other decreases. The span between the two new frequencies is directly related to the suspension parameters. Compared with the case of rigid suspension, the other modal frequencies are all lower, and the change is not so obvious. Rigid suspension is equivalent to increasing the body weight, and also changes the mass center of the carbody. The equipment moves together with the carbody, and hence there is only one vertical bending and one lateral bending frequency. From the above analysis, we can see that the key to the modal matching analysis is the influence of suspension parameters on the first bending mode; the other modes can be ignored.

To further research the influence of equipment suspension parameters on the carbody bending mode, the next section studies the effects of changing the equipment mass, suspension position, and suspension stiffness on the carbody mode. The influence of the location of the suspended equipment on carbody vibration is shown in Fig. 5. From these results, we can see that the equipment has the least influence on the carbody mode when the suspension position is about 7.5 m from the carbody center, which is close to the bogie. Between the bogie and the carbody center, a greater longitudinal distance from equipment to carbody center implies the carbody bending frequency will be closer to the bending frequency without suspended equipment. The low-order coupled bending frequency of the carbody obviously increases as the longitudinal distance to the suspended equipment increases, whereas the coupled frequency caused by the carbody decreases. However, there is only one bending frequency in the rigid suspension case. We can also see that, when the equipment is suspended above the bogie, the mode of the carbody is reversed. From the above analysis, we can conclude that elastically suspended equipment should be positioned away from the bogie to improve the first-order bending frequency of the carbody. There is 3.5 m from the bogie to the end of the car, and 7.5 m from bogie to car center, which is sufficient longitudinal distance to improve the bending frequency. Thus, elastically suspended equipment should be suspended near the center of the carbody.

The next section examines the relationship between the modal frequency and transmissibility. The input excitation is a sinusoidal signal. The influence of equipment mass on carbody modes is shown in Fig. 6. In Fig. 6a, the suspension frequency ($f = (2\pi)^{-1}(m^{-1}k)^{-0.5}$) is set to 12 Hz, and we can see that the span between high and low frequencies becomes wider as the equipment mass increases. This increase in equipment mass causes the transmissibility of high frequencies to gradually decrease, while that of low frequencies increases. In Fig. 6b, the suspension stiffness is maintained at 33.6 MN/m (when the equipment mass is

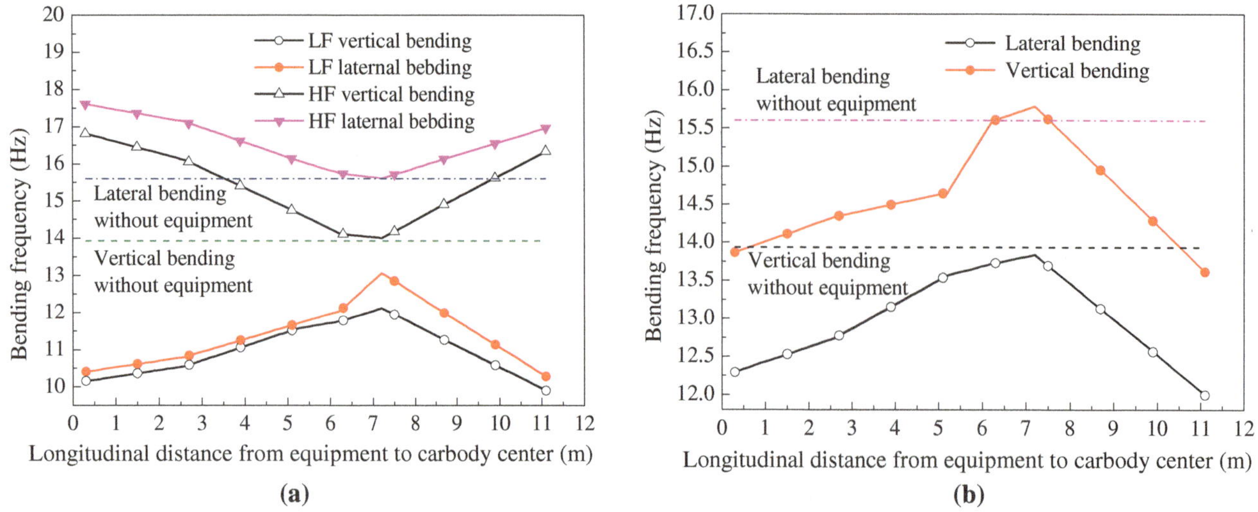

Fig. 5 Relationship between the location of suspended equipment and carbody mode. **a** Elastic suspension. **b** Rigid suspension

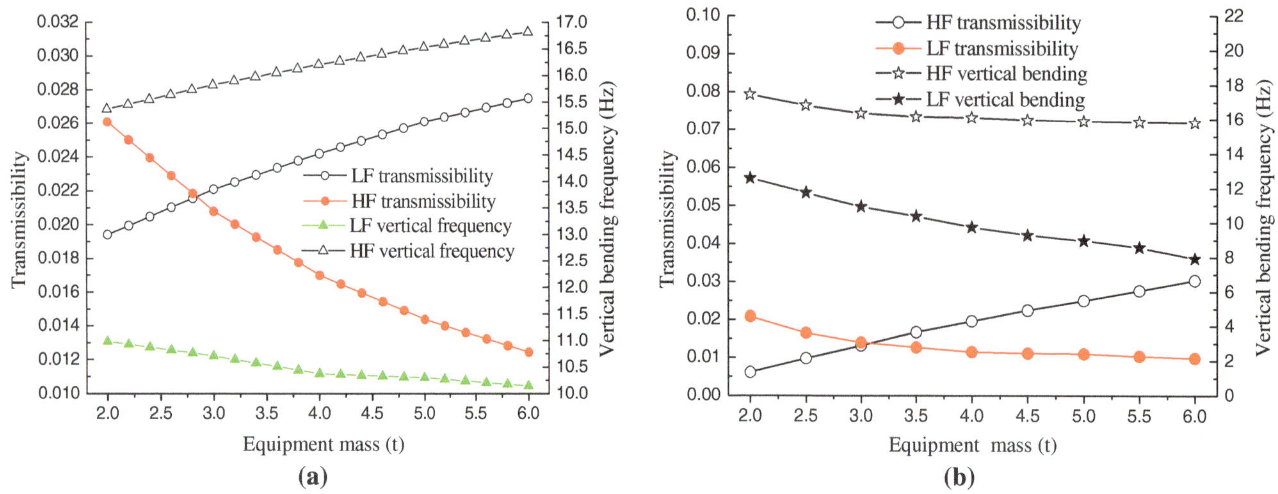

Fig. 6 Relationship between elastic suspension weight and carbody mode. **a** Suspension frequency of 12 Hz. **b** Suspension stiffness of 33.6 MN/m

6,000 kg, the suspension frequency is 12.45 Hz, which is close to the carbody bending frequency). With an increase in mass, the bending frequency of the carbody gradually decreases. The change in the low-frequency component is more obvious than that of the high frequency. At the same time, the high-frequency transmissibility increases and the low-frequency transmissibility is reduced, which can reduce the sensitive frequency felt by passengers and improve the ride quality. Therefore, the heavier the equipment, the smaller its suspension frequency should be to reduce the suspension stiffness of the equipment and increase the transmissibility of high frequencies. However, we must be careful to avoid bogie resonance. As shown in Fig. 7, the rigid suspension of the equipment is equivalent to increasing the mass of the carbody without changing its stiffness. This serves to decrease the carbody first-order bending frequency and increase the transmissibility. However, these changes are not so obvious.

The relationship between the carbody bending mode and suspension frequency is depicted in Fig. 8. An increase in the suspension frequency causes the bending frequency of the carbody to increase rapidly, and the low-order bending frequency becomes close to that in the absence of suspended equipment. The high-order bending frequency tends to infinity, and is eventually equivalent to the carbody bending frequency under rigid suspension. However, as the suspension frequency changes, the carbody transmissibility also varies. The increase in suspension frequency leads to an increase in low-frequency transmissibility, and a reduced high-frequency transmissibility. When the suspension frequency reaches a certain value (about 10.75 Hz), the transmissibility of high and low frequencies is equal. At this point,

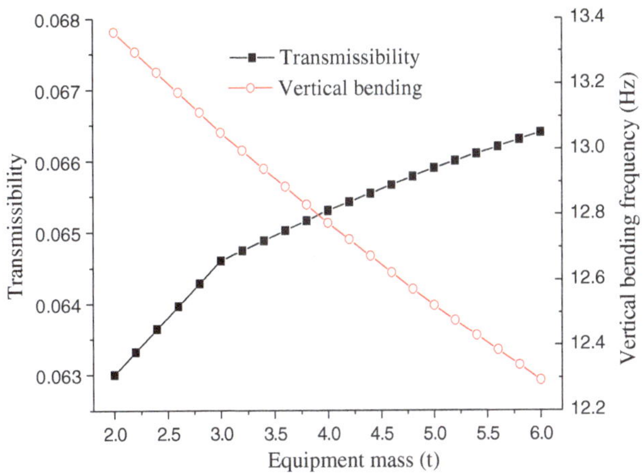

Fig. 7 Influence of rigid suspension equipment on carbody mode

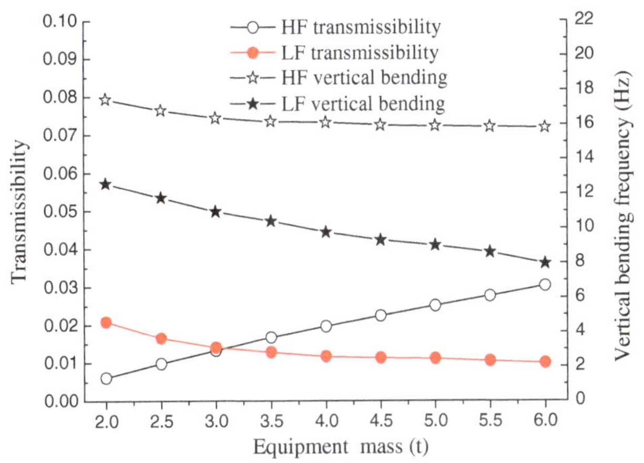

Fig. 8 Influence of equipment suspension stiffness on carbody mode

the vibration is minimized. To improve the ride quality, the first-order bending frequency should be as high as possible, and the transmissibility should be as low as possible. To meet these two conditions simultaneously, the frequency of suspended equipment should be less than the carbody first-order bending frequency under rigid suspension, so that the carbody transmissibility can be optimized.

5 Vehicle dynamic system modeling and vibration response analysis

5.1 Dynamic rigid–flexible coupled theory and modeling

Elastic vibration of the carbody cannot be properly reflected in the traditional rigid body model, and our track test shows that the equipment clearly influences the carbody modal frequency. To properly reflect the coupling between suspension equipment and carbody elastic vibration, we require a rigid–flexible coupled dynamic model. In

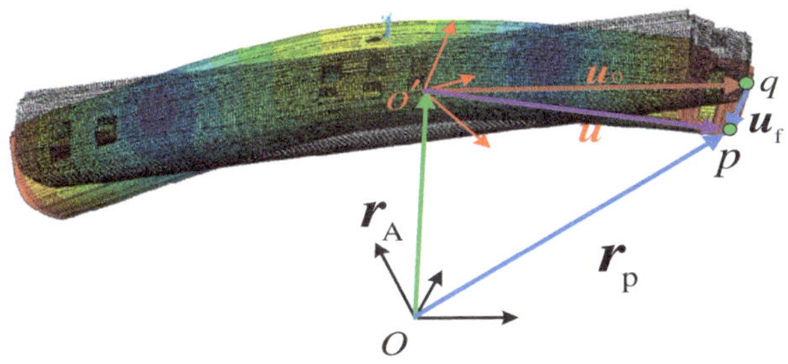

Fig. 9 Deformable carbody coordinates

this section, we discuss the rigid–flexible theory. Consider an undeformed carbody in which the location of each point q relative to a local fixed reference system O is represented by the vector $\boldsymbol{r}_A$ (see Fig. 9).

The point q moves to a new location p under elastic deformation, so that the position vector relative to the global coordinate system is

$$\boldsymbol{r}_p = \boldsymbol{r}_A + \boldsymbol{A}(\boldsymbol{u}_0 + \boldsymbol{u}_f), \tag{4}$$

where $\boldsymbol{A}$ is the transformation matrix, and $\boldsymbol{r}_A$ is the location vector of the fixed reference system relative to the global system. The global location vector of an arbitrary point on the flexible carbody is $\boldsymbol{r}_p$, $\boldsymbol{u}_0$ is the position of point q in the undeformed state, and $\boldsymbol{u}_f(t)$ denotes the deformed vector. The displacement vector field due to the structure flexibility is expressed as

$$\boldsymbol{u}_f(t) = \sum_{j=1}^{N} \boldsymbol{n}_j \eta_j(t), \tag{5}$$

where $\boldsymbol{n}_j$ represents the deformation shape j and $\eta_j(t)$ is the shape amplitude.

Differentiating the deformation equation with respect to time yields

$$\dot{\boldsymbol{r}}_p = \dot{\boldsymbol{r}}_A + \dot{\boldsymbol{A}}\boldsymbol{u}_f + \boldsymbol{A}\dot{\boldsymbol{u}}_f. \tag{6}$$

The kinetic energy of the deformable body can be defined as

$$T = \frac{1}{2}\int_V \rho \dot{\boldsymbol{r}}_p^T \boldsymbol{r}_p \mathrm{d}V - \frac{1}{2}\dot{\boldsymbol{\eta}}^T \boldsymbol{m}\dot{\boldsymbol{\eta}}, \tag{7}$$

where $\boldsymbol{m}$ is the mass of the flexible body and η the vector of shape amplitude. In the multi-body dynamics software Simpack, we can use virtual work theory to describe a generalized elastic force. The dynamic equations of the system are

$$\begin{cases} \boldsymbol{M}_e\ddot{\boldsymbol{\eta}} + \boldsymbol{D}_e\dot{\boldsymbol{\eta}} + \boldsymbol{K}_e\boldsymbol{\eta} = \boldsymbol{f}_e \\ \boldsymbol{C}_e(\dot{\boldsymbol{\eta}}, t) = 0 \end{cases}, \tag{8}$$

where $\boldsymbol{M}_e$, $\boldsymbol{D}_e$, $\boldsymbol{K}_e$, $\boldsymbol{C}_e$ denote the general mass matrix, general damping matrix, general stiffness matrix, and constraint matrix, respectively; $\boldsymbol{f}_e$ denotes a general force

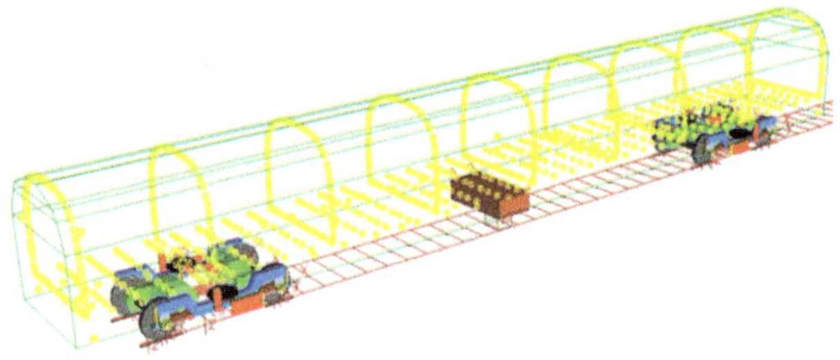

Fig. 10 Dynamical model of vehicle system

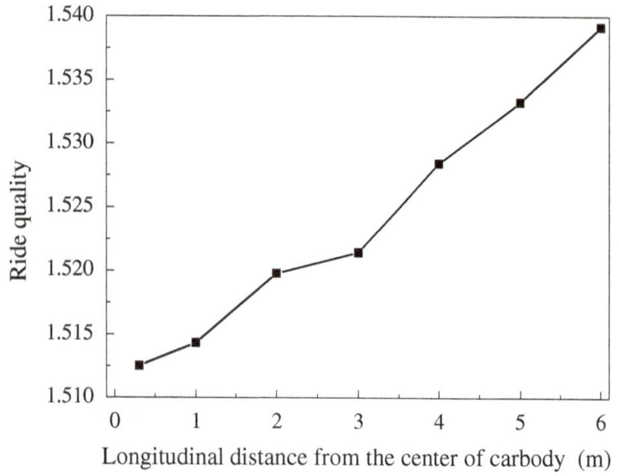

Fig. 11 Equipment position and ride quality

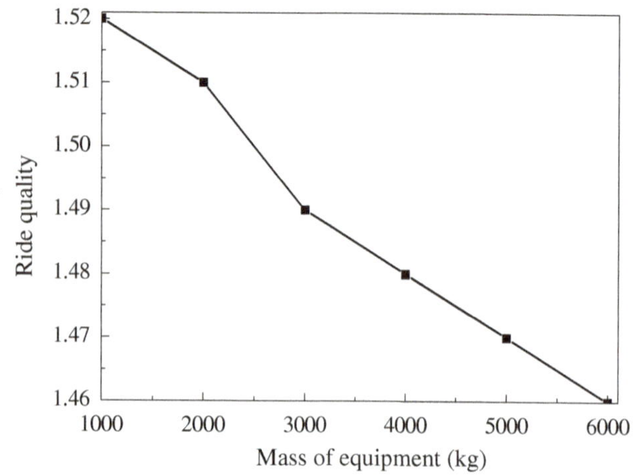

Fig. 12 Equipment mass and ride quality

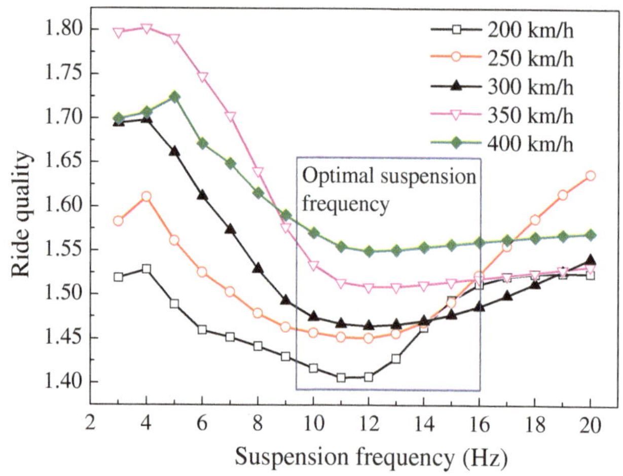

Fig. 13 Equipment frequency and ride quality

consisting of force elements, gravity, external force, and general inertia force.

In engineering practice, the finite element method is often used to model flexible carbodies because of its versatility. First, we extract the carbody substructure SUB file and the vibration modal RST file through finite element analysis. Then, a SID file is generated by the pre-processing module of Simpack. By considering the wheelset and bogie to be a rigid structure, we fix the suspended equipment to a beam of the carbody underframe by four hangers. The suspension equipment has a mass of 6,000 kg, and the carbody with its accessory structures has a mass of 28,000 kg. The model of this vehicle system is shown in Fig. 10, which contains 34 rigid degrees of freedom and 20-order eigenmodes.

5.2 Analysis of vibration response

The vehicle running quality is seriously affected by the suspension parameters of the equipment. It is difficult to study the influence of these parameters experimentally, but we can simulate the vibration characteristics of the carbody under different combinations of suspension parameters. We can then analyze the matching relationship between the equipment and the modal frequency. The following is based on the rigid–flexible coupled dynamics model. The measured irregularity excitation is used to determine the input response, allowing the matching principles between the equipment suspension parameters and carbody mode to be identified. Figures 11, 12 and 13 show the vertical ride quality of the central carbody for various locations, mass, and frequencies of the suspended equipment. From Figs. 11 and 12 we can see that the ride quality index is significantly improved by the use of elastic suspension. The following conclusions can be drawn from an analysis of the equipment location. First, a suspension position close to the center of the carbody is favorable in terms of ride quality. Second, heavier equipment improves ride quality. From these two observations, we can conclude that heavy equipment should be suspended as close as possible to the center of the carbody to improve ride comfort. Figure 13 shows the influence of suspension stiffness on the ride quality index under elastic suspension. According to Fig. 8, we can see that vibration is not adequately reduced when the frequency of suspended equipment is too high or too low. As the suspended frequency increases, the high-frequency part of the carbody coupled bending vibration will decrease. However, when the suspension frequency is higher than the carbody bending frequency under rigid suspension, the low-frequency bending vibration plays a more important role. As a result, the vibration is aggravated, and ride quality will worsen. A suspension frequency that is too low will lead to serious equipment vibration and a worsening of the carbody stability. However, when the frequency is too high, the connection becomes similar to that of rigid suspension, and the carbody vibration will be more serious. Therefore, the suspension frequency of the equipment should be close to, and preferably smaller than, the carbody bending frequency in the rigid case.

6 Conclusions

Using finite element simulations and analyzing test data, we verified that our modeling approach was correct and provided a valid foundation for further calculations. We then analyzed the relationship between carbody modal frequencies and equipment suspended from the car. The resulting modal matching principles can be summarized as follows:

- To improve the frequency of carbody high-frequency bending and reduce energy transfer, equipment with a large mass should be suspended toward the middle of the carbody.
- The weight of the equipment strongly affects the first bending frequency and energy transfer of the carbody. The frequency of heavy suspended equipment should be sufficiently low to increase the transmissibility of high frequencies and improve the vibration characteristics of the carbody.
- The bending frequency of the carbody can be effectively improved by increasing the suspension stiffness. However, this will also increase low-frequency energy transfer. When the suspension frequency is lower than the first-order bending frequency of the carbody under rigid suspension, high-frequency energy transfer plays a leading role. Thus, the suspension frequency of the equipment should be slightly lower than the carbody bending frequency.

Acknowledgments This work was financially supported by the National Natural Science Foundation of China (No. U1334206) and the National Basic Research Program of China (No. 2011CB711106).

References

1. Zeng J, Luo R (2007) Vibration analysis of railway passenger car systems by considering flexible carbody effect. J China Railw Soc 29(6):19–25
2. Wu PB, Xue SH, Yang CH (2005) Dynamic response of high-speed passenger car based on flexible carbody model. J Traffic Transp Eng 5(2):5–8
3. Wang D, Lee YS et al (2009) Mitigating the effect of impact loading on a vehicle using an essentially nonlinear absorber. J Veh Syst Dyn 4(10):1183–1204
4. Zhou JS, Zhang W et al (2009) Vibration reduction analysis of the dynamic vibration absorber on the flexible carbody of railway vehicles. J China Railw Sci 30(3):86–90
5. Takigami T, Tomioka T (2008) Bending vibration suppression of railway vehicle carbody with piezoelectric elements. J Mech Syst Transp Logist 1(1):111–121
6. Kozek M, Benatzky C, Schirrer A (2008) Vibration damping of a flexible carbody structure using piezo-stack actuators. J Int Fed Autom Control Seoul Korea 7:6–11
7. Huang CH, Zeng J (2010) Flexural vibration suppression of carbody for high2speed passenger car based on constrained damping layers. J Traffic Transp Eng 10(1):36–41
8. Holst C (1998) Active damping of carbody vibraton. M Master thesis, Department of Mechanics, Royal Institute of Technology (KTH)
9. Braghin F, Cheli F, Melzi S, Negrini S et al (2013) Experimental modal analysis and numerical modelling of agricultural vehicles. J Conf Proc Soc Exp Mech Ser 5(11):213–220
10. Lethonen TJ, Juhala M (2005) Predicting the ride behavior of a suspended agricultural tractor. J Int J Heavy Veh Syst 1:131–142
11. Xiang YC, Xiang WL (2012) Modal analysis of hanging equipment carbody of CRH5 EMU. J Mach Build Autom 41(4):43–46
12. Kang HG, Zeng J et al (2011) Coupled vibration analysis between carbody and underframe of high-speed intergrated test train. J Beijing Jiaotong Univ 6(35):62–66
13. Zhao YY, Liu SY, Zhang W (2012) The influence of hanging way and the flexible suspension mass on the body modal frequency. J Jiamusi Univ (Natural Science Edition) 30(3):326–333
14. Zhao HW, Tian AQ, Wang WJ et al (2011) Simulation and consistence test analysis of carbody structure vibration modal. J Dalian Jiaotong Univ (Natural Science Edition) 32(6):26–29
15. Lu ZJ, Tian HQ, Zhou D (2005) Modal analysis of 270 km/h high-speed motor car. J China Railw Sci 26(6):18–23
16. Hao LB, Hu QN, Li G (2004) Disscussion of finite element analysis on modal condition of passenger cars in the preparation conditions. J Roll Stock 42(11):4–7
17. Tomioka T, Takigami T, Ichiro K (2009) Modal analysis of railway vehicle carbodies using a linear prediction model. J Syst Des Dyn 32(6):918–931
18. Luo GB, Zeng J (2013) Study on carbody flexible vibrations with considering underframe suspended equipment. J Mod Manuf Eng 5:1–6 (in Chinese)
19. Wu HC, Wu PB (2012) Effect of equipment suspension stiffness on riding quality E. J Noise Vib Control 32(8):73–77
20. Sun WJ, Gong D et al (2012) Influences of suspended equipment under carbody on high speed train ride quality. J Procedia Eng 16:812–817

23

DC and AC transport properties on $La_{0.8}Sr_{0.2}MnO_3$

**Zhantao Wei · Xinsheng Yang · Li Lv ·
Min Zhang · Yong Zhang**

Abstract Magnetoresistive sensor can be widely used in modern transportation field, such as the vehicle positioning and navigation system, vehicle detection system, and intelligent transportation system. In order to improve the efficiency of magnetoresistive sensor, we synthesized $La_{0.8}Sr_{0.2}MnO_3$ polycrystalline bulks at different sintering temperatures and investigated their DC and AC transport properties in this work. As a result, all samples showed insulator–metal (I–M) phase transition, and the transition temperature ($T_{I\text{–}M}$) shifted to higher temperature with the increase of sintering temperature. The $T_{I\text{–}M}$ measured at different AC frequencies was smaller than that measured at DC condition, which implied that the I–M phase transition was suppressed at AC frequencies. The resistivity measured at high AC frequencies was larger than that measured at low AC frequencies, which could be attributed to the change of the magnetic penetration depth (δ). However, the room-temperature AC-magnetoresistance (MR) at low frequencies was much larger than that at high frequencies and room-temperature DC-MR. These findings demonstrate that reducing the AC frequency is an effective way for enhancing the room-temperature MR, which can be used to promote the efficiency of magnetoresistive sensor.

Keywords Magnetoresistance · Perovskite manganese oxides · DC and AC transport properties

Z. Wei · X. Yang · L. Lv · M. Zhang · Y. Zhang
Key Laboratory of Magnetic Levitation Technologies and Maglev Trains, Ministry of Education of China, Chengdu 610031, China

Z. Wei · X. Yang (✉) · L. Lv · M. Zhang · Y. Zhang
Superconductivity and New Energy R&D Center, Southwest Jiaotong University, Chengdu 610031, China
e-mail: xsyang@swjtu.edu.cn

1 Introduction

Magnetoresistance sensor, as one of the most active magnetic sensor technologies, which is widely applied in the vehicle positioning and navigation system, vehicle detection system, green transport systems, and so on [1]. Therefore, how to improve the efficiency of magnetoresistance sensor has been one of hot topics in these fields.

Since the discovery of colossal magnetoresistance (CMR) in perovskite manganese oxides with formula $L_{1-x}A_xMnO_3$ (where L is a trivalent rare earth, and A is a divalent alkaline earth ion), many efforts have been devoted to integrate its transport properties and physical mechanisms, such as Jahn–Teller effect [2], double exchange [2, 3], phase separation [4], and charge order [5]. Meanwhile, lots of novel physical phenomena, such as spin glass behavior [6, 7] and exchange bias [8], were discovered. These findings indicated that $L_{1-x}A_xMnO_3$ can be used in magnetoresistive sensor for improving the efficiency [9]. However, CMR effect is sensitive to temperature and only appears in the vicinity of the Curie temperature in a relative high magnetic field. This seriously limits the efficiency and application of magnetoresistive sensor. Thus, finding new ways to improve the room-temperature magnetoresistance (MR) in a low magnetic field become a hot topic, and many studies have been focused on it. It was found that introducing a second phase into the matrix is an effective way to improve magnetoresistance [10]. Gupta et al. [11] found that using Mo to substitute the Mn of $La_{0.67}Sr_{0.33}MnO_3$ led to an enhancement of room-temperature MR. Many attempts have been made to study its AC transport properties which were measured with radio frequency ($f > 1$ MHz). Rebello and Mahendiran [12] showed that the value of the AC-MR at room temperature in $La_{0.7}Sr_{0.3}MnO_3$ is extremely high

($\approx$47 % in a magnetic field where the magnetic induction intensity $B = 0.1$ T and $f = 3$–5 MHz). Aparnadevi and Mahendiran [13] revealed that the frequency dependent magnetotransport properties are dominated by the field and frequency dependence of AC permeability. However, the detailed AC transport properties of $La_{0.8}Sr_{0.2}MnO_3$ at frequency less than 1 MHz have not been investigated.

In this work, a systematic study of AC electrical transport in the frequency range $f = 0$–14 kHz based on the $La_{0.8}Sr_{0.2}MnO_3$ was made. Their DC and AC transport properties were also presented. The room-temperature AC-MR at low frequencies is much larger than that at high frequencies and room-temperature DC-MR, which can be used to promote the efficiency of magnetoresistive sensor.

2 Experimental

$La_{0.8}Sr_{0.2}MnO_3$ polycrystalline bulks were prepared by sol–gel method, as elaborated in [10]. The obtained powers were carefully ground and pressed into wafers. Then, the wafers were sintered at various temperatures of 600, 800, 1,000, and 1,200 °C.

The phase structure and morphology were analyzed by X-ray diffractometer (XRD, X'Pert Panalytical) and field emission scanning electron microscope (FESEM, JSM-7001F), respectively. The DC and AC transport measurements were performed using electrical properties measurement system (ET-9000) with a lock-in amplifier (SR830) by the standard four-probe method. To insure the constant current measurement, a large resistor with a resistance 1,000 times higher than the value of samples was applied in the AC transport measurement.

3 Results and discussion

Figure 1 shows XRD patterns of all samples. No peaks of other impurities are found under instrumental resolution, confirming that the product is composed of pure $La_{0.8}Sr_{0.2}MnO_3$. The strength of the major peak increases gradually as the sintering temperature rises. It is known that the average grain size of unit cell is characterized by Scherrer equation [8]: $D \sim K\lambda/\beta\cos\theta$, where K is the shape factor, λ is the wavelength of X-ray, β is the full width at half maximum (FWHM), and θ is the diffraction angle. The average grain size, lattice parameters (a, c), and cell volume are calculated, as shown in Table 1. Obviously, with the sintering temperature increasing, the lattice parameter a decreases gradually, while the lattice parameter c increases. Moreover, the average grain size also increases with the increase of sintering temperature, which could be explained by the cluster effect [14]. Typical FESEM images of all samples with different sintering temperatures are shown in Fig. 2. It can be seen that the grain size increases with the sintering temperature.

The DC resistivity was measured as a function of temperature ($\rho-T$), as shown in Fig. 3. All samples exhibit the insulation-metal (I-M) phase transition, and the transition temperature ($T_{\text{I-M}}$) shifts to high temperature regions as the sintering temperature increases. However, when a low magnetic field ($B = 0.3$ T) is applied, the $T_{\text{I-M}}$ shifts to low temperature regions, which could be confirmed by Fig. 5a. The resistivity decreases monotonously as the sintering temperature increases because of the decrease of grain boundaries [15]. It is obvious that the resistivity of the samples sintered at 600 °C is larger than that of other samples. The reason may lie in that there are a large number of magnetic disorders and amorphous grain boundaries in the samples sintered at 600 °C, which lead to the electrons receiving more scattering than that in other samples [16]. However, it is found that the room-temperature MR at 0.3 T is small, which limits its application in magnetoresistive sensor.

Furthermore, the $\rho-T$ behaviors of $La_{0.8}Sr_{0.2}MnO_3$ polycrystalline samples at different AC frequencies were measured, as shown in Fig. 4. All samples exhibit a similar trend as the temperature increases at AC frequencies varying from 0.133 to 13.333 kHz. All samples also have the I–M phase transition. It is noted that all curves show a flat range in the low frequency regions (133, 1,333 Hz) and then a gentle range in high frequency regions (3,333, 7,777, 13,333 Hz), which can be observed more clearly in samples with higher sintering temperature. It could be considered as an outcome of the classical skin effect. As the frequency of the alternative current increases, the flow of current is restricted to a specified thickness, which results in the increase of resistivity [17].

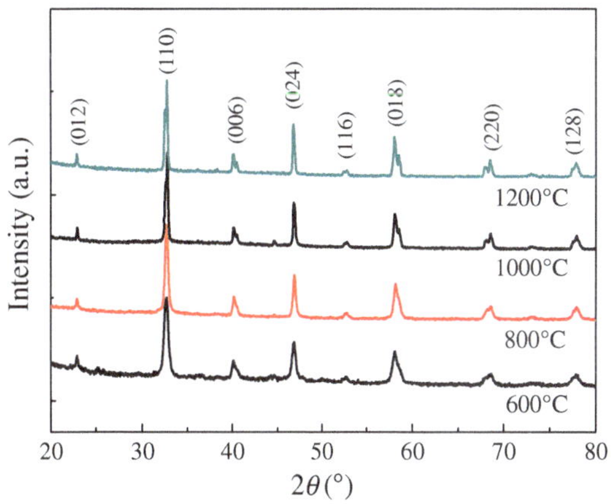

Fig. 1 X-ray diffraction patterns of $La_{0.8}Sr_{0.2}MnO_3$ polycrystalline bulks sintered at various temperatures

Table 1 Structure parameters of samples sintered at different temperatures

Sintering temperature (°C)	600	800	1,000	1,200
Average grain size (nm)	20	27	54	75
Lattice constant (Å)				
a	5.4970	5.4916	5.4749	5.4512
c	13.3739	13.3943	13.4299	13.4586
Cell volume (Å^3)	349.967	349.605	348.613	347.082

Figure 5a shows the sintering temperature dependent $T_{I\text{–}M}$ of samples measured at DC frequency in different applied magnetic fields. Obviously, the applied magnetic field leads to the decrease of $T_{I\text{–}M}$, and the $T_{I\text{–}M}$ increases as the sintering temperature increases under 0 and 0.3 T. According to Dutta et al. [16], the sample with lower sintering temperature has a higher density of grain-boundary, which could scatter the elections and then cause the increase of resistivity. Naturally, lower value of $T_{I\text{–}M}$ appears due to a stronger inhibitory effect from grain-boundary in the samples with lower sintering temperature. The frequency dependent $T_{I\text{–}M}$ of samples at AC frequency in zero magnetic fields is shown in Fig. 5b. With frequency increasing, $T_{I\text{–}M}$ first increases, reaching a maximum, and then decreases. The $T_{I\text{–}M}$ measured at different AC frequencies is smaller than that measured at DC condition, which implies that the I–M phase transition is suppressed at AC frequencies.

In order to study in detail the influence of frequency on resistivity, we define the resistance change rate at different frequencies as $\Delta\rho/\rho(0) = (\rho(f) - \rho(0))/\rho(0)$, where the $\rho(f)$ is the resistance obtained at different AC frequencies in zero DC magnetic field, and $\rho(0)$ is the resistance obtained at $f = 0$ Hz. Figure 6 shows the temperature dependent $\Delta\rho/\rho(0)$ of all samples measured at low frequencies ($f = 133$, 1,333 Hz) in zero DC magnetic field. Obviously, all samples show a peak near the $T_{I\text{–}M}$ in $\Delta\rho/\rho(0)$ versus T curve. This phenomenon is similar to the

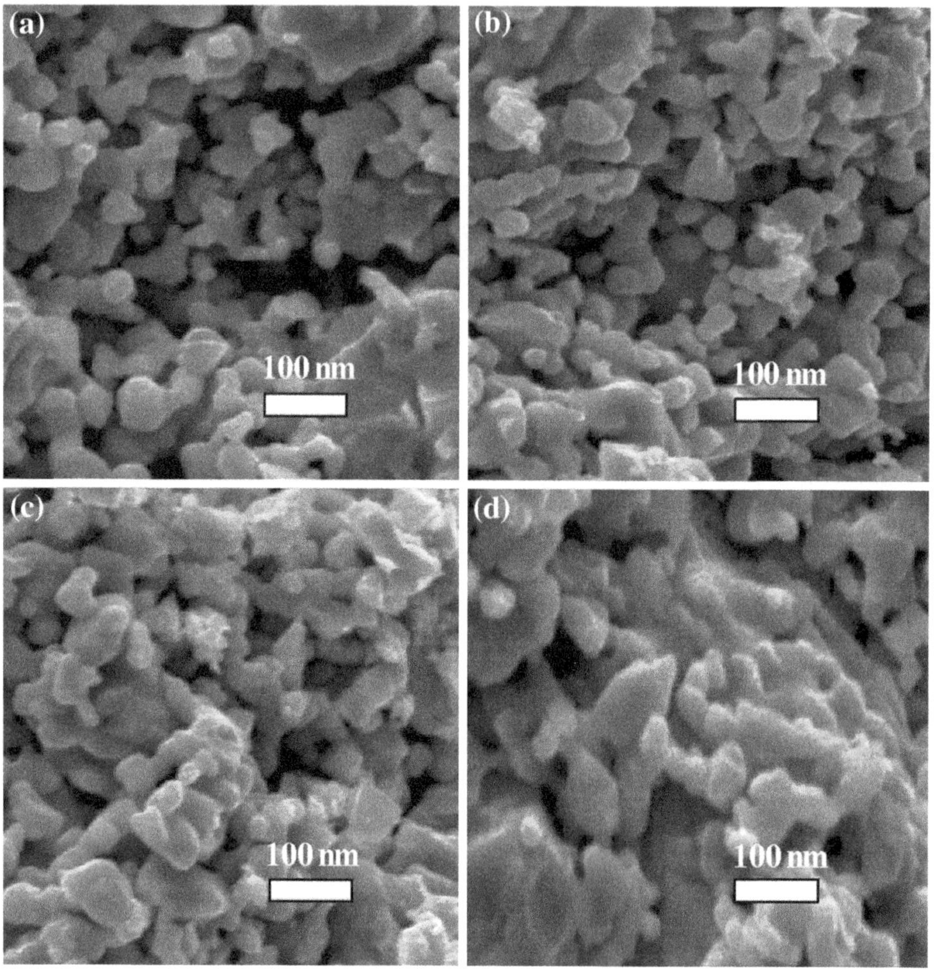

Fig. 2 Typical FE–SEM images of samples with different sintering temperature. **a** 600 °C. **b** 800 °C. **c** 1,000 °C. **d** 1,200 °C

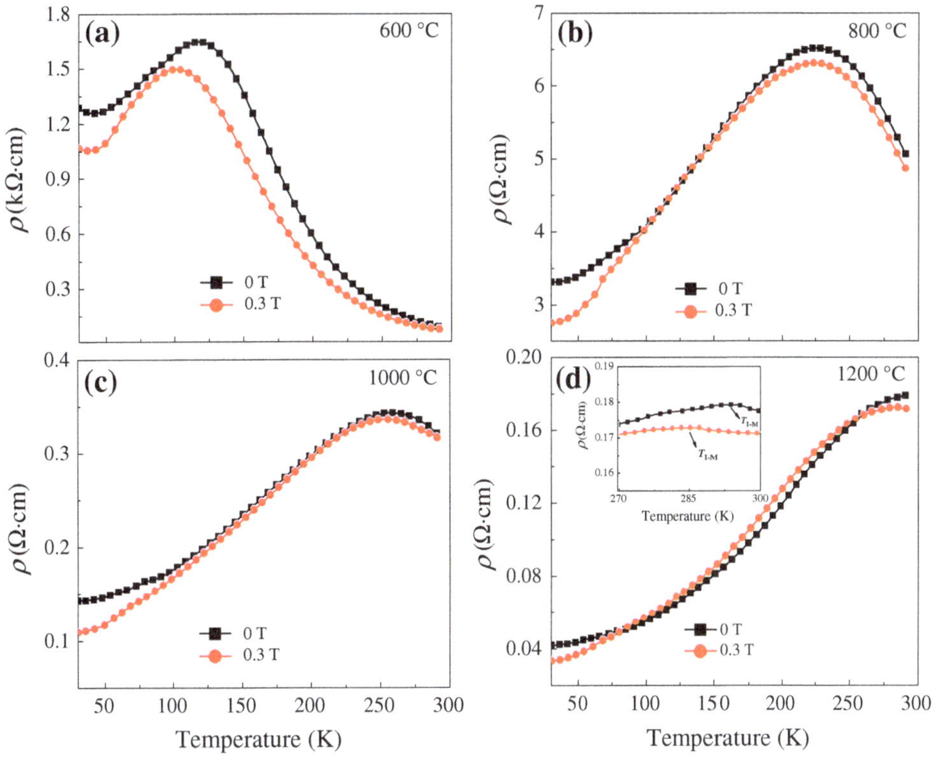

Fig. 3 Temperature dependent resistivity in different applied magnetic fields for samples sintered at different temperatures. **a** 600 °C. **b** 800 °C. **c** 1,000 °C. **d** 1,200 °C. *Inset* in (**d**) transition temperature T_{I-M} of sample sintered at 1,200 °C

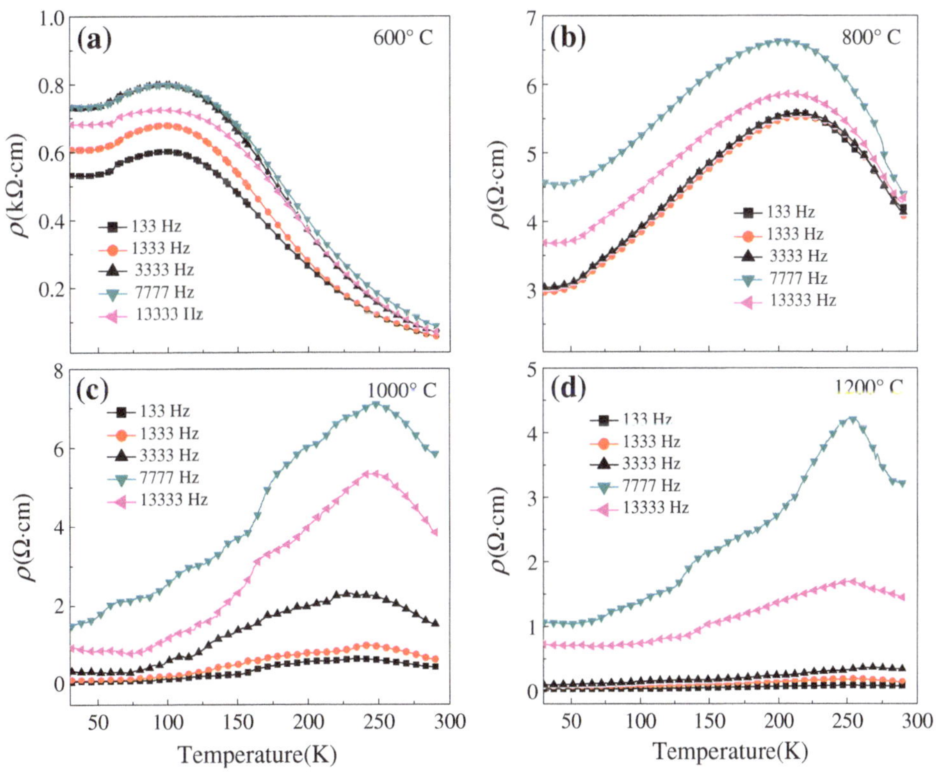

Fig. 4 Temperature dependent resistivity at different AC frequencies in zero magnetic field for all samples sintered at different temperatures. **a** 600 °C. **b** 800 °C. **c** 1,000 °C. **d** 1,200 °C

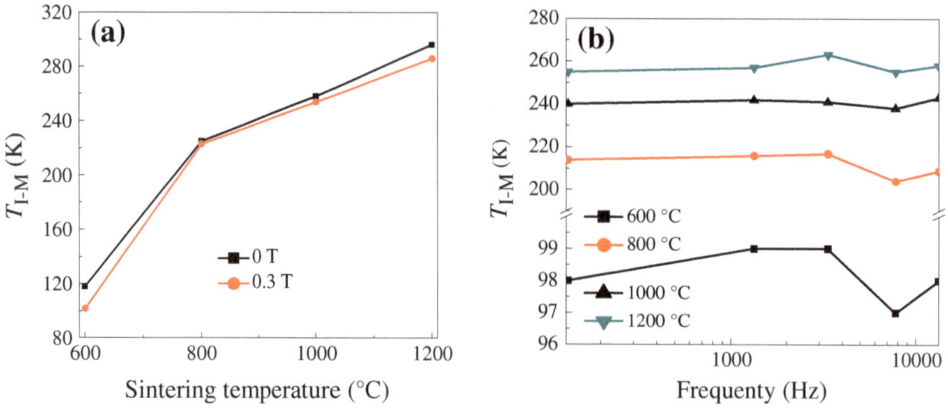

Fig. 5 Sintering temperature dependent T_{I-M} of samples at DC frequency in different applied magnetic fields (**a**) and frequency dependent T_{I-M} of samples at AC frequency in zero applied magnetic field (**b**)

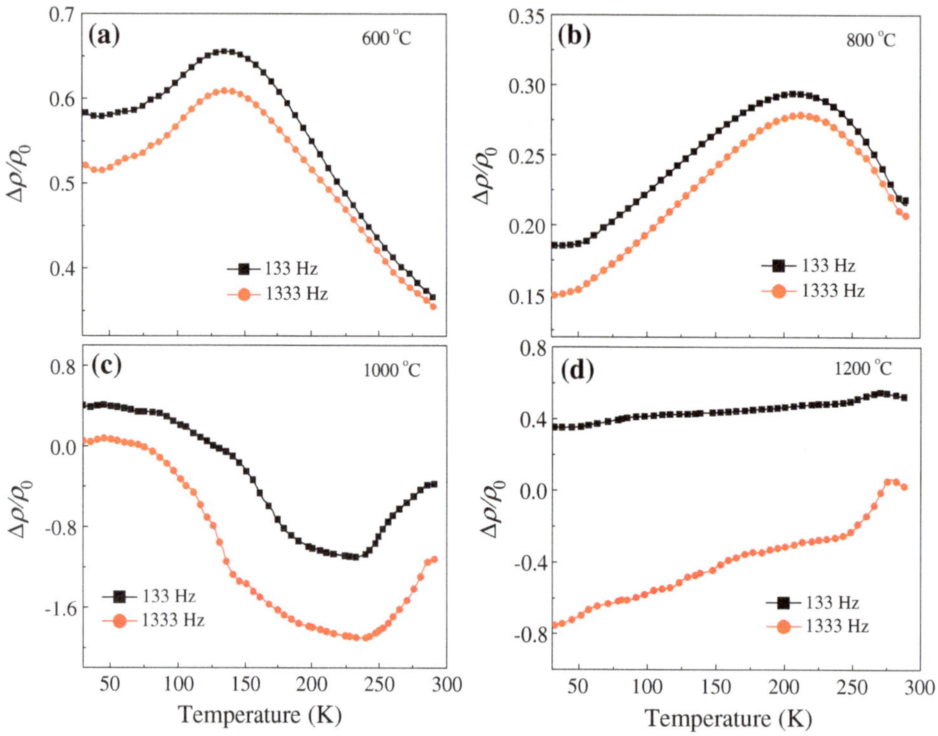

Fig. 6 Temperature dependent $\Delta\rho/\rho(0)$ at low frequencies ($f = 133$ Hz, 1,333 Hz) in zero DC magnetic field for samples sintered at different temperatures. **a** 600 °C. **b** 800 °C. **c** 1,000 °C. **d** 1,200 °C

intrinsic magneto-resistance peak, which could be due to the magnetic domains and strain free structure [18]. The features may be explained by the fact that the change of AC frequency causes a induce magnetic field, and thus leads to the change of resistivity. Further theoretical research is required to fully understand its physical mechanism.

Figure 7 shows the AC-MR $((\rho(H) - \rho(0))/\rho(0) \times 100\ \%)$ as a function of frequency at 290 K under 0.3 T. All samples exhibit a similar MR-f behavior that is MR first increases, reaching a maximum, and then decreases. The MR measured at low frequencies is obviously larger than that measured at high frequencies, especially for the samples with higher sintering temperature. Obviously, the room-temperature AC-MR is largely enhanced compared with the room-temperature DC-MR (inset in Fig. 7). It provides a novel way to promote the sensitivity and efficiency of magnetoresistive sensor.

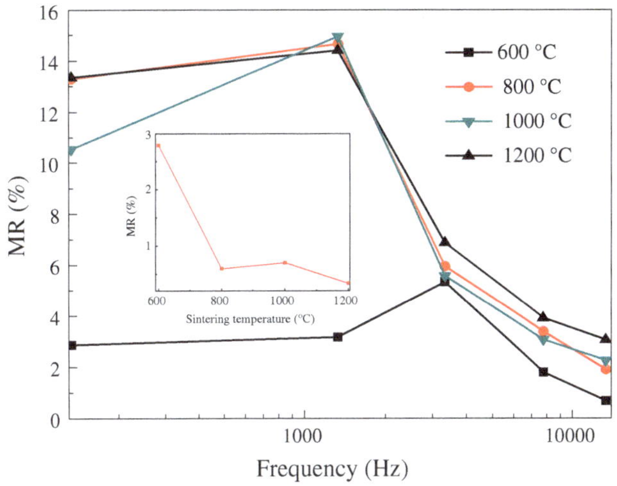

Fig. 7 Frequency dependence of AC-MR of all samples under 0.3 T at 290 K. *Inset* DC-MR at 290 K as a function of sintering temperature

4 Conclusion

The detailed AC and DC transport properties of $La_{0.8}Sr_{0.2}MnO_3$ polycrystalline samples have been studied. All samples show the I–M phase transition in both DC and AC measurements. The $T_{I\text{–}M}$ measured at different AC frequencies is smaller than that measured at DC condition, which implies that the I–M phase transition is suppressed by AC frequency. The resistivity at high AC frequencies is obviously larger than that at low AC frequencies. However, the room-temperature AC-MR at low frequencies is much larger than that at high frequencies and room-temperature DC-MR. These findings provide an alternative strategy to probe the spin-charge coupling in manganites and open a huge potential application prospect in modern intelligent transportation field.

Acknowledgments This work was supported by the National Magnetic Confinement Fusion Science Program (No. 2011GB112001), Program of International S&T Cooperation (No. 2013DFA51050), the National Natural Science Foundation of China (Nos. 51002125, 11104224, 11004162, and 51271155), the Research Fund for the Doctoral Program of Higher Education of China (No. 20110184120029), the Fundamental Research Funds for the Central Universities (Nos. 2682013ZT16, SWJTU12CX18, SWJTU11ZT31, 2682013CX004, and SWJTU11BR063), and the Science Foundation of Sichuan Province (Nos. 2011JY0031 and 2011JY0130).

References

1. Taghvaeeyan S, Rajamani R (2011) Use of vehicle magnetic signatures for position estimation. Appl Phys Lett 99(13):134101-134101-3
2. Hrabovský D, Caicedo JM, Herranz G et al (2009) Jahn–Teller contribution to the magneto-optical effect in thin-film ferromagnetic manganites. Phys Rev B 79(5):052401
3. Gu M, Song C, Yang F et al (2012) Tuning magnetic and transport properties through strain engineering in $La_{0.7}Sr_{0.3}MnO_3$/$La_{0.5}Sr_{0.5}TiO_3$ superlattices. J Appl Phys 111(8):084906-084906-6
4. Saber MM, Egilmez M, Schoofs F et al (2012) Strain, spin disorder, and thickness dependence of magneto-transport in $Sm_{0.55}Sr_{0.45}MnO_3$ films. Appl Phys Lett 100(25):252408-252408-5
5. Hu L, Sheng Z, Hu X et al (2012) Control of the charge/orbital ordering transition in epitaxial $La_{7/8}Sr_{1/8}MnO3$ thin films through compressive strain. J Phys D Appl Phys 45(17):175002
6. Wang KF, Wang Y, Wang LF et al (2006) Ferromagnetic metal to cluster-glass insulator transition induced by A-site disorder in manganites. Appl Phys Lett 88(16):152505-152505-3
7. Rivadulla F, Lopez-Quintela MA, Rivas J (2004) Origin of the glassy magnetic interactions in $GdMe_xMn_{1-x}O_3$ perovskite (Me = transition metal). Phys Rev Lett 3:167206-167206-4
8. Qian T, Li G, Zhang T et al (2007) Exchange bias tuned by cooling field in phase separated $Y_{0.2}Ca_{0.8}MnO_3$. Appl Phys Lett 90(1):012503-012503-3
9. Daubaras A, Zilys M (2012) Vehicle detection based on magneto-resistive magnetic field sensor. Electron Electric Eng 118(2):27–32
10. Yang L, Yang X, Cheng C et al (2012) Structure and electrical properties in polymer bonded $La_{0.7}Sr_{0.3}MnO_3$ composites. Phys Procedia 27:92–95
11. Gupta S, Ranjit R, Mitra C et al (2001) Enhanced room-temperature magnetoresistance in $La_{0.7}Sr_{0.3}MnO_3$-glass composites. Appl Phys Lett 78(3):362–364
12. Rebello A, Mahendiran R (2010) Influence of length and measurement geometry on magnetoimpedance in $La_{0.7}Sr_{0.3}MnO_3$. Appl Phys Lett 96(3):032502-032502-3
13. Aparnadevi M, Mahendiran R (2013) Alternating current magnetotransport in $Sm_{0.1}La_{0.6}Sr_{0.3}MnO_3$. AIP Adv 3(1):012114-012114-8
14. Djerdj I, Tonejc AM (2006) Structural investigations of nanocrystalline TiO_2 samples. J Alloys Compd 1:159–174
15. Vazquez-Vazquez C, Blanco MC, Lopez-Quintela MA et al (1998) Characterization of $La_{0.67}Ca_{0.33}MnO_3 \pm \delta$ particles prepared by the sol–gel route. J Mater Chem 8(4):991–1000
16. Anulekha Dutta, Gayathri N, Ranganathan R (2003) Effect of particle size on the magnetic and transport properties of $La_{0.875}Sr_{0.125}MnO_3$. Phys Rev B 68(5):054432
17. Naik VB, Rebello A, Mahendiran R (2009) A large magnetoinductance effect in $La_{0.67}Ba_{0.33}MnO_3$. Appl Phys Lett 95(8):082503-082503-3
18. Hwang HY, Cheong S-W, Ong NP et al (1996) Spin-polarized intergrain tunneling in $La_{2/3}Sr_{1/3}MnO_3$. Phys Rev Lett 77:2041–2044

24

Transport properties and microstructure of $La_{0.7}Sr_{0.3}MnO_3$ nanocrystalline thin films grown by polymer-assisted chemical solution deposition

Min Zhang · Li Lv · Zhantao Wei · Xin Sheng Yang · Xin Zhang

Abstract Perovskite-based materials can be widely used in the aerospace and transportation field. Perovskite manganese oxides $La_{0.7}Sr_{0.3}MnO_3$ (LSMO) thin films were grown on $LaAlO_3$ (100) and Si (100) single crystal substrates by the polymer-assisted chemical solution deposition (PACSD) method. Electronic transport behavior, microstructure, and magnetoresistance (MR) of LSMO thin films on different substrates were investigated. The resistance of LSMO films fabricated on $LaAlO_3$ substrates is smaller than that on the Si substrates. The magnetic field reduces resistance of LSMO films both on Si and LAO in the wide temperature region, when the insulator-metal transition temperature shifts to higher temperature. The low-field magnetoresistance of LSMO films on Si in low temperature range at 1 T is larger than that of LSMO films on LAO. However, the MR of LSMO film on LAO films at room-temperature is about 5.17 %. The thin films are smooth and dense with uniform nanocrystal size grain. These results demonstrate that PACSD is an effective technique for producing high quality LSMO films, which is significant to improve the magnetic properties and the application of automotive sensor.

Keywords Polymer-assisted chemical solution deposition (PACSD) · $La_{0.7}Sr_{0.3}MnO_3$ (LSMO) thin films · Transport properties and microstructure

M. Zhang · L. Lv · Z. Wei · X. S. Yang (✉) · X. Zhang
Key Laboratory of Magnetic Levitation Technologies and Maglev Trains, Ministry of Education of China, and Superconductivity and New Energy R&D Center, Southwest Jiaotong University, Chengdu 610031, China
e-mail: xsyang@swjtu.edu.cn

1 Introduction

Perovskite-based functional materials have a wide application prospect due to colossal magnetoresistance (CMR) and magnetic refrigeration effects. They can be used to make the magnetic head, magnetic resistance memory, and motor transducer. The transducer can be used to measure and control current, position, angle etc. Moreover, they can be used in aerospace, automotive industry as magnetic refrigeration material. In recent years, as the developments of anisotropic MR effect and CMR in ferromagnetic/non-magnetic metal multilayer films, the thin film MR sensor has become one of the most active magnetic sensor technologies, which are widely used in mineral exploration, underground drilling, position detection, navigation etc.

CMR materials such as $(La,Ca)MnO_3$, $(La,Sr)MnO_3$, and $(La,Ca,Sr)MnO_3$ have been extensively investigated on polycrystal, single crystal, and thin films [1–3], due to theirs potential applications in devices such as memory devices, magnetic field sensors, infrared detectors, and spin-based devices [4–8].

Shim et al. [9] reported that the maximum low-field magnetoresistance (LFMR) of LSMO film, was 0.68 %, obtained at 300 K in the 450 Oe field region. Liu et al. [10], also obtained that the LFMR of LSMO film on Si (100) substrate were as high as 2.54 % at $H = 0.05$ T and 300 K. Different from the intrinsic double-exchange-type CMR, induced by a strong magnetic field in the tesla range, the LFMR effect is attributed to spin dependent tunneling across the grain boundaries (GBs).

Many techniques have been applied to preparing $La_{0.7}Sr_{0.3}MnO_3$ epitaxial and polycrystalline films on different substrates including pulsed laser deposition (PLD) [11, 12], sputtering [13, 14], metal organic chemical vapor deposition (MOCVD) [15, 16], molecular beam epitaxy

(MBE) [17], chemical solution deposition (CSD) [18] etc. Among these methods, the CSD method has some advantages for the films deposition and growth such as low cost, high stability, and simple process. As for the CSD technique for the preparation of LSMO films, many researches about epitaxy, concentration of precursor solution, annealing, LFMR as well as the GBs on substrates of MgO, $LaAlO_3$, Si, $SrTiO_3$, and GaAs have been reported [18, 19].

In this article, LSMO films were synthesized on $LaAlO_3$ (100) and Si (100) by polymer-assisted chemical solution deposited (PACSD). The smooth and dense films can be fabricated via this method. Preparation and transport properties on $LaAlO_3$ (100) and Si (100) were investigated because silicon is a good candidate for spin injection and the lattice mismatch of LAO is small enough to induce epitaxy. The microstructure, the surface, and the electronic transport behavior and magnetoresistance (MR) properties of LSMO thin films were presented.

2 Experimental

The nanocrystalline thin films of $La_{0.7}Sr_{0.3}MnO_3$ were grown by PACSD method on (100) $LaAlO_3$ and Si (100) single crystal substrates. For preparation of LSMO thin films, Si (1 0 0), and $LaAlO_3$ (1 0 0) single crystal substrates were cleaned. First, analytical reagents $La(NO_3)_3 \cdot 6H_2O$, $Sr(NO_3)_2$ and $Mn(NO_3)_3 \cdot 6H_2O$ were weighed accurately and dissolved in DMF (*N*,*N*-Dimethylformamide) to obtain anhydrous clear solution with a final solution concentration of 0.3 mol/L. Then, Polyvinylpyrrolidone K30 (PVP) was used to increase the viscosity of solution. Deposition of LSMO was carried out by a spin coater at 500 rpm for 10 s, followed by 4,000 rpm for 50 s. After deposition, the wet films were dried at 80 °C for 10 min. Then the films were pyrolized at temperature 200 °C for 10 min and 500 °C for 20 min. The dried films were finally crystallized at 830–870 °C for 1–2 h in air ambience.

The phases were determined by X-ray diffraction using the Philips X-ray diffractometer X'Pert MRD diffractometer (Cu-K α radiation, $\lambda = 1.54$ Å), the microstructures of the films were also observed using field emission scanning electron microscope (FESEM, JSM-7001F), and the transport properties and magnetic properties were measured by Physical Property Measurement System (PPMS, Quantum Design 9 T).

3 Results and discussion

Figure 1 shows the typical θ–2θ X-ray diffraction spectra of LSMO film grown on LAO (100) substrates. Only (100) and (200) peaks are observed without any impurity peaks, indicating highly *c*-axis oriented growth with pseudo-cubic (100) orientation. From the data, the calculated lattice constants are $a = 3.876$ Å for LSMO and $a = 3.796$ Å for LAO. The lattice mismatch between LSMO (100) films and LAO (100) substrates is 2.1 %, which are evaluated via the definition of

$$\frac{a_f - a_s}{a_s} \times 100\ \%. \tag{1}$$

XRD pattern of LSMO film on Si (100) substrate is presented in Fig. 2. All the main diffraction peaks belong to LSMO phase, implying that the LSMO film on Si is of single phase with perovskite structure. It is obviously different from the film fabricated on the single crystal substrates of $LaAlO_3$, which are advantageous for the growth of (100) orientation films. The film growth on Si (100) substrate resulted polycrystalline with pseudo-cubic orientation due to the large lattice mismatch degree between LSMO (100) films and Si (100) ($a = 0.539$ nm) substrates. Nevertheless, the small mismatch between LSMO (110) and Si (100) was reported to be 1.8 % in [10]. Therefore, LSMO grains grow via a self-textured growth mode and are randomly oriented.

In order to evaluate the quality of LSMO thin films from the PACSD method, the microstructures of the films were also examined. Figure 3 shows the FE-SEM images of LSMO films on LAO and LSMO films on Si, which displays that the films are smooth and have high crystal quality. It is clear from Fig. 3b that the LSMO films on Si consist of nanometer polycrystalline. Typical grain sizes are in the range of 50–70 nm with some cavities, which may be related to the volatilization of organic precursors. This is in agreement with the results in [9]. Comparing the morphology of the LSMO films on Si (Fig. 3a) and LAO (Fig. 3b), we find that the latter seems to have denser microstructure and smaller crystallites size of 15 nm or so. It can be seen that LSMO film on LAO is the columnar structure with a thickness of 400 nm. The bulk of film is

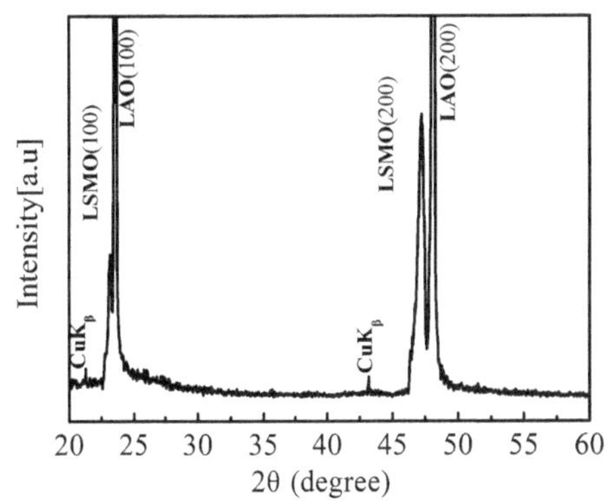

Fig. 1 X-ray diffraction of $La_{0.7}Sr_{0.3}MnO_3$ deposited on LAO (100) substrate

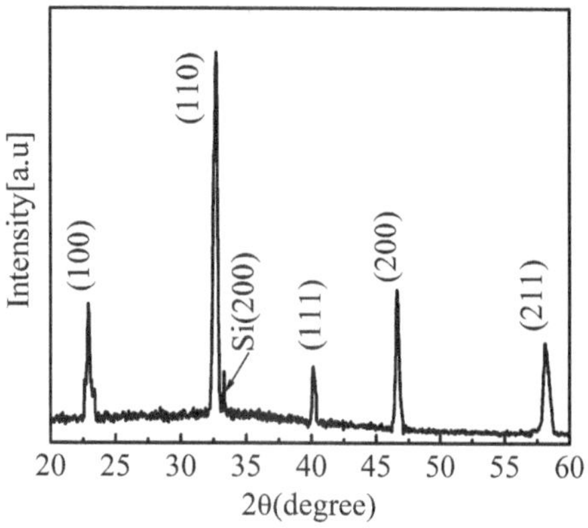

Fig. 2 X-ray diffraction of $La_{0.7}Sr_{0.3}MnO_3$ deposited on Si (100) substrate

epitaxial but the surface exhibits some polycrystalline structure.

These results show that substrate has a significant effect on the crystallographic properties of the resulting thin film.

Figure 4 shows temperature-dependent electrical resistivity ($\rho - T$) for LSMO films on LAO from 360 to 5 K under 0 and 1 T. Note that metal-insulator transition is at 349 K for $H = 0$ T corresponding to bulk samples temperature around 360 K, which proves a good quality of the film. The film exhibits systematic increase in metal-insulator transition temperature (T_{MI}) with an increase of applied magnetic field. Due to limitations of the measurement equipment, insulator-metal transition is not observed in the film under applied magnetic field $H = 1$ T.

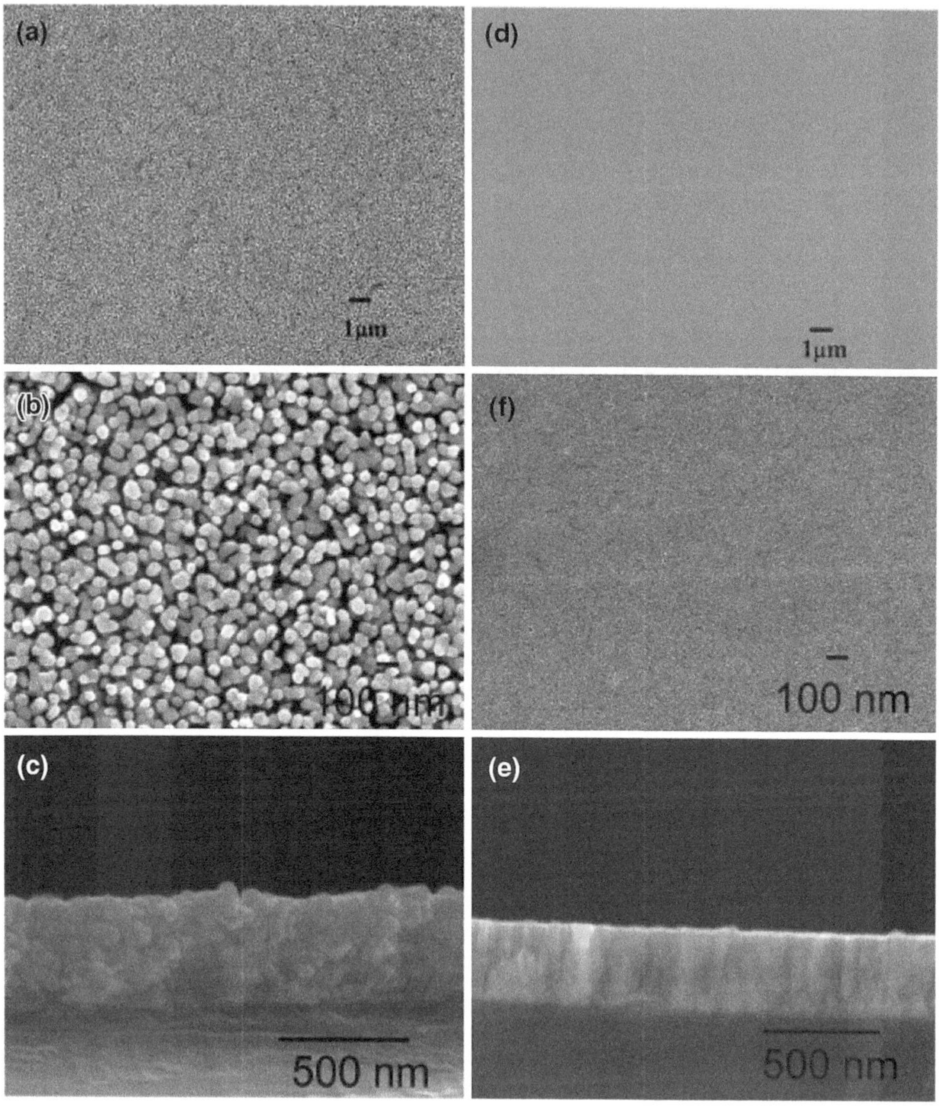

Fig. 3 Typical FE-SEM images for **a** LSMO film on Si and **b** LSMO film on Si collected at ×5,000 and ×50,000 magnification, respectively. **c** The cross-sectional morphology of LSMO film on Si. **d** LSMO film on LAO film and **e** LSMO film on LAO collected at ×5,000 and ×50,000 magnification, respectively. **f** The cross-sectional morphology of LSMO film on LAO

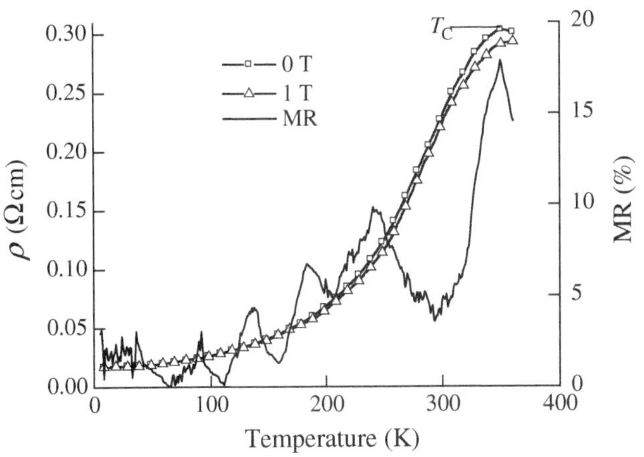

Fig. 4 Temperature dependence of resistivity for LSMO film on LAO (100) substrate. Solid lines are experimental data, and symbols represent the MR curve for LSMO film

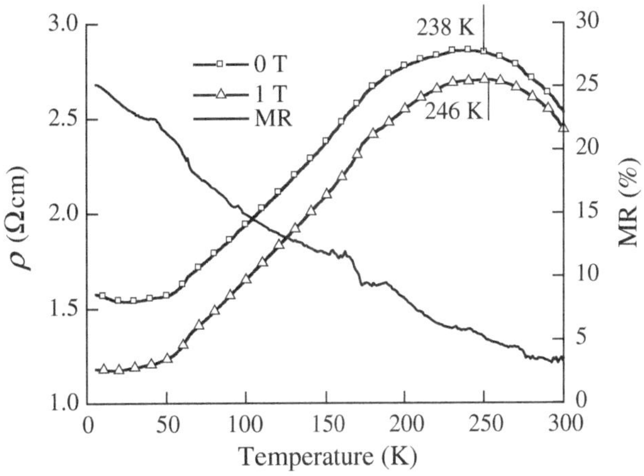

Fig. 5 Temperature dependence of resistivity for LSMO films on Si (100) substrate. Solid lines are experimental data, and symbols represent the MR curve for LSMO film

The resistance is reduced under applied magnetic field 1 T. Within double exchange model, the applied magnetic field can align these disordered spins between parallel arrangements Mn^{3+}–Mn^{4+} of LSMO, increase the transition probability of itinerant electrons, and reduce the spin disorder scattering; thereby the resistance is decreased effectively.

The curve of MR versus temperature is shown in Fig. 4, which has an obvious MR peak near T_C and the MR fluctuates in whole temperature range. The MR ratio here is illustrated as MR $= (\rho_O - \rho_H)/\rho_O \times 100$ %, where ρ_O is the zero field resistivity and ρ_H is the resistivity in the applied field H. With decreasing temperature, MR shows an overall decrease with undulant pattern. The MR of the LSMO films on LAO mainly results from the suppression of spin fluctuation by magnetic field within grains and the contribution of the GBs to MR is inconsiderable [20]. The maximum value of MR is 2.67 % at 0.6 T (not shown here) and 5.17 % at 1 T when $T = 300$ K.

The temperature dependence of the resistance at different magnetic field (0 and 1 T) from 300 to 5 K and MR curve for LSMO films on Si are shown in Fig. 5. Compared with LSMO films on LAO, the transition temperatures of LSMO films on Si are lowered down to 238 K. Under the applied magnetic fields of 1 T, the resistance is remarkably suppressed in the wide temperature range and the T_{MI} shifts to a higher temperature (246 K), which is dominated by the spin-polarized tunneling at the GBs [21]. The external magnetic increases the carrier capacity and leads to a change in resistivity ρ. In addition, resistance gradually decreases at lower temperatures for LSMO films on LAO, but rises from 50 to 5 K for LSMO films on Si. The different transport behaviors are assumedly due to the magnetic domain walls at GBs and the reentrant effect in the polycrystalline films [20].

For the LSMO films on Si, the MR increases with decreasing temperature. It is suggested that magnetic external field prevents spin-dependent scattering of polarized electrons at the GBs and results in a large MR.

Obviously, the MR behavior of LSMO films on Si is different from LSMO films on LAO. A similar phenomenon was also observed by Liu et al. [22]. Single crystal and textured films have an MR peak in the vicinity of T_{MI}, while the polycrystalline LSMO films do not show the peak behavior. Thus, the crystalline orientation of the films plays an important role in governing the MR characterization.

4 Conclusions

Polycrystal LSMO thin films were synthesized on $LaAlO_3$ (100) and Si (100) single crystal substrates through spin-coating method by utilizing PACSD with PVP. XRD data shows evidence that the LSMO films on LAO is a—axis oriented and LSMO films on Si is polycrystalline, implying that the orientation of films depend on lattice mismatch degree between substrates and LSMO films. The resistance of LSMO films fabricated on $LaAlO_3$ substrates is smaller than that on the Si substrates. The magnetic field leads to reduced resistance of LSMO films both on LAO and Si in a wide temperature region, when the insulator-metal T_{MI} is shifted to a higher temperature. The LFMR of LSMO films on Si in low temperature range at 1 T is larger than that of LSMO films on LAO. The MR of LSMO films on LAO displays large values of MR as high as 5.17 % at room temperature. The MR of LSMO films on Si increases with decreasing temperature. The change in MR is related to GBs effect. Both films are smooth and dense with uniform nanocrystal size grain. The grain size of LSMO films on Si is larger than LSMO films on LAO. From the analysis, it is found that crystalline orientation of substrates plays an important role in affecting grain size, structure, transport properties, and the MR characterization. These results demonstrate that PACSD is a viable method for growing high quality LSMO films.

Acknowledgments The research was supported by the Program of International S&T Cooperation 2013DFA51050, National Magnetic Confinement Fusion Science Program (2011GB112001), Science Foundation of Sichuan Province (2011JY0031, 2011JY0130), the financial support of the National Natural Science Foundation of China (No. 51271155, No. 51002125), and the Fundamental Research Funds for the Central Universities (SWJTU12CX018).

References

1. Urushibara A, Moritomo Y, Arima T et al (1995) Insulator-metal transition and giant magnetoresistance in $La_{1-x}Sr_xMnO_3$. Phys Rev B 51:14103–14109
2. Huang Y-H, Xu Z-G, Yan C-H et al (2000) Soft chemical synthesis and transport properties of $La_{0.7}Sr_{0.3}MnO_3$ granular perovskites. Solid State Commun 114:43–47
3. Rajeswari M, Shreekala R, Goyal A et al (1998) Correlation between magnetic homogeneity, oxygen content, and electrical and magnetic properties of perovskite manganite thin films. Appl Phys Lett 73:2634–2672
4. Thomas KA, de Silva PSIPN, Cohen LF et al (1998) Influence of strain and microstructure on magnetotransport in $La_{0.7}Ca_{0.3}MnO_3$ thin films. J Appl Phys 84:3939–3948
5. Lee KR, Chung YJ, Lee JH et al (2003) Electrical and magnetic properties of $La_{0.7}Ca_{0.3}MnO_3$ films prepared by RF magnetron sputtering method for colossal magnetoresistance applications. Thin Solid Films 426(1):205–210
6. Gao J, Shen SQ, Li TK et al (2003) Current-induced effect on the resistivity of epitaxial thin films of $La_{0.7}Ca_{0.3}MnO_3$ and $La_{0.85}Ba_{0.15}MnO_3$. Appl Phys Lett 82:4732–4734
7. Kang Y-M, Kim H-J, Yoo S-I et al (2009) Excellent low field magnetoresistance properties of the $La_{0.7}Sr_{0.3}Mn_{1+d}O_3$-manganese oxide composites. Appl Phys Lett 95:052510–052512
8. Viret M, Drouet M, Nassar J et al (1997) Low-field colossal magnetoresistance in manganite tunnel spin valves. Europhys Lett 39:545–549
9. Shim IB, Lee HM, Park KT et al (2002) Low-field magnetoresistance in sol–gel derived $La_{2/3}Sr_{1/3}MnO_{3-\delta}$ thick films. J Magn Magn Mater 242–245:1169–1171
10. Liu SM, Zhu XB, Yang J et al (2004) Fabrication of polycrystalline $La_{0.7}Sr_{0.3}MnO_3$ thin films on Si (1 0 0) substrates by chemical solution deposition. Phys B 353:238–241
11. Krishnan KM, Modak AR, Lucas CA et al (1996) Role of epitaxy and polycrystallinity in the magnetoresistance and magnetization of La0.8Sr0.2MnO3 thin films. J Appl Phys 79(8):5169–5171
12. Satoh I, Oniduka M, Kobayashi T (2002) PLD growth of $La_{0.7}Sr_{0.3}MnO_3$ tilted nanocolumn boundaries on constricted step-edged GaAs substrates and MR properties. Appl Surf Sci 197–198:527–531
13. Hawley MF, Adams CD, Arendt PN et al (1996) CMR films structure as a function of growth and processing. J Cryst Growth 174:455–463
14. Park S-I, Kim YH, Cho YS et al (2001) Characterization of La–Sr–Mn–O Films on Si(100) Grown by Using an rf–Sputtering Process under Different O_2 Partial Pressures. J Korean Phys Soc 38(1):38–41
15. Balevičius S, Cimmperman P, Petrauskas V et al (2005) Two phase structure of ultra-thin La–Sr–MnO films. Thin Solid Films 515:691–694
16. Toro RG, Fiorito DMR, Fragalà ME et al (2010) A novel MOCVD strategy for the fabrication of cathode in a solid oxide fuel cell: synthesis of $La_{0.8}Sr_{0.2}MnO_3$ films on YSZ electrolyte pellets. Mater Chem Phys 124:1015–1021
17. Méchin L, Adamo C, Wu S et al (2012) Epitaxial $La_{0.7}Sr_{0.3}MnO_3$ thin films grown on $SrTiO_3$ buffered silicon substrates by reactive molecular beam epitaxy. Phys Status Solidi A 209(6):1090–1095
18. Hasenkox U, Mitz C, Waser R (1999) Chemical Solution Deposition of Epitaxial $La_{1-x}(Ca, Sr)_x$ MnO_3.Thin Films. J Electroceram 3(3):255–260
19. Kartopu G, Yalçin O, Demiray AS et al (2011) Magnetic and transport properties of chemical solution deposited (100)—textured La0.7Sr0.3MnO3and La0.7Ca0.3MnO3 nanocrystalline thin films. Phys Scr 83:015701.
20. Yang SY, Kuang WL, Liou Y et al (2004) Growth and characterization of $La_{0.7}Sr_{0.3}MnO_3$ films on various substrates. J Magn Magn Mater 268:326–331
21. Chen Y, Wang G, Zhang S et al (2011) Magnetocapacitance effects of $Pb_{0.7}Sr_{0.3}TiO_3/La_{0.7}Sr_{0.3}MnO_3$ thin film on Si substrate. Appl Phys Lett 95:052910–052912
22. Liu SM, Zhu XB, Yang J et al (2006) The effect of grain boundary on the properties of $La_{0.7}Sr_{0.3}MnO_3$ thin films prepared by chemical solution deposition. Ceram Int 32:157–162

25

Experimental study of temperature gradient in track slab under outdoor conditions in Chengdu area

Zhao Pingrui · Liu Xueyi · Liu Guan

Abstract Temperature is one of the important loads for designing slab track. The characteristic of slab track temperature varies greatly with different regional climates. In this work, a bi-block slab track model was built under outdoor conditions in Chengdu area; the statistical characteristic of temperature gradient in track slab and the relationship between temperature gradient and surface air temperature were tested and analyzed. The results show that the track slab temperature gradient will vary periodically according to the surface air temperature, and show a clear nonlinearity along the height direction. The temperature gradient distribution is extremely uneven: the temperature gradient in the top part of the track slab is larger than that in the bottom part; the most frequently occurring temperature gradient of the track slab is around −3.5 °C/m and more than 75 % locates in the level −10 to 10 °C/m; concrete with a relatively good heat exchange condition with the surrounding air has a narrower band distribution. In addition, the frequency distribution histogram should exclude the time zone from 00:00 to 06:00 because there is almost no traffic in this period. The amplitude of track slab temperature variation is obviously lower than that of the air temperature variation, and the former is approximately linear with the latter.

Keywords Slab track · Track slab · Temperature gradient · Temperature field · Surface air temperature

With the rapid development of high-speed railway, as a new type of track structure, slab track was widely used in China, especially for the lines for which the maximum operation speed exceeds 300 km/h. Compared to the traditional ballasted track, concrete or asphalt material was used in slab track instead of discrete ballast which will deform easily due to the flow and wear under high-speed train-induced vibration. This can give the slab track a guaranteed ability to maintain long-term and high-precision track geometry. At present, most of slab tracks in China's high-speed railway are reinforced concrete structure—including CRTS I, CRTS II, and CRTS III slab track; and CRTS I and CRTS II bi-block slab track [1].

Temperature in slab track will change with the environmental temperature, especially for the top layer of the slab track system—track slab. Due to the thermal expansion and contraction of the concrete, the temperature variation will lead to thermal stress in slab track. Therefore, thermal stress is a very important factor to design the slab track. Normally, the thermal load can be divided into three parts (Fig. 1): the overall sectional temperature change, temperature gradient, and self-constrained temperature, which will cause expansion stress, warping stress, and inner stress, respectively [2]. For the continuous slab track, the expansion stress will control the design; while for the unit slab track, the warping stress can be the control factor of the design [3].

It is convenient to measure and make statistics of the environmental temperature. If there is a relationship between the temperature in slab track and the ambient air temperature, the statistical characteristic of the slab track temperature can be extracted from the statistical characteristic of the air temperature. In China slab track design, the maximum temperature gradient for design is set from pavement design [4]. Many researchers have conducted tests to measure the temperature field in concrete pavement [5, 6]. However, because of the large dimension of the

Z. Pingrui · L. Xueyi · L. Guan (✉)
MOE Key Laboratory of High-speed Railway Engineering, Southwest Jiaotong University, Chengdu 610031, China
e-mail: 75994776@qq.com

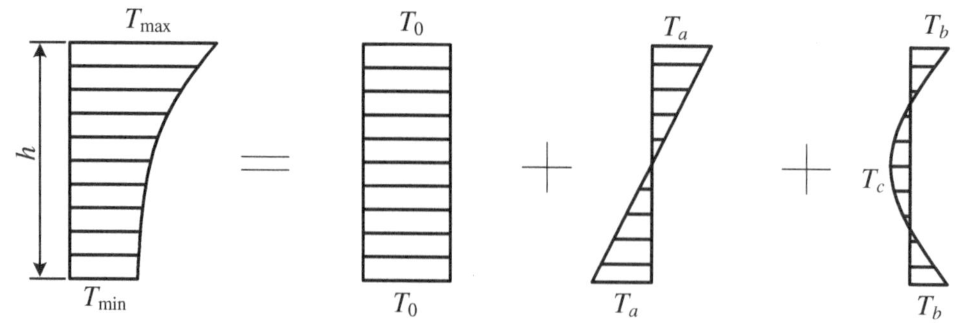

Fig. 1 Typical temperature distribution in track slab

pavement and different heat exchange conditions of pavements, the data measured from concrete pavement cannot be used directly for design of slab track. Slab track is a structure that bears long-term temperature load and repeated train load [7]. In order to study the fatigue performance of the slab track, we must take into account the coupling action of the train load and temperature load [8, 9]. The maximum temperature value cannot reflect the fatigue performance of the slab track completely. This paper will focus on the test of temperature field in slab track to establish the connection between slab track temperature gradient and air temperature.

1 Test plan

The temperature change in track slab is related to the sun heat radiation and cooling condition [10]. According to the structural dimension and characteristic of the CRTS I bi-block slab track, the slab was selected. A test was conducted from the end of May to the end of July, covering a period of two months. In the experiment, platinum resistance temperature sensors were used to capture the temperature of various kinds of track structure outdoors. The temperature sensors and the track slab for this test are shown in Fig. 2.

Ignoring the local influence of the bi-block sleeper, a track slab model with a dimension of 1 m × 1 m × 0.3 m was designed and constructed. At the middle, edge, and corner parts of the track slab (see Fig. 3a), there is one test section. At each of the test section three temperature sensors with a 100 mm distance were set (Fig. 3b). Above the track slab (2 cm above the surface) there is one sensor to measure the surface air temperature. Totally 10 temperature sensors are placed.

Because of the poor heat conductivity of the concrete [11], there will be differences between each sensor's data at the same section. We define the average of the data from the three sensors in the same section as the overall sectional temperature, and use the temperature difference between sensors at two different sections to calculate the temperature gradient.

The DH5975 temperature measuring system was used to collect and record the slab track temperature and surface air temperature automatically. The collection interval was 30 min. But for the first 7 days, the interval was set to 15 min to reflect the influence of the cement hydration [12].

Fig. 2 The temperature sensors and the track slab for this test

2 Time history of overall sectional temperature change in track slab

Taking the average of measurements from the three sensors in each section as the overall sectional temperature in track slab, the relationship between the sectional temperature and the surface air temperature is shown in Fig. 4.

The varying tendency of the temperature at different parts of the track slab is similar, and the moment of reaching their extreme value is almost the same. At

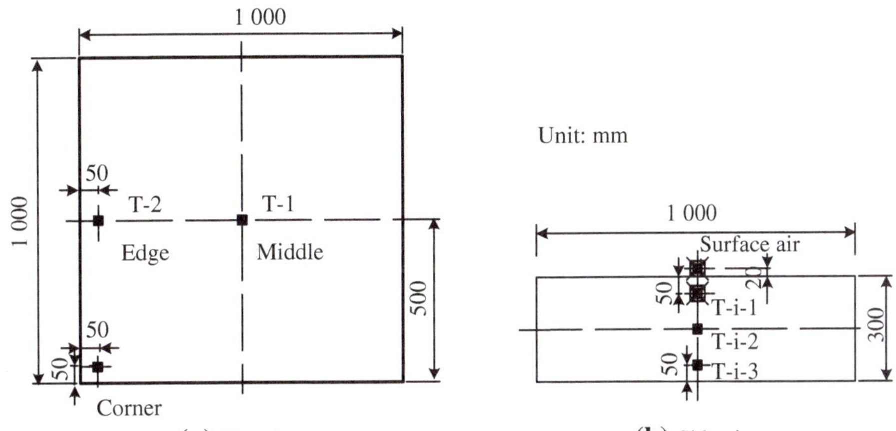

(a) Top view (b) Side view

Fig. 3 Schematic arrangement of the temperature field test

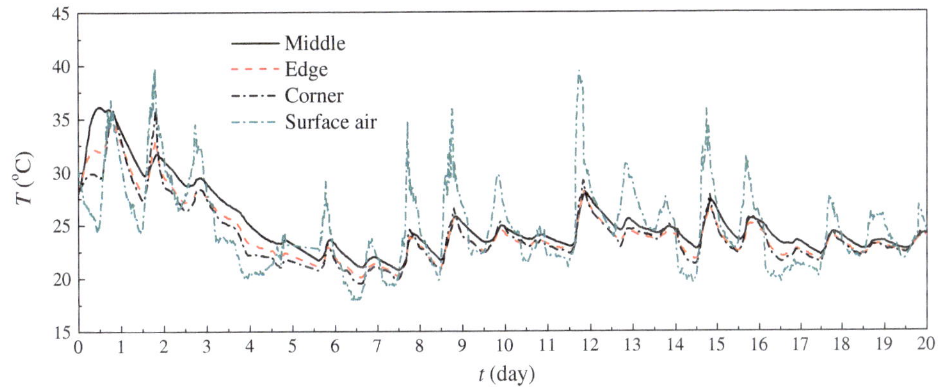

(a) 0 to 20 days

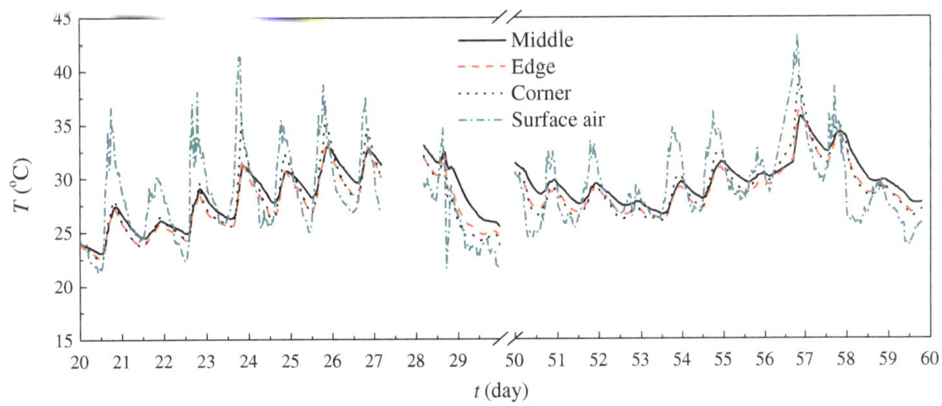

(b) 20 to 60 days

Fig. 4 Time history of track slab temperature and surface air temperature

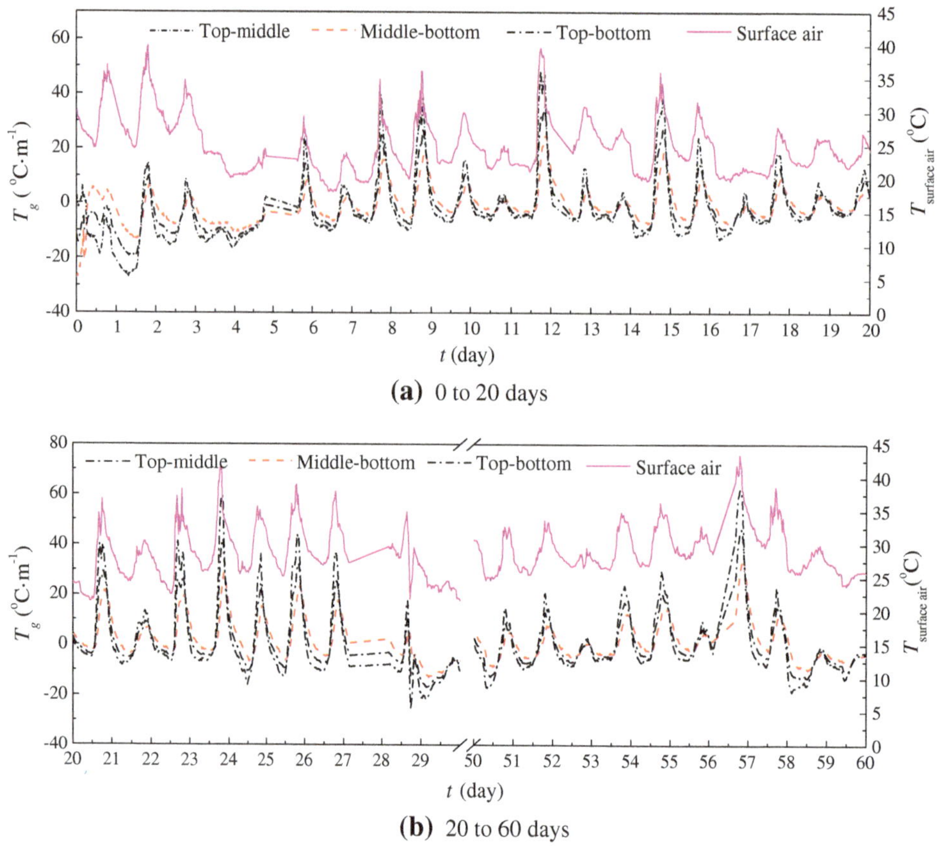

Fig. 5 Time history of track slab temperature gradient and surface air temperature (*middle part*)

different parts of the concrete, the heat exchange conditions are different because of the different contact areas with the surrounding air, causing a difference in temperature measurements. When the temperature rises during the daytime, the temperature increment at the corner and edge parts is a bit larger than the one at the middle part, and vice versa. The amplitude of temperature changes is in a descending order from the corner, edge to middle.

The track slab temperature varies periodically with the air temperature—rising during the daytime and dropping at nighttime. The higher the air temperature, the higher the track slab temperature. The time for track slab temperature to reach its extreme value is 2 h later than that for the air temperature for a 30-cm-thick track slab in Chengdu area.

3 The time history of temperature gradient in track slab

Using the acquired data from the three temperature sensors in the same test section of the track slab, we can calculate the temperature gradient of the track slab through the following formula:

$$T_g = (T_i - T_j)/(H_i - H_j), \tag{1}$$

where T_g is temperature gradient (°C/m); T_i and T_j are temperatures measured by two sensors at the top, middle, or bottom layer of the track slab (°C); and H_i and H_j are the heights of the two sensors (m).

The time history of temperature gradient at the middle part of the track slab is shown in Fig. 5. In order to facilitate comparison, the air temperature time history is also shown in the figure (the purple line). The top–middle, middle–bottom, and top–bottom represent the temperature gradients between top and middle layers, middle and bottom layers, and top and bottom layers of the concrete, respectively.

As can be seen in Fig. 5, the temperature gradient in track slab changes almost simultaneously with the surface air temperature. The temperature gradient amplitude goes high when the surface air temperature rises. The temperature gradient calculated using the top and middle sensors is clearly larger than that calculated using the middle and bottom sensors, which shows a nonlinear distribution along the height direction. The temperature gradient calculated using the top and bottom sensors is located between the above two. In designing the slab track, the nonlinear part is generally ignored and it is assumed that there is a linear distribution along the whole section. The nonlinear part will cause inner stress within the section.

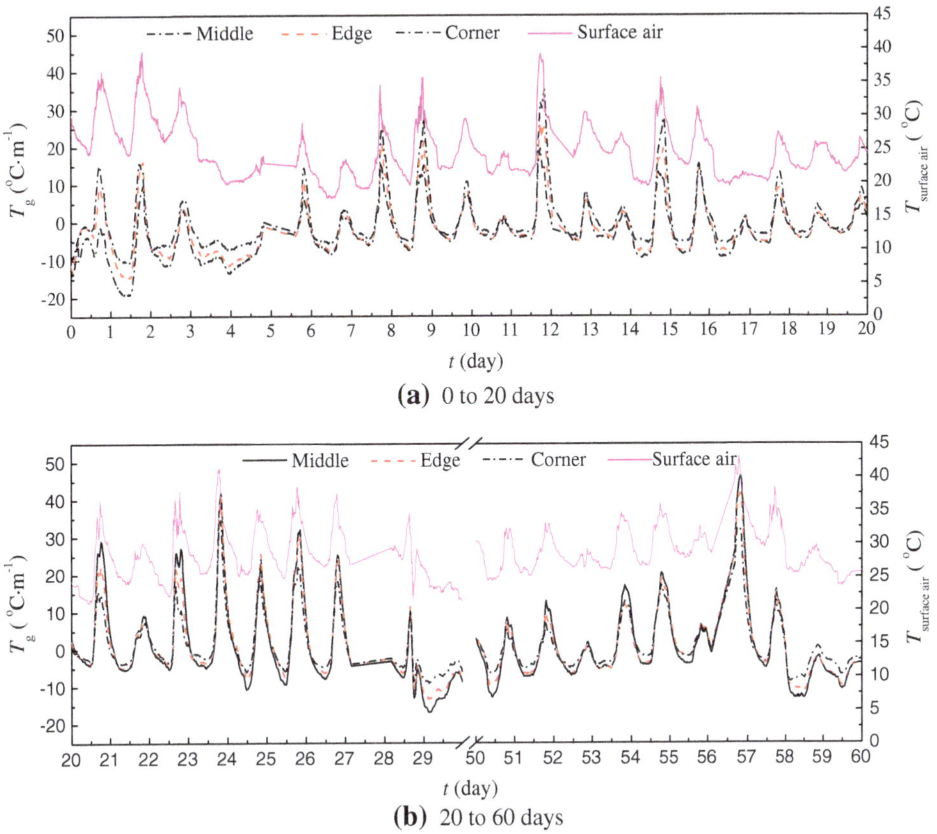

Fig. 6 Time history of track slab temperature gradient and surface air temperature (different part)

During the cement hydration phase, due to the hydration heat, the temperature at the bottom and middle layer of track slab is higher than that at the top, where heat can radiate to the outside easily, causing a large negative temperature gradient of several days.

Figure 6 shows the time history of the temperature gradients at different parts of the track slab, where the temperature gradients are calculated from the top and bottom sensors. The total trend of the temperature gradient at different locations of the track slab is the same as the surface air temperature simultaneously. Because of the different heat exchange condition with surface air, the temperature gradient at the middle part is larger than that at the edge, and both larger than that at the corner. During the hydration phase, a verse discipline can be found. During the design of slab track, the temperature gradient from the middle part should be selected as the design parameter, which can guarantee a safety design of slab track.

4 The distribution of temperature gradient in track slab

Excluding the first 7 days affected by hydration heat, a frequency account analysis was carried out using the time history of the track slab temperature gradient, and the frequency distribution histogram is obtained as shown in Fig. 7 (top row, from left to right is for the middle part, edge part, and corner part, respectively), where the ordinate p denotes the proportion of the temperature distribution in these tested days. The temperature gradient ranges from −20 to 50 °C/m, but the distribution is extremely uneven. The most frequently occurring temperature gradient of the track slab is around −3.5 °C/m. Compared with the middle part, the corner part of the track slab has a narrower band distribution. This phenomenon should be attributed to the three directions of heat exchange with the surrounding air, while at the edge only two directions, and at the middle just one direction.

In top row of Fig. 7, all the data throughout the whole day are used, but in real operation, the period from midnight till 4 o'clock is the so-called vertical "window" time, when all the high-speed trains will stop for the daily maintenance of engineering structures. Therefore, when we do the statistical analysis for the fatigue analysis under the coupling action of train load and temperature load, the "window" time should be excluded. Also in the present operation, after the "window" time till 6 o'clock, only a few track inspection vehicles will run, so the statistics should exclude the data from 00:00 to 06:00. The statistical

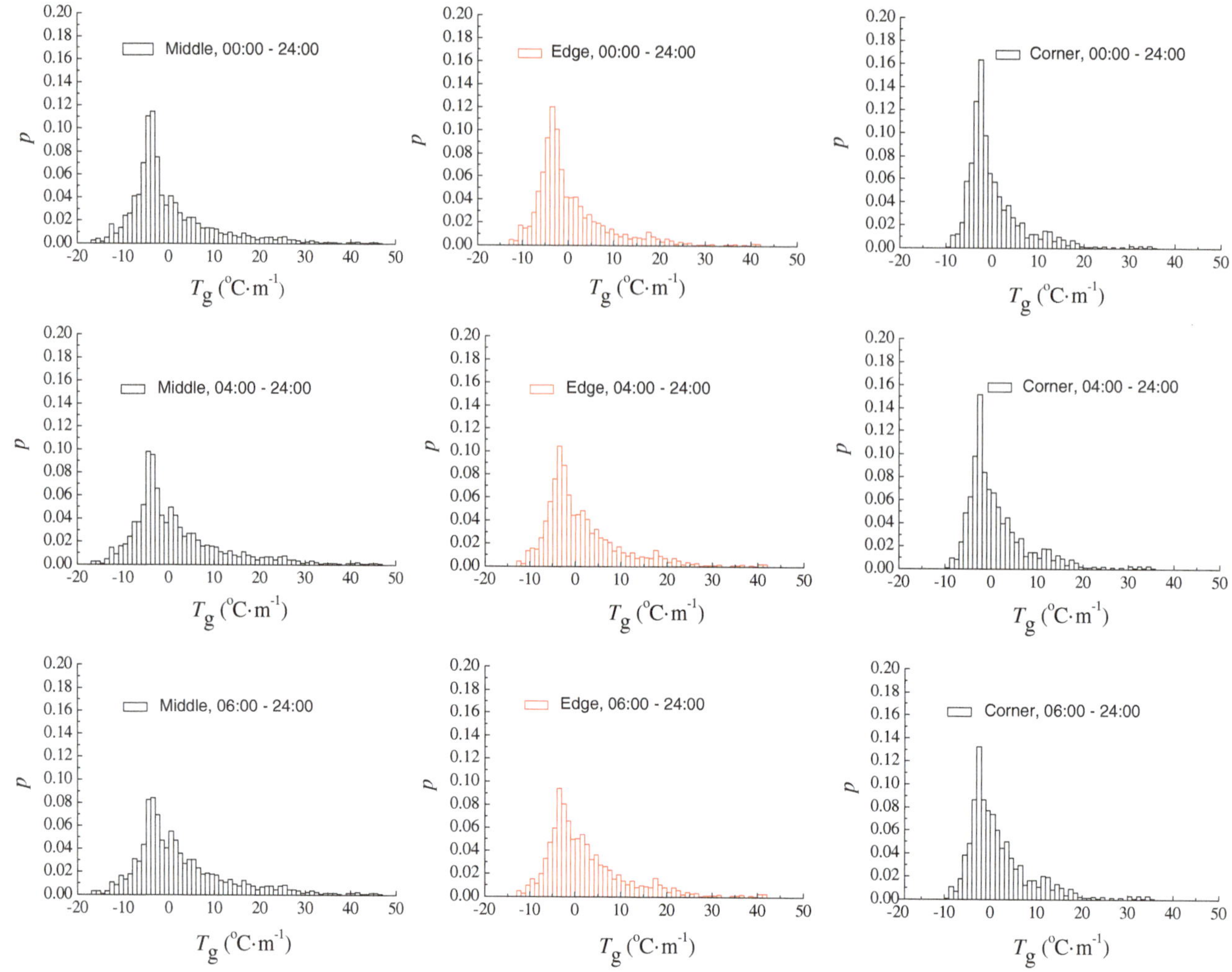

Fig. 7 Track slab temperature gradient distribution

distribution of temperature gradient in track slab using different time zone data is shown in Fig. 7 (middle row is for time zone [4,24], and bottom row is for time zone [6,24]). 00:00 to 06:00 is the coldest time zone of a day, when concrete track slab is in a cooling stage that shows negative temperature gradients. It will decrease the probability of the most frequently occurring negative temperature gradient if data are excluded in this time zone. The frequency distribution histogram using the data excluding the time zone from 00:00 to 06:00 should be used, or adjusted according to the real "window" time when we do the coupling fatigue analysis of track slab under train load and temperature gradient.

Using 10 °C/m as step, the percentage of different levels of temperature gradient is calculated and shown in Table 1. The most frequently occurring temperature gradient appears in −10 to 0 °C/m, 57.9 % for the middle part, 59.6 % for the edge part, and 63.1 % for the corner. The second frequently occurring temperature gradient is in level 0–10 °C/m, with the percentage of 22.7 %, 25.4 %, and 27.4 % for the middle, edge, and corner, respectively. Little differences exist among the different parts. When checking the fatigue of the track slab, the data acquired from the middle part of the track slab from the safety viewpoint are available. Taking account of the "window" time and the low traffic operation after maintenance work, the percentage of level −10 to 0 °C/m drops to 50.6 % and 45.9 %, and the percentage of level 0–10 °C/m rises to 27.3 % and 30.5 % accordingly. More than 80 % of the temperature gradient is between −10 and 10 °C/m throughout the whole day. When the coupling action with train load is considered as an effect factor, there is still more than 75 % period located in the level from −10 to 10 °C/m.

5 Relationship between track slab temperature amplitude and air temperature

Figure 8 shows the amplitude of daily temperature variation. The highest amplitude of the surface air temperature

Table 1 Percentage of different levels of temperature gradient (%)

Time zone	Level of temperature gradient (°C/m)	−20 to −10	−10 to 0	0–10	10–20	20–30	30–40	40–50
00:00–24:00	Middle	5.3	57.9	22.7	7.8	4.3	1.2	0.7
	Edge	2.7	59.6	25.4	8.3	2.8	0.9	0.5
	Corner	0.0	63.1	27.4	7.8	0.9	0.9	0.0
04:00–24:00	Middle	5.1	50.6	27.3	9.4	5.2	1.5	0.8
	Edge	2.2	52.9	29.9	10.0	3.4	1.0	0.6
	Corner	0.0	55.7	32.8	9.4	1.0	1.0	0.0
06:00–24:00	Middle	4.6	45.9	30.5	10.5	5.9	1.6	0.9
	Edge	1.9	48.0	33.4	11.2	3.7	1.2	0.6
	Corner	0.0	50.6	36.6	10.5	1.2	1.2	0.0

variation is close to 20 °C and the amplitude of track slab temperature variation is close to 10 °C. The amplitude of track slab temperature variation rises with that of the surface air, which indicates that the track slab temperature variation has a great correlation with the change of the surface air temperature, as shown in Fig. 9. The amplitudes of temperature variation from more to less are in the order of the edge, the corner, and the middle area.

A fitting formula is given as follows:

$$\Delta T_{\text{slab,daily}} = K\Delta T_{\text{surfaceair,daily}}, \tag{2}$$

where $\Delta T_{\text{slab,daily}}$ and $\Delta T_{\text{surface air,daily}}$ are track slab and surface air temperature variations, respectively (°C), and K is the scale factor. For this test, which was taken in an outdoor environment in Chengdu, the scale factor K for the middle, edge, and corner areas are 0.275, 0.307, and 0.419, respectively; and the goodness of fit of the formula for the above three areas are 0.859, 0.877, and 0.862, respectively.

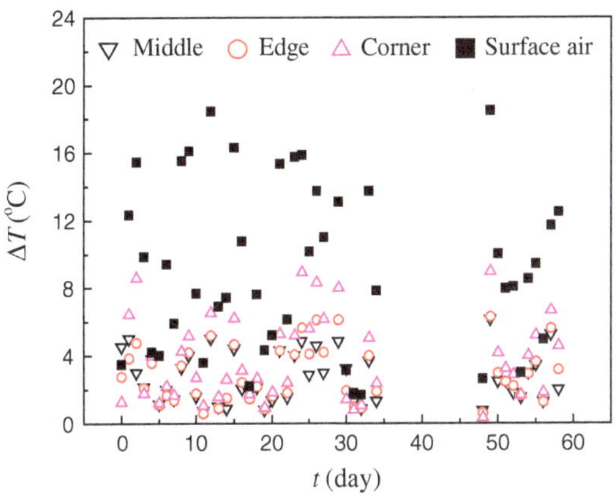

Fig. 8 Distribution of temperature variation amplitude

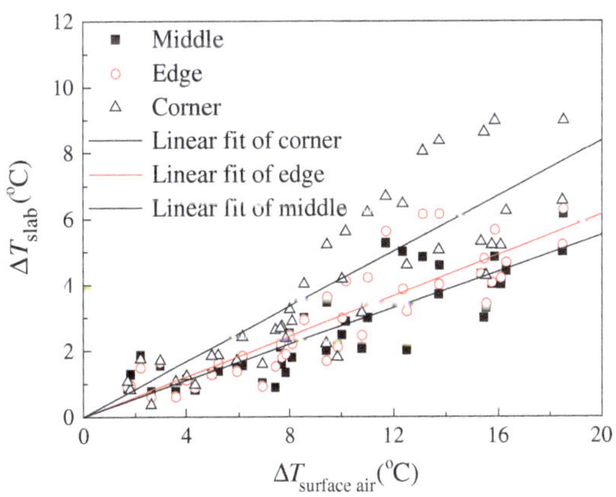

Fig. 9 Relationship between temperature variation of track slab and surface air

6 Conclusion and recommendation

This test was taken in Chengdu outdoor environment and has a significant meaning of pinpointing the correlation between the environmental temperature and providing a statistical method for analysis of track slab's temperature load.

A bi-block track slab model was constructed, and the temperature field of the track slab together with the surface air temperature was recorded and analyzed. Some conclusions can be made as follows, which can provide a reference for slab track design and its long-term service performance study.

(1) Track slab temperature and surface air temperature reveal a periodical phenomenon with a day period, which shows a strong connection between them. The temperature gradient in track slab changes almost simultaneously with the surface air temperature, and shows a clear nonlinearity along the height direction. The temperature gradient in top part is larger than that in the bottom part.

(2) The frequency account analysis shows that the temperature gradient distribution is extremely uneven. The most frequently occurring temperature

gradient of the track slab is around −3.5 °C/m. In more than 75 % period of a day, the temperature gradient is located in the level −10 to 10 °C/m. The concrete part with a better heat exchange condition with the surrounding air (e.g., the corner part of the slab track) has a narrower band distribution.

(3) At present, due to the fact that there is no traffic or low traffic during the "window" period, the data for frequency distribution histogram should exclude the time zone from 00:00 to 06:00, or should be adjusted according to the real "window" time when we do the coupling fatigue analysis of track slab under train load and temperature gradient.

(4) The amplitude of track slab temperature variation is obviously lower than that of the air temperature variation, and there is a proportionality between them. Because of the differences of heat transfer conditions, the results of the test are different: The amplitudes of temperature variation from more to less are in the order of the edge, the corner, and the middle area.

(5) Limited by the test conditions, only a 2-month-long test was carried out. More tests should be done using this method to get more precise coefficient and convincing conclusion.

Acknowledgments This work was supported by the National Key Basic Research Program of China (973 Program) (2013CB036202), the National Natural Science Foundation of China (51008258), and Fundamental Research Funds for the Central Universities (SWJTU12CX065).

References

1. He HW (2006) The ballastless track technology. China Railway Publishing House, Beijing (in Chinese)
2. Liu XY, Zhao PR, Yang RS et al (2010) The design theory and method for ballastless track of passenger dedicated railway line. Southwest Jiaotong University Press, Chengdu (in Chinese)
3. Zhao PR, Liu XY (2008) The temperature-stress calculation method and parameter analysis of continuous track slab. Railw Eng 11:81–85 (in Chinese)
4. Cong ZH, Zheng NX, Jin Y (2012) Temperature field division at Liupanshan area of Ningxia for asphalt pavement performance. Adv Mater Res 511:12–15
5. Zhimin T, Zhengjun M, Zou X (2013) Pavement temperature estimation model based on field temperature data. Tongji Daxue Xuebao/J Tongji Univ 41:700–704 (in Chinese)
6. Haigui K, Yuanxun Z, Yingchun C, Ya L (2007) Regression analysis of actual measurement of temperature field distribution rules of asphalt pavement. Chin J Highw Transp 20(6):13–18
7. Rostásy FS, Budelmann H (2007) Verification of thermal restraint of a railways trough structure by long-term monitoring. Struct Infrastruct Eng 3(3):237–244
8. Shao P, Li H, Wu S, Wang T, Liu Z, Li H (2013) Measurement and research on temperature warping of CRTS I track slab and crack between track slab and cement asphalt mortar cushion. Zhongguo Tiedao Kexue 34(2):18–22
9. Zhao C, Song X, Zhu X (2013) Temperature deformation analysis of CRTS II ballastless slab track. In 2013 joint rail conference (pp. V001T01A016-V001T01A016). American Society of Mechanical Engineers
10. Wang SR, Sun L, Li Q (2009) Temperature measurement and temperature stress analysis of ballastless track slab. J Railw Eng Soc 2:52–55
11. Coleri E, Popescu L, Signore JM, Wu R, Harvey JT (2012) PCC slab temperature gradients as a function of structure and environment experience from the SHRP II R21 composite pavement test track. In: 10th international conference on concrete pavements
12. Yang S (1996) A temperature prediction model in new concrete pavement and a new test method for concrete fracture parameters. Doctoral dissertation, Texas A&M University

26

Modal analysis of prestressed draft pad of freight wagons using finite element method

Sachin Sudhakar Harak · Satish Chand Sharma · Suraj Prakash Harsha

Abstract The present work intends to investigate dynamic behaviour of draft gear using finite element method. The longitudinal force that the draft gear absorbs usually leads to the failure of its components, especially, the load bearing draft pads. Dynamic behaviour of an individual draft pad and a draft gear is determined and characterized with exciting frequencies and corresponding mode shapes. The effect of compressive prestress load on the dynamic behaviour of an individual draft pad is also determined as the draft pads in assembled state are under constant axial compressive force in the draft gear. The vibration characteristics of individual draft pad are compared with draft pads that are part of draft gear. The modal analysis gives us a basis for subjecting a draft pad to higher frequency loading for determining its fatigue behaviour.

Keywords Draft gear · Draft pad · Prestress · Natural frequency · Mode shapes · Finite element method

1 Introduction

In longitudinal train dynamics, draft gear is the most important component of the autocoupler. The autocoupler with friction type draft gear is the most common wagon connection in the Australian and North American freight system [1]. Similar autocouplers with draft gear are also used in the Indian freight system and their adoption in the freight system can be traced back in the early 1960s. Mark-50 was adopted in 1963, RF-361 was adopted in 1965 and SL-76 was adopted in 1977. Being an important part of the autocoupler system, the draft gear serves to absorb energy generated during the normal starting and braking of the freight railway wagon system and also during different train handling conditions, thus avoiding damage and failure of the coupling system. It thus functions as a passive vibration isolator to control and absorb high amplitude forces thereby avoiding damage to laden goods. The frequency of this longitudinal force depends on the train running condition and varies continuously.

The longitudinal force that the draft gear absorbs usually leads to the failure of its components, especially, the draft pads. Garg and Dukkipati [2] specified the problems caused by longitudinal forces and attributed the failure of draft gears due to excessive longitudinal draft forces. Initial work on longitudinal train dynamics was aimed towards reducing longitudinal oscillations in passenger train. Work done in this context involved measurement and simulation of in-train forces to reduce longitudinal oscillations [3]. The relation between lateral coupler force components and wheel unloading were studied by El-Sibaie [4] and a new method was proposed for evaluating the curving stability of any freight railway car under buff and draw conditions. Chen [5] proposed a mathematical model to calculate transient responses of the coupler, whereas McClanachan et al. [6] conducted experiments to determine the occurrences of coupler impacts combined with pitching motions in the wagon body. Then these interactions were simulated using NUCARS, ADAMS/Rail and a train-wagon interaction model. Cole and Sun [7] evaluated fatigue life of autocoupler systems with and without self-locking features of the draft gear for three types of wagon connection coupling systems. Nasr and Mohammadi [8] studied the effect of train braking delay time on train longitudinal dynamics. They carried out simulations for three different

S. S. Harak (✉) · S. C. Sharma · S. P. Harsha
Vibration and Noise Control Lab., Mechanical and Industrial Engineering Department, Indian Institute of Technology, Roorkee 247667, Uttarakhand, India
e-mail: sachin.s.harak@gmail.com

cases of long, medium and short brake application times for three different train forward velocities.

In this article, the dynamic behaviour of draft pad and draft gear is investigated using finite element method. The draft gear consists of six draft pads which are assembled in series (one above another) and are always under a constant axial compressive prestress in unloaded condition. Prestress influences the values of the stiffness matrix by causing stress stiffening and affects the response frequencies of the system thus impacting its modal responses [9]. The draft pad consists of a rubber compound/elastomer along with two steel plates. The modal behaviour of draft pad and draft gear assumes significance as it is necessary to identify frequencies at which fatigue failure of draft pads can occur. The vibration characteristics of the draft pad and the draft gear for the first 15 modes have been determined using finite element method. The vibration excitation of an individual draft pad has been compared with the dynamics of draft pads in the draft gear for identifying various frequencies for a given mode shape.

2 Dynamic analysis of draft pad

2.1 FE analysis

For a dynamic system, the general form of equation of motion is

$$m\ddot{x} + c\dot{x} + kx = f(t).$$

Modal analysis determines the natural frequencies and mode shapes of a continuous structure. Thus, modal analysis helps in determining vibration characteristics of any system. In finite element solution, the differential equation of motion is discretized into a number of finite element equations which form a system of algebraic equations to be solved: $[\mathbf{K}]\{\mathbf{u}\} = \{\mathbf{F}\}$, where $[\mathbf{K}]$ is the stiffness matrix, $\{\mathbf{u}\}$ is the nodal displacement vector and $\{\mathbf{F}\}$ is the applied load vector. These equations are solved by finite element package ANSYS [10].

For a free vibration analysis, the natural frequencies (ω_i) and the mode shapes (ϕ_i) are determined using following relation: $([\mathbf{K}] - \omega_i^2[\mathbf{M}])\{\phi_i\} = 0$ [11]. For the prestressed modal analysis, the stress field is taken into account by modifying the stiffness matrix $[\mathbf{K}]$. While evaluating the mode shapes, the material is assumed to behave linearly elastic neglecting non-linearities, damping is not included, and no excitation of the structure is assumed.

2.2 Model for modal analysis

For vibration-based analysis, which is model dependent, an accurate model is important. FEM has been used for determining the mode shapes and natural frequency of continuous system like the freight wagon [12]. Three-dimensional models of draft pad and draft gear are created using *Solidworks* and exported to ANSYS environment for numerical analysis.

2.2.1 Individual draft pad

The draft pad (RF-8) is illustrated through Fig. 1. Such pads serve as rubber springs in "RF-361" model of draft gears manufactured by Miner Enterprises Inc., USA, and are used in freight wagons autocoupler of Indian Railways. Each RF-8 pad has ultra-high capacity natural rubber spring package to absorb the high longitudinal forces arising due to operations like acceleration and/or braking and also due to changes in track topography. This rubber compound/elastomer is sandwiched between two steel plates and perfectly bonded to them. The material properties of steel are—modulus of elasticity: 200×10^9 N/m^2, Poisson's ratio: 0.30 and material density: 7,850 kg/m^3. The rubber compound/elastomer has modulus of elasticity: 21.946×10^6 N/m^2, Poisson's ratio: 0.4970 and material density: 1,220 kg/m^3. The mechanical properties such as modulus of elasticity and Poisson's ratio of rubber compound/elastomer are obtained from experimental data. Experimental data refers to stress–strain plot obtained for an individual draft pad from compression test conducted in-house. MCalibration software has been used to derive the mechanical properties from this experimental plot. In general, the elastomer being used is vulcanized natural rubber with a hardness value of 75 on the Shore hardness scale (A scale). The rubber compound/elastomer under consideration is also described by its tensile strength and tear strength. Typical values are 180 and 40 kg/cm^2, respectively. A maximum compression set of 25 % at 70 °C for 24 h is also taken into consideration for accommodating geometric changes. The boundary conditions are: all nodes at the outer surface of bottom plate are fixed and the prestress pressure load is applied at outer surface of top plate.

2.2.2 Draft gear

Figure 2a illustrates three-dimensional view and Fig. 2b presents sectional view of draft gear. The various components of the draft gear (RF-361) under consideration are draft pads (six in numbers) with top follower, shoes (three in numbers)

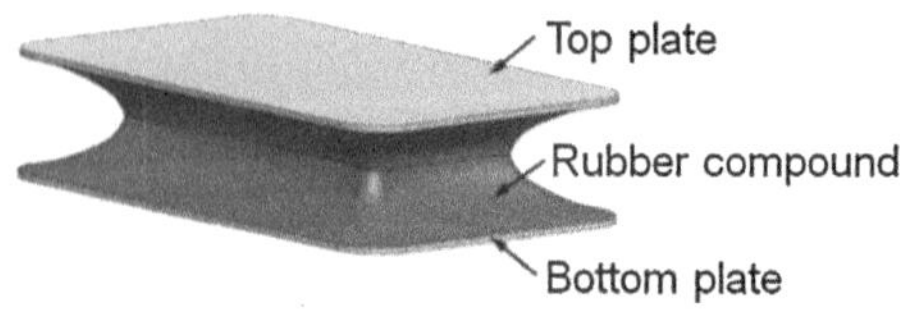

Fig. 1 Three-dimensional representation of typical draft pad (RF-8)

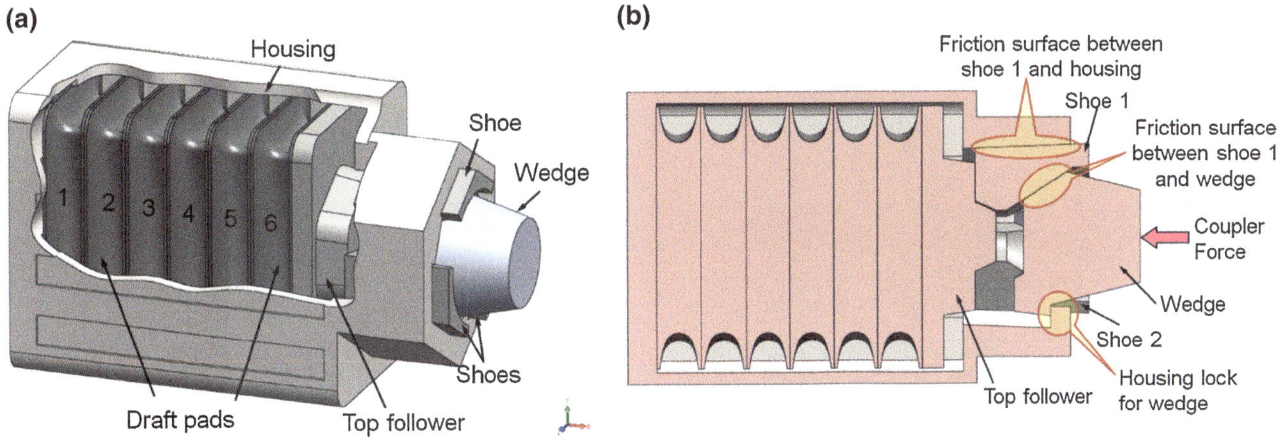

Fig. 2 Draft gear unit **a** three-dimensional representation and **b** sectional representation

and a wedge. All these components are assembled in a housing with the three shoes arranged circumferentially around the wedge. The rubber springs absorb the longitudinal (coupler) forces applied on the draft gear. The wedge and shoes operate between the top follower and the draft gear housing. The applied forces reach the rubber springs/draft pads only after overcoming the friction in two stages. In stage one, friction is provided by the surface contact between the wedge and the shoes. In stage two, friction is provided by surface contact between shoes and draft gear housing as illustrated in Fig. 2b. The top follower provides support to the shoes. In both the stages, the friction between the components, i.e. wedge and shoes; shoes and housing, damps out the shock before the rubber springs/draft pads actually absorb the longitudinal force(s).

It means that the wedge and shoes act as shock damper and the applied force is then transferred to the rubber springs/draft pads. Each rubber spring/draft pad consists of elastomeric material sandwiched between two metal plates. Advantage of using draft gear having such damping effect is that impact energy on the couplers can gradually be absorbed by all the couplers. This results in lower coupler force in leading wagons. The height of a single draft pad prior to assembly is 67.2 mm. In assembled condition the effective height of a single draft pad is 50 mm. The material properties taken are same as those in Sect 2.2.1. The boundary conditions are taken as: all nodes at the bottom surface of the housing are fixed and the prestress pressure load is applied at outer surface of the wedge.

3 Results

The natural frequencies of an individual draft pad as illustrated in Fig. 3 for the first five modes are obtained using finite element package (ANSYS) and presented in Table 1. As the first five frequencies are observed to be dominating, hence they are considered. These values are obtained under free stress condition.

These first five mode shapes of an individual draft pad are presented in Fig. 3. The draft pad as illustrated in Fig. 1 consists of three components, viz. top plate, rubber compound and bottom plate. All these three components are bonded together and behave as a single unit. Since the bottom plate is fixed, the motion of rubber compound and top plate describes each mode shape. The origin of the coordinate system is located at the geometric centre of the pad. The triad helps in visualizing the axes orientations. Accordingly, it is seen that

- The first mode shape shown in Fig. 3a describes oscillatory (angular) motion of the rubber compound and top plate about x-axis of the draft pad.
- The second mode shape illustrated in Fig. 3b shows oscillation of the rubber compound and top plate about y-axis. However, the y-axis about which the oscillation occurs does not coincide with the draft pad y-axis but passes through the base plate of draft pad.
- The third mode shape in Fig. 3c describes oscillation of the rubber compound and top plate about z-axis. The z-axis in this case also does not coincide with the draft pad z-axis, but passes through the base plate of draft pad.
- The fourth mode shape shown in Fig. 3d shows oscillation of the rubber compound and top plate about y-axis. The y-axis in this case is contained by the top plate as is evident from the colour code.
- The fifth mode shape illustrated in Fig. 3e indicates oscillatory motion of the rubber compound and top plate about z-axis. This z-axis passes through the top plate as is evident from the colour code.

The natural frequencies of the draft gear are also evaluated in a similar manner. These natural frequencies are

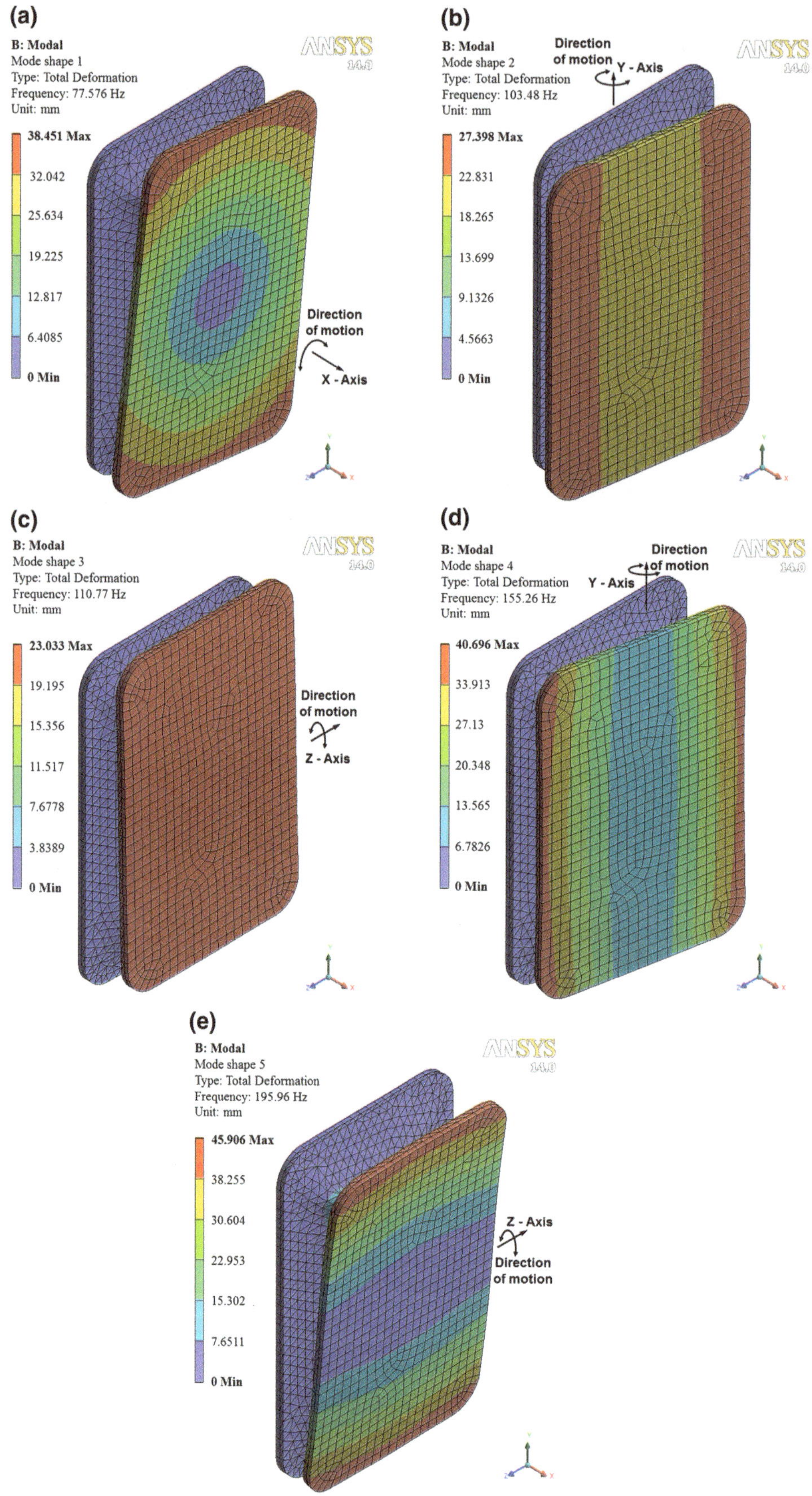

Fig. 3 Mode shapes of draft pad for **a** first, **b** second, **c** third, **d** fourth and **e** fifth natural frequency

Table 1 Natural frequencies of an individual draft pad in Hz

Mode	1	2	3	4	5
Frequency	77.57	103.48	110.77	155.26	195.96

obtained under prestress condition and are presented in Table 2.

A comparison of the mode shapes of each draft pad in the draft gear with mode shapes of an individual draft pad is done. It is observed that first five mode shapes of the individual draft pad are dominating the behaviour of each draft pad in the draft gear for all 15 mode shapes of the draft gear. This is because the draft pads housed in the draft gear have their own individual behaviour which cumulatively results in different mode shapes of the draft gear. These mode shapes of draft gear as such have no resemblance with mode shapes of an individual draft pad.

Figure 4 describes the first mode shape of a draft gear with prestress condition. The draft gear housing is very stiff as compared to draft pad and thus has a very poor modal response. Hence it has been hidden in this view. Modal behaviour of each draft pad is presented separately. It is seen that all the draft pads oscillate about the longitudinal (x) axis of draft gear. Each draft pad exhibit modal behaviour close to its first modal shape. Since, the modal behaviour is symmetrical at the interface of third and fourth pad, hence only modal behaviour of the first three draft pads is presented in the illustration.

A general description of all the remaining mode shapes of draft gear is also necessary to establish the correlation between the mode shapes of draft pad and draft gear. For the second mode of the draft gear, linear displacement of the interface between pad 3 and pad 4 in lateral (z) direction is observed causing the draft pads to exhibit modal behaviour close to its second mode shape. The third mode shows linear displacement of the interface between pad 3 and pad 4 in vertical (y) direction causing the draft pads to exhibit modal behaviour close to its third mode shape. For the fourth mode shape, the interface between pad 3 and pad 4 remains fixed and the interfaces between pads 1 and 2, 2 and 3 and interfaces between pads 4 and 5, 5 and 6 oscillate about longitudinal (x) axis in opposite sense. As a result, modal behaviour corresponding to the first mode shape of

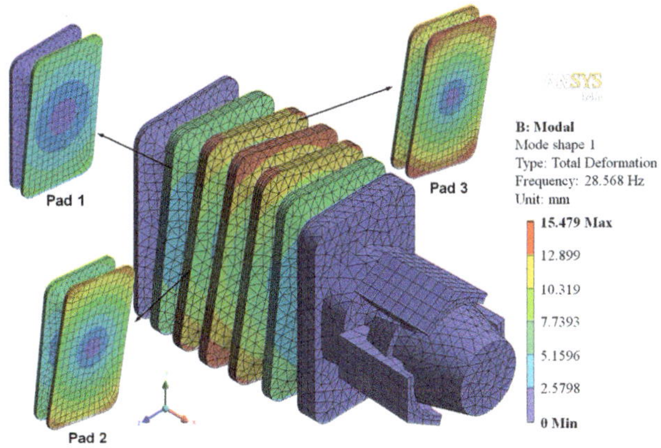

Fig. 4 First mode shape of a draft gear

draft pad is observed in some draft pads. Fifth mode shape shows oscillation of the interface between pad 3 and pad 4 about vertical (y) axis. The results is that some draft pads exhibit second modal behaviour while some exhibit fourth modal behaviour of individual draft pad.

Sixth mode shape shows oscillation of the interface between pad 3 and pad 4 about lateral (z) axis. As a result some draft pads exhibit third modal behaviour of individual draft pad. In the seventh mode shape, interfaces between pads 2 and 3, 4 and 5 remains fixed and interfaces between pads 1 and 2, 3 and 4, and 5 and 6 oscillate about longitudinal (x) axis causing every single pad to execute motion in accordance to first mode shape. The eight mode shape of draft gear is oscillation of all draft pad interfaces about lateral (z) axis. This oscillation results in fifth mode shape for all pads. For the ninth mode shape of draft gear, the interface between pads 3 and 4 remains fixed while the adjacent interfaces (i.e. interface between pads 1 and 2, 2 and 3 and interfaces between pads 4 and 5, 5 and 6) oscillate in opposite direction. This mode thus causes all pads to behave in accordance to first mode shape. For the tenth mode shape of draft gear, interface between pads 2 and 3 and pads 4 and 5 oscillate between about vertical (y) axis in opposite phase. The resulting effect is that all pads execute second modal behaviour.

For the eleventh mode shape, all pad interfaces oscillate about vertical (y) axis in same phase and the resulting motion is in confirmation to the fourth modal shape. For

Table 2 Natural frequencies of draft gear under prestress load

Mode no.	1	2	3	4	5	6	7	8
Frequency (Hz)	28.56	34.53	38.97	55.58	65.33	72.81	79.36	96.47

Mode no.	9	10	11	12	13	14	15
Frequency (Hz)	98.16	100.53	101.58	110.38	110.82	121.99	135.08

the twelfth mode shape all adjacent pad interfaces oscillate about longitudinal (x) axis in opposite direction causing each pad to follow first modal behaviour. For the thirteenth mode shape the interface between pads 2 and 3 and pads 4 and 5 oscillates about the lateral (z) axis. As a result all pads execute motion in accordance to the third modal shape. For the fourteenth mode shape, all pad interfaces move linearly in the longitudinal direction. This is the only mode in which no pad behaviour confirms to any of the first five modes. In the fifteenth mode shape the interface between pads 3 and 4 oscillates about the vertical (y) axis while interfaces between pads 2 and 3 and pads 4 and 5 move linearly in opposite phase in the lateral (z) direction.

Table 3 Correlation between mode shapes and natural frequencies of an individual draft pad and draft gear

Individual draft pad	Draft gear		
Mode no.	Pad no.	Similar mode no.	Corresponding frequency (Hz)
1	1, 6	1, 4, 7, 9, 12	28.56, 55.58, 79.36, 98.16, 110.38
	2, 5	1, 7, 9, 12	28.56, 79.36, 98.16, 110.38
	3, 4	4, 7, 9, 12	55.58, 79.36, 98.16, 110.38
2	1, 6	2, 5, 10, 15	34.53, 65.33, 100.53, 135.08
	2, 5	10, 15	100.53, 135.08
	3, 4	2, 5, 10, 15	34.53, 65.33, 100.53, 135.08
3	1, 6	3, 6, 13,	28.56, 72.81, 110.82
	2, 5	3, 13	28.56, 110.82
	3, 4	3, 13	28.56, 110.82
4	1, 6	11	101.58
	2, 5	5, 11	65.33, 101.58
	3, 4	11	101.58
5	1, 6	8	96.47
	2, 5	8	96.47
	3, 4	8	96.47

The resulting effect of this motion causes all draft pads to behave in accordance to second mode shape.

The correlation between the first five mode shapes of an individual draft pad with the first 15 mode shapes and frequencies of the draft gear is summarized in Table 3. It is seen that the first five modes of an individual draft pad have different frequencies when this draft pad coexists with other draft pads in a draft gear. As an example, for an individual draft pad at mode shape 1 the frequency is 77.57 Hz. Column 2 in Table 3 identifies which pads in the draft gear exhibit a similar mode shape, column 3 identifies during which mode shapes of the draft gear this behaviour is observed and column 4 conveys the corresponding frequency value of the draft pads.

Further, the individual draft pad is also subjected to prestress condition before evaluating its modal behaviour. The draft pads are assembled in the draft gear in compressed state, i.e. any draft pad will have a height of 50 mm which is 17.2 mm less than the free height of 67.2 mm. The equivalent value of compressive load corresponding to this deflection is estimated to be around 24×10^3 N. The prestress condition is defined in the form of static load applied on the top plate of the draft pad. These compressive loads give rise to stress field. Once the static analysis establishes the stress field; it can be applied to the draft pad as prestress through options provided in the analysis package. This procedure was repeated for eight different compressive loads till the actual working load was reached. The results for various prestress loads have been summarized and presented in Table 4.

The percent variation in the frequencies for prestress compressive load of the draft pad is plotted in Fig. 5. For an individual draft pad with increase in compressive prestress load the frequency decreases for the first three modes but for the fourth and fifth mode, increase in the compressive prestress load causes increase in the frequency values. The behaviour of the first three modes is in conformance to observations by [9], i.e. with increase in compressive prestress load the frequency decreases, but for the fourth and fifth mode the behaviour is in contract to the observations by [9].

Table 4 Frequency variation of first five modes of draft pad for prestress compressive load

Mode no.	Modal frequency (Hz) for compressive load							
	3 kN	6 kN	9 kN	12 kN	15 kN	18 kN	21 kN	24 kN
1	76.78	75.84	74.80	73.70	72.53	71.30	69.99	68.59
2	101.98	100.49	98.77	96.95	95.02	92.99	90.84	88.56
3	110.08	109.26	108.30	107.29	106.23	105.11	103.95	102.72
4	156.60	157.98	158.94	159.73	160.38	160.90	161.30	161.58
5	198.05	200.68	202.69	204.50	206.18	207.76	209.24	210.63

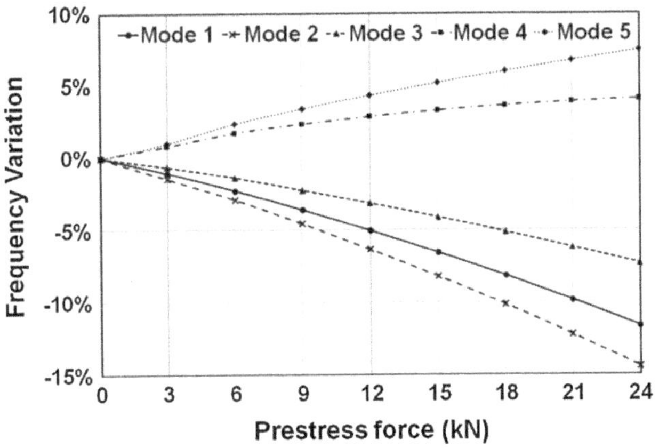

Fig. 5 Frequency variation due to prestress load for various mode shapes

4 Conclusions

The mode shapes and frequencies for an individual draft pad and a draft gear have been obtained using finite element approach. It is seen that the mode shapes of draft pads in a draft gear are dominated by the first five modes (Fig. 3) of an individual draft pad. For these dominant mode shapes of an individual draft pad, different values of frequencies (Table 3) of the draft gear have been identified. These values are different than the various values of frequency obtained for an individual pad. These different values of frequency assumes significances as the draft gear is subjected to variable loading frequency due to acceleration and braking of the locomotive. Hence, a range of frequency values have been identified against which a draft pad must be tested for fatigue response. As per the schedule of technical requirements by Indian Railways [13], a draft pad is subjected to cyclic loading of 10 Hz for determining its fatigue behaviour. The values identified (Table 3) are greater than the existing value but have significance as the possibility that the frequency of cyclic load acting on the draft gear in actual operation will not exceed this specified value cannot be ruled out. The modal analysis thus gives us a basis for subjecting a draft pad to higher frequency loading for determining its fatigue behaviour.

Increase in the value of compressive load on the draft pad shows that not all mode shapes have decreasing values of frequency, due to geometric non-linearity. Of the five modes, only the first three mode shapes show such behaviour. Mode shapes 4 and 5 show increase in frequency with increase in compressive preload. This behaviour is in contrast to the vibrational behaviour exhibited by the first three modes. Since, material non-linearity is neglected during the analyses, such behaviour can be attributed only to two parameters; either the geometric non-linearity or the high Poisson's ratio of rubber compound. It can thus be concluded that geometric non-linearity and/or high Poisson's ratio can cause the natural frequency of draft pads to increase with increasing compressive preload in case of draft pads (which is a elastomeric material).

References

1. Polach O, Berg M, Iwnicki S (2006) Simulation. In: Iwnicki S (ed) Handbook of rail vehicle dynamics, 1st edn. CRC Press, Boca Raton, pp 359–422
2. Garg VK, Dukkipati RV (1984) Dynamics of railway vehicle systems. Academic Press, Toronto
3. Duncan IB, Webb PA (1989) The longitudinal behaviour of heavy haul trains using remote locomotives. In: Proceedings of fourth international heavy haul conference, Brisbane, pp 587–590
4. El-Sibaie M (1993) Recent advancements in buff and draft testing techniques. In: Proceedings of IEEE/ASME joint railroad conference, Pittsburgh, pp 115–119
5. Chen O (1998) Dynamic simulation of Taipei EMU train. Veh Syst Dyn 30:143–167
6. McClanachan M, Cole C, Roach D, Scown B (1999) An investigation of the effect of bogie and wagon pitch associated with longitudinal train dynamics. Veh Syst Dyn 33:374–385
7. Cole C, Sun YQ (2006) Simulated comparisons of wagon coupler systems in heavy haul trains. Proc Inst Mech Eng Part F: J Rail Rapid Transit 220:247–256
8. Nasr A, Mohammadi S (2010) The effects of train brake delay time on in-train forces. Proc Inst Mech Eng Part F: J Rail Rapid Transit 224:523–534
9. Bedri R, Al-Nais MO (2005) Prestressed modal analysis using finite element package ANSYS. In: Li Z, Vulkov L, Wásniewski J (eds) Lecture Notes in Computer Science. Springer, New York, pp 171–175
10. Lu M, Geng H, Yang B, Yu L (2010) Finite element method for disc-rotor dynamic characteristics analysis of gas turbine rotor considering contact effects and rod preload. In: Proceedings of IEEE international conference on mechatronics automation, Xi'an, pp 1179–1183
11. Rao SS (2009) Mechanical vibrations. Pearson Education, New Delhi
12. Harak SS, Sharma SC, Harsha SP (2014) Structural dynamic analysis of freight railway wagon using finite element method. Proc Mater Sci 6:1891–1898
13. Indian Railways Schedule of Technical Requirement No. 55-BD-90 for rubber springs of high capacity draft gear for centre buffer couplers of BG Bogie Wagons having 625.5 mm draft gear pocket, Amendment No. 3, September 1993, pp 1–11

27

Preparation of Nb_3Al superconductor by powder metallurgy and effect of mechanical alloying on the phase formation

Zhao Liu · Yongliang Chen · Lupeng Du · Pingyuan Li · Yajing Cui · Xifeng Pan · Guo Yan

Abstract Adoption of powder-in-tube method to fabricate superconducting wire can realize a large application of Nb_3Al prepared by powder metallurgy. Powder metallurgy was used to synthesize Nb_3Al under various heat-treatment conditions, annealing temperature was varied from 700 to 1,000 °C and heating time was varied from 10 to 50 h. X-ray diffraction patterns reveal that a reaction between Nb and Al took place and formed $NbAl_3$ phase. Under current heat-treatment conditions (annealing temperature was varied from 700 to 1,000 °C and heating time was varied from 10 to 50 h), $NbAl_3$ was so stable that it did not further react with the unreacted Nb and was not sensitive to the heat-treatment condition. By mechanical alloying, adoption of high-energy ball milling significantly decreases particle size and enhances surface free energy, which promotes the formation of Nb_3Al phase. X-ray diffraction patterns indicate that relatively pure Nb_3Al phase was obtained under the same heat-treatment condition. Energy-dispersive X-ray analysis measurement demonstrates that the obtained samples were close to the right stoichiometry of A15 structure Nb_3Al.

Z. Liu · Y. Chen (✉) · L. Du · P. Li · Y. Cui
Key Laboratory of Magnetic Levitation Technologies and Maglev Trains, Ministry of Education of China, and Superconductivity and New Energy R&D Center, Southwest Jiaotong University, Chengdu 610031, People's Republic of China
e-mail: llchen@home.swjtu.edu.cn

X. Pan · G. Yan
Western Superconducting Technologies Co., Ltd., Xi'an 710018, People's Republic of China

Keywords Nb_3Al · Superconductivity · Powder metallurgy · Mechanical alloying

1 Introduction

With development of modern transportation and pursuit of high speed in various vehicles, superconducting maglev is promising in taking an important role in the future transportation system. Selection of a proper superconducting wire is vital in fabrication of superconducting magnet. Nb_3Al, due to its excellent superconducting properties, has drawn comprehensive attention since 1980s [1, 2]. A15 structure materials are famous for their superconductivities and Nb_3Al has the higher critical temperature among all the A15 structure materials [3, 4]. Furthermore, high value of critical current density (J_c) under high applied field makes it a promising substitution material for the fabrication of high-field magnet. The current existing fabrication method, powder-in-tube (PIT), makes it possible for a large-scale application of Nb_3Al prepared by powder metallurgy.

In International Thermonuclear Experimental Reactor (ITER) [5], compared with the well-developed Nb_3Sn, Nb_3Al possesses better stress/strain tolerance. This makes it an ideal choice for the fabrication of the toroidal field (TF) coil in terms of the large size of the D-shape TF [6, 7]. At the moment, the relatively successful methods to prepare Nb_3Al wire are restricted to high-temperature methods, because of the difficulty of the phase formation. Therefore, in order to acquire Nb_3Al in the right stoichiometry, high-temperature generation specialized instruments are needed, such as RHQT [8]. In this work, powder metallurgy was used to synthesize Nb_3Al and the effect of mechanical alloying on the phase formation was investigated.

2 Methods

The experimental process includes into two parts. In the first part, normal hand grinding method was used to mill the Nb, Al powder. The mole ratio of Nb/Al = 3/1 was weighed, placed in an agate mortar and pestle, and mixed by gentle hand grinding for 0.5 h. The mixtures were pressed into round pellet with a diameter of 11.5 mm and a thickness of 1.5 mm using hardened steel die with uniaxial 12 MPa pressure. Then, the pellet was sealed into a vacuum-fused quartz tube with a vacuum degree of 5.5×10^{-3} Pa. Finally, the samples were heat treated in a tube furnace and argon was vented to avoid permeability or leakage.

In the second part, mechanical alloying was adopted and a SPEX [9] mixer 8000 high-energy mill was used. All other experimental procedures were the same.

The crystal structure was studied by powder X-ray diffraction (XRD) using an X'Pert MRD diffractometer with Cu $K\alpha$ radiation. All observed reflections were indexed. Lattice constants were determined from LeBail refinements. Microstructure and composition of the sample were analyzed using a field emission scanning electron microscope (FESEM) equipped with an energy-dispersive X-ray analysis (EDX). DC magnetization was measured with a SQUID magnetometer (MPMS, Quantum Design) and resistivity measurements were performed with a physical property measurement system (PPMS, Quantum Design).

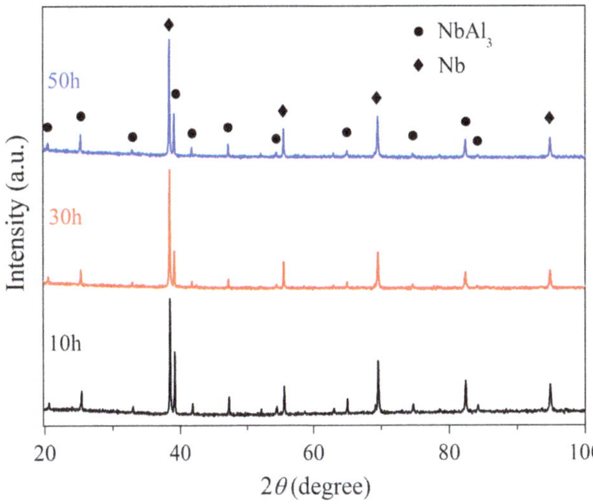

Fig. 1 XRD patterns of samples sintered at 800 °C

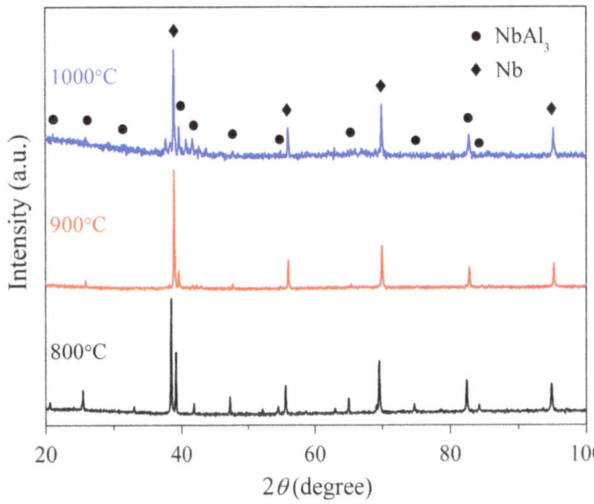

Fig. 2 XRD pattern of samples sintered at 10 h for different temperatures

3 Results and discussion

Figure 1 shows the XRD patterns of Nb/Al reaction samples sintered at 800 °C for different times. It can be observed that the main phases can be indexed as $NbAl_3$ and Nb. According to the phase diagram, Nb is inferred as the α-Nb which is the Nb-rich solid solution of aluminum in niobium. Figure 2 shows the XRD patterns of the samples sintered for 10 h at different sintering temperatures. Compared with the data of Fig. 1, it is obvious that both the changes of temperature and extension of heating duration make no difference to the resultant phase formation. Figure 2 shows the XRD patterns of the samples sintered for 10 h at different sintering temperatures. By comparison, it is apparent that either the change of temperature or extension of heating duration makes no difference to the resultant phase formation.

Thus, it is concluded that the reaction between Nb and Al results in the formation of $NbAl_3$ under the selected heat-treatment conditions. Furthermore, $NbAl_3$ is the most thermodynamically stable of the Nb–Al intermediate phases and so the kinetics of conversion to Nb_3Al by reaction with Nb can be expected to be slog. Consequently, it is concluded that, in the range of 800–1,000 °C, the temperature and duration of heat treatment are not the controlling factors of the formation of Nb_3Al.

The SEM image of sample sintered at 800 °C for 50 h is displayed in Fig. 3. It can be seen from the picture that the sample is very porous and the size of the biggest grain in this field of view is more than 10 μm. The very coarse morphology of the sample would give the diffusion reaction a very long diffusion distance. Considering the very high formation enthalpy [9] of Nb_3Al, the long diffusion distance between the two grains makes it impossible for the formation of Nb_3Al using hand milling. The EDX data on grains (see Table 1) support the conclusion that the big grains are $NbAl_3$.

Figure 4 shows the XRD patterns of Nb_3Al samples prepared for different spex milling and sintering times at

Fig. 3 SEM image for the sample sintered at 800 °C for 50 h

Table 1 EDX results of a random point on the grain

Element	Weight (%)	Atom (%)
Al	44.0	73.0
Nb	56.0	27.0

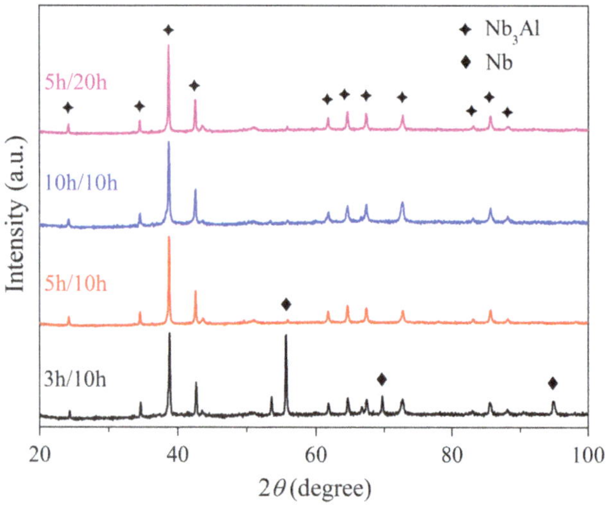

Fig. 4 XRD patterns of the samples prepared for different milling and sintering times at 900 °C

900 °C. The main phase can be indexed as the A15 structure Nb_3Al. The high-purity Nb_3Al sample was obtained when the spex milling time reached to 5 h, and spex milling for 10 h and sintering for 10 h gave the optimal sample. The sample that was spex milled for only 3 h contained a great quantity of Nb. This can be explained by short spex milling time; the Nb/Al powders lacked enough extra driving forces for diffusion reaction. Furthermore, it can be seen from all the four patterns that no $NbAl_3$ phase was found, thus indicating that spex milling is effective in increasing the activity of the two powders by improving the mixing and reducing the diffusion distance.

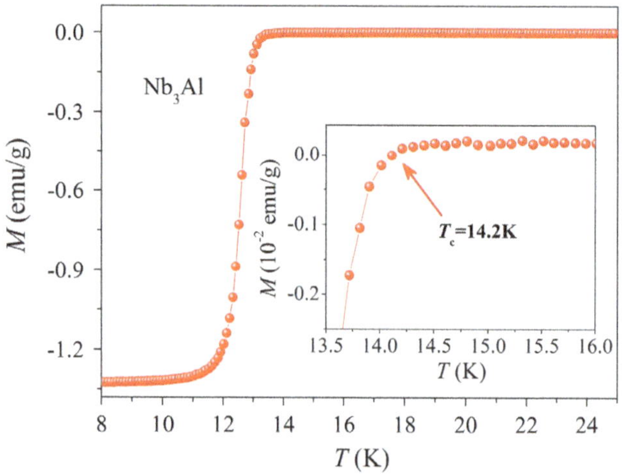

Fig. 5 The temperature dependence of magnetization for the Nb_3Al sample, and *inset* shows the enlarged view of the superconducting transition

Figure 5 shows the temperature dependence of magnetization for the sample spex milled for 5 h and sintered for 20 h, and the critical temperature (T_{conset}) is 14.2 K which is 4 K lower than the optimal value. The impurity phases and the porous microstructure may give rise to the poor superconducting property, although little impurity phase was indexed from the XRD patterns. It should be noted that no superconducting transition of Nb was detected, which further confirms that the Nb has sufficiently reacted. However, it is important to realize that the Nb_3Al single phase range is relatively wide according to the phase diagram. In addition, the heat-treatment temperature selected in this work can only give an Nb_3Al sample with the stoichiometry of only 20 at.% Al. Therefore, the deviation from the right stoichiometric concentration of Aluminum may be another reason that accounts for the low T_c of the sample, though the sample is an almost-pure single phase as the XRD reflects.

4 Conclusions

1. The composition of Nb/Al powder used in the experiment is not in the Nb_3Al solid solution homogeneity region in phase diagram of Nb–Al binary alloy, but the as-grown sample appears to be single phase.
2. 800 °C is not adequate in terms of kinetics for the equilibration of Nb_3Al phase formation.
3. Spex milling is effective in increasing activity of the powders by improving the mixing of the powders, simultaneously reducing the diffusion distance. By adoption of spex milling and the conventional PIT

method, fabrication of Nb_3Al magnet under low temperature can be realized.

4. The impurities and the deviation of the right stoichiometric concentration of aluminum in Nb_3Al are probably responsible for the poor quality of superconducting transition.
5. Combination of spex milling and the PIT method may be one of the solutions to prepare Nb_3Al wire under low temperature in the future.

Acknowledgments This work was supported by the National Magnetic Confinement Fusion Science Program (Grant No. 2011 GB112001), the Program of International S&T Cooperation (Grant No. 2013DFA51050), the National Natural Science Foundation of China (Grant Nos. 11104224, 11004162, 51377138, and 51302224), the Research Fund for the Doctoral Program of Higher Education of China (Grant No. 20110184120029), the Fundamental Research Funds for the Central Universities (Grant Nos. 2682013ZT16, SWJTU11ZT31, 2682013CX004, and SWJTU11BR063), and the Science Foundation of Sichuan Province (Grant Nos. 2011JY0031 and 2011JY0130).

References

1. Yamada Y, Ayai N (1999) Development of Nb_3Al superconductors for International Thermonuclear Experimental Reactor (ITER). Cryogenics 39:115–122
2. Koizumi N, Takeuchi T, Okuno K (2005) Development of advanced Nb_3Al superconductors for a fusion demo plant. Nucl Fusion 45:431–438
3. Takeuchi T (2000) Nb_3Al conductors for high-field applications. Supercond Sci Technol 13:R101–R119
4. Okuno K, Shikov A, Koizumi N (2004) Superconducting magnet system in a fusion reactor. J Nucl Mater 329–333:141–147
5. Seeber B (2007) Critical current of a Nb_3Sn bronze route conductor under uniaxial tensile and transverse compressive stress. Supercond Sci Technol 20:S184–S188
6. Ando T (1996) Design consideration of the ITER-TF coil with a react-and-wind technique using Nb_3Al conductor. IEEE Trans Magn 32:2324–2327
7. Fukuzaki T, Takeuchi T, Banno N (2002) Effect of rapid heating, quenching and transformation conditions on the superconducting properties and microstructure of Jelly-Roll processed Nb_3Al superconductors. Supercond Sci Technol 15:1404–1409
8. Koizumi N, Ando T, Sugimoto M et al (1998) Development of a 13-T and 40-kA Nb_3Al conductor for toroidal coils of fusion reactors. Fusion Eng Des 41:277–281
9. Calka A, Wexler D, Zhou J et al (1998) Mechano-synthesis of niobium–aluminum superconductors for the high field applications. Mater Sci Forum 269–272:1043–1048

28

Obtaining vehicle parameters from bridge dynamic response: a combined semi-analytical and particle filtering approach

R. Lalthlamuana · S. Talukdar

Abstract Dynamic load imposed on the bridge by moving vehicle depends on several bridge–vehicle parameters with various uncertainties. In the present paper, particle filter technique based on conditional probability has been used to identify vehicle mass, suspension stiffness, and damping including tyre parameters from simulated bridge accelerations at different locations. A closed-form expression is derived to generate independent response samples for the idealized bridge–vehicle coupled system considering spatially non-homogeneous pavement unevenness. Thereafter, it is interfaced with the iterative process of particle filtering algorithm. The generated response samples are contaminated by adding artificial noise in order to reflect field condition. The mean acceleration time history is utilized in particle filtering technique. The vehicle-imposed dynamic load is reconstructed with the identified parameters and compared with the simulated results. The present identification technique is examined in the presence of different levels of artificial noise with bridge response simulated at different locations. The effect of vehicle velocity, bridge surface roughness, and choice of prior probability density parameters on the efficiency of the method is discussed.

Keywords Dynamic load · Particle filter · Forward solution · Spatially non-homogeneous · Conditional probability

R. Lalthlamuana · S. Talukdar (✉)
Department of Civil Engineering, Indian Institute of Technology Guwahati, Guwahati 781039, India
e-mail: staluk@iitg.ernet.in

1 Introduction

Every bridge has certain restriction for the vehicle load and length. When the limit exceeds, permit has to be sought from the competent authority to pass the vehicle through the bridge. Weigh bridges are installed at important sections of highways to restrict the overloaded vehicles to enter the bridge. Presently, weigh-in-motion system in use can estimate the axle loads. But it incurs high cost of installation and maintenance. Accuracy is also affected by the speed of vehicle and unevenness of the pavement. It is to be mentioned that vehicle and bridge are integral parts of transportation system. The performance of one is affected by the performance of other and vice versa. Their behavior is coupled due to forces at contact points. For given structural configuration, construction materials, and road surface condition, the physical parameters of vehicle also play a significant role in bridge dynamic behavior. Traditionally, the bridge design load is calculated by magnifying static live load with impact factor. Days are not far when complete moving load time history would be necessary to check the design of long-span bridges. Recognizing the practical significance of the research on moving load identification, efforts have been made to estimate vehicle load from the bridge dynamic response using various techniques to improve the accuracy.

For the determination of axle load, weigh-in-motion system using instrumented bridge was developed by Moses [1]. Clayton and Peter [2] investigated truck weights from the perspective of regulatory limits. Accurate knowledge of dynamic forces acting on the built-up bridges is important to know the remaining service life. The interaction of moving vehicles with bridge has attracted the attention of many researchers for predicting their behavior by analytical and numerical techniques. However, it is difficult to measure the interaction force directly as their dynamics are coupled with both temporal and spatial variation. Theoretical model of bridge–vehicle interaction was developed by Green and Cebon [3] and Yang and Yau [4] within the limitations of linear dynamics. The determination of vehicle parameters from bridge response measurement requires a development of forward and an inverse scheme. The inverse solution is usually iterative in nature, whereas forward scheme may adopt an analytical or numerical approach depending on the bridge–vehicle model and degree of complexity involved. Yu and Chan [5] reviewed the recent works on moving load identification on bridges. Connor and Chan [6] employed least square method to estimate equivalent static load and their dynamic variation with time based on bridge response measurement. Vehicle–bridge interaction was ignored in the system model. An interpretive method was developed by Law et al. [7] where bridge has been modeled as an assembly of lumped masses, and Chan et al. [8] improved the model using Euler Bernouli continuous system. He also conducted laboratory experiments for the identification of moving mass from measured strain [9]. Later on, Law and Fang [10] proposed a theoretical optimal state estimation with the use of dynamic programming, by which moving load could be identified, overcoming the difficulties of ill conditioning of state matrix in time and frequency domain approach mentioned in Refs [11, 12]. Moving load identification in multi-span beams was also reported in Refs. [13, 14], where the effect of noise, number of vibration modes, and effect of support flexibility for non-rigid bearings were considered. Recently, Wu and Law [15] adopted stochastic finite-element-based method for moving load identification and further, they presented a new approach of vehicle axle load identification using Karhunen–Loeve expansion of stochastic process with irregular road surface [16]. In the last two decades, system identification using various numerical and experimental techniques has been tried in the field of structural engineering. Ghanem and Shinozuka [17] reviewed different identification algorithms and studied their applications to structures subjected to earthquake excitation. Shinozuka and Ghanem [18] verified different identification algorithms based on data obtained during controlled experiments on physical model. Deng and Cai [19] used genetic algorithm to identify parameters of vehicles moving on bridges. Development of Bayesian state estimation methodologies has added a new dimension in system identification involving various uncertainties [20]. Most important Bayesian estimation is Kalman filtering which is applicable to linear models and Gaussian type of uncertainties, and can be regarded as the stepping stone for the development of particle filtering method. In the last decade, the particle filter method based on conditional probability theory has attracted many researchers in the field of communication engineering, robotics, image processing, and ecology to estimate state of the system and hidden parameters from noisy signals/data [21–23]. Weerts et al. [24] applied particle filtering for state updating in rainfall–runoff models in hydrological applications. Schon et al. [25] reported on the computational complexity that increases with state dimension and suggested a marginalization technique to improve particle filtering which was then successfully used to an integrated navigation system of Swedish aircraft. Chatzi and Smyth [26] considered sensor heterogeneity for system identification of three degree of freedom system using unscented Kalman filter and particle filter technique. Narsella and Manohar [27] employed finite-element-based particle filter method for system identification with multiple sensor data for structural health monitoring purpose. Use of particle filter technique to estimate damage in vibrating beams without prior information of undamaged state has been documented by Pokale and Gupta [28]. Ching et al. [29] and Nasrella and Manohar [30] applied a particle filter method for state as well as system parameter estimation of dynamic systems implementing finite-element model in forward scheme with deterministic excitation. It was concluded that particle filter method can provide consistent result for non-linear models also.

Literatures available on the identification of moving load parameters on bridges using particle filter technique are scanty. Further researches are necessary to establish the popularity of the particle filter technique in vehicle–bridge coupled dynamics in the presence of stochasticity of contact forces. It was realized by researchers that particle filtering technique is computationally expensive, as a large number of response samples are required to obtain converged results. Although particle filter technique may take care of model inaccuracies, forward solution of mathematical model of the dynamic system with iterative or numerical schemes consumes lot of time to detect hidden parameters in the signal. In view of this fact, it would be advantageous if an idealized but quite accurate physical model is evolved which enables closed-form solutions to be interfaced with the iterative process of the particle filtering technique. In the present paper, a forward solution of the bridge–vehicle coupled system for non-homogeneous pavement input has been obtained in closed form to enable rapid generation of samples, required in the iterative algorithm to overcome the demerits of the existing particle filter method. Since accelerometers are the most practically used sensors, the simulated acceleration time history has been used to illustrate the approach with idealized model of bridge and vehicle. It may be noted that most of the authors limited their study to the estimation of gross vehicle weight from the bridge dynamic response. The effect of vehicle suspensions on the bridge dynamic response has been recognized by Green et al. [31], where they observed that use of soft and well-damped suspension produces smaller dynamic wheel loads on the bridge pavement. Moreover, estimation of vehicle suspension and tyre parameters, in indirect way, is also helpful to vehicle owners since a suitable maintenance policy may be worked out from the knowledge of current status. The present study formulates a coupled bridge–vehicle system using continuum approach, where parameters of bridge and vehicle in combined way influence the system output and, therefore, governs the moving force time history on the bridge. Particle filtering technique along with semi-analytical scheme has been used to estimate not only vehicle body mass but also tyre mass, suspension stiffness, suspension damping, tyre stiffness, and tyre damping. The identified vehicle parameters have been used to reconstruct interaction force time history and compared with the true value. The algorithm used in the present study along with semi-analytical method of sample generation has been tested by comparing the results with theoretical and experimental results available in published literatures. The identification technique used in the paper has also been examined in the presence of noise with response samples at different locations, vehicle velocity, and bridge surface roughness. The effect of range of the parameters to construct prior probability density function on the efficiency of the method is also discussed.

2 Bridge–vehicle dynamic interaction

2.1 Bridge–vehicle system equations

In particle filtering technique, it is necessary that forward problem of the bridge–vehicle model can be solved by adopting suitable scheme. The choice of model depends on the purpose of analysis, computational cost, and desired accuracy. Usually, transverse dimension of bridge deck is small compared to its span, and therefore a simplified beam model is a suitable option to establish a clear connection between bridge response and other influencing parameters [32], especially in case of implementing a computationally intensive method for the estimation of system parameters using large number of response samples in stochastic dynamics. The simpler model when tuned to fundamental natural frequency of the real bridge is expected to predict dynamic behavior similar to real structure. In the present study, a single-span bridge has been idealized as Euler–Bernoulli beam of uniform cross section and damping properties. Vehicle has been modeled as a lumped sprung mass m_v, and the mass of wheel, tyre, and part of the suspension is referred to as the unsprung mass m_w. The characteristics of the vehicle suspension system have been assumed to be linear. The bridge–vehicle model is shown in Fig. 1.

All translatory motions are assumed to be positive in upward direction. The sprung mass m_v is subjected to heave motion z_1 in vertical direction, and the vertical displacement of unsprung mass m_w is subjected to z_2. The vehicle body and the unsprung mass are connected by suspension system comprising spring element of stiffness k_v and dashpot with damping constant c_v, respectively. Tyre stiffness and damping are k_t and c_t, respectively. It may be noted that the vertical degrees of freedom z_1, z_2 and bridge deflection $y(x,t)$ are measured with reference to their respective static equilibrium position at any time instant t. The equation of motion for the vehicle is coupled with the bridge equation of motion through the interaction force existing at the contact point of the two systems.

The governing differential equations of motion of the two lumped masses can be written as

$$m_v\ddot{Z}_1(t) + c_v\{\dot{Z}_1(t) - \dot{Z}_2(t)\} + k_v\{Z_1(t) - Z_2(t)\} = 0, \tag{1}$$

$$\begin{aligned} m_w\ddot{Z}_2(t) &+ k_t\{Z_2(t) - y(x_c,t) - h(x_c)\} \\ &+ k_v\{Z_2(t) - Z_1(t)\} + c_v\{\dot{Z}_2(t) - \dot{Z}_1(t)\} \\ &+ c_t[\dot{Z}_2(t) - \{\dot{y}(x_c,t) + \dot{h}(x_c)\}] \\ &= 0, \end{aligned} \tag{2}$$

where $h(x)$ is the pavement roughness at a distance x from the reference station and x_c denotes the location of wheel contact form the same reference station. The governing equation of transverse motion of the beam can be written as

$$\mathrm{EI}\frac{\partial^4 y(x,t)}{\partial x^4} + m_b\frac{\partial^2 y(x,t)}{\partial t^2} + c_b\frac{\partial y(x,t)}{\partial t} = f_c(x,t), \tag{3}$$

where m_b, EI, and c_b represent the mass per unit length, flexural rigidity of bridge, and viscous damping per unit length of bridge, respectively. Assuming tyre remains in contact with the bridge at all times, interaction force ($f_c \geq 0$) in space and time variable can be expressed as

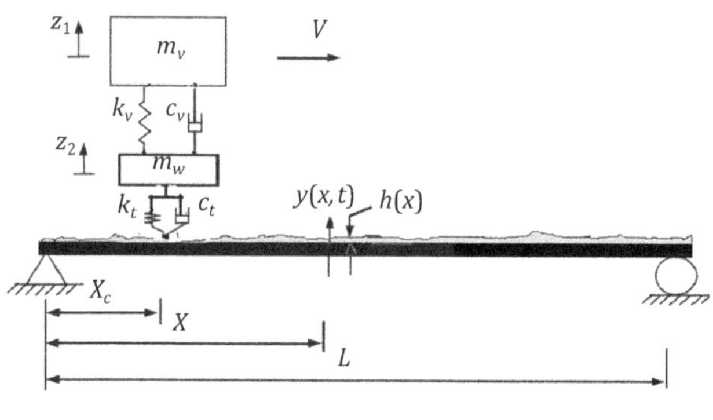

Fig. 1 Bridge-vehicle Model

$$f_c(x,t) = -[k_t\{z_2(t) - y(x,t) - h(x)\} + c_t\{\dot{z}_2(t) - \dot{y}(x,t) - \dot{h}(x)\}]\,\delta(x - x_c) - \{m_w + m_v\}g\delta(x - x_c), \tag{4}$$

where δ is the Dirac delta function having the following property

$$\int_{-\infty}^{\infty} f(x)\,\delta(x - x_c)\mathrm{d}x = f(x_c). \tag{5}$$

2.2 Bridge deck roughness

In the present study, we introduce a roughness, which is non-homogeneous in space even though vehicle velocity is constant, by adopting the following equation:

$$h(x) = h_m(x) + \sum_{s=1}^{N} \varsigma_s \cos(2\pi\,\Omega_s x + \theta_s), \tag{6}$$

where $h(x)$ is the deck surface unevenness which includes two parts. The first part $h_m(x)$ is a deterministic mean which may represent construction defects, pot holes, approach slab settlement, expansion joints, development of corrugation, etc. The second part of Eq. (6) is a Gaussian process [33] with a random phase angle θ_s uniformly distributed from 0 to 2π. N is the number of terms used to build up the road surface roughness,ς_s is the amplitude of cosine wave, Ω_s is the spatial frequency (rad/m) within the interval [Ω_L, Ω_U] where Ω_L, Ω_U are lower and upper cut-off frequencies of spatial unevenness, respectively. The parameters ς_s and Ω_s are computed as

$$\varsigma_s = \sqrt{2S(\Omega_s)\Delta\Omega}\ ;\Omega_s = \Omega_L + \left(s - \frac{1}{2}\right)\Delta\Omega;\ \Delta\Omega = \frac{(\Omega_U - \Omega_L)}{N}. \tag{7}$$

In Eq. (7), $S(.)$ is the power spectral density function (m^3/rad) at spatial frequency Ω_r of road surface roughness. The index 's' refers to a discrete point inside the frequency range Ω_L to Ω_U, where power spectral density of road roughness is to be known. In the present study, power spectral density function has been expressed as [34]

$$S(\Omega_s) = S(\Omega_0) \times \left(\frac{\Omega_s}{\Omega_0}\right)^{-2}, \tag{8}$$

where Ω_0 is referred as discontinuity frequency and is taken as $1/2\pi$ (rad/m).

On examination of Eq. (8), it is revealed that at very low spatial frequency ($\Omega_s \to 0$), the power spectral density becomes unbounded, i.e., $S(\Omega_s) \to \infty$. In view of this, Yin et al. [35] suggested an improved equation as follows:

$$S(\Omega_s) = S(\Omega_0) \times \frac{\Omega_0^2}{\Omega_s^2 + \Omega_L^2}. \tag{9}$$

The PSD function given by Eq. (9) has been adopted in the present study.

2.3 Discretization of bridge equations of motion

Using mode superposition technique, the bridge deflection in flexure has been shown by Meriovitch [36] as

$$y(x,t) = \sum_{k=1}^{\infty} \varphi_k(x) q_k(t), \tag{10}$$

where $\phi_k(x)$ is the flexural mode of the beam for simply supported boundary condition and q_k(t) is the generalized co-ordinates in the kth mode. The natural frequencies of simply supported bridge can be obtained as [36]

$$\omega_k = \sqrt{\frac{\mathrm{EI}}{m_b}\left(\frac{k\pi}{L}\right)^4}, \tag{11}$$

where ω_k is the bridge natural frequency in the kth mode, and EI, m_b, and L represent flexural rigidity, mass per unit length, and span of the bridge, respectively.

The mass normalized mode shape is given by [36]

$$\varphi_k(x) = \sqrt{\frac{2}{m_b L}} \sin\frac{k\pi x}{L}. \tag{12}$$

Now, substituting Eq. (10) in Eq. (3) and multiplying both sides of the equation by $\phi_j(x)$ and then integrating with respect to x from 0 to L with the use of orthogonality conditions, the equation of motion can be discretized in normal co-ordinates as

$$\ddot{q}_k(t) + 2\xi_k\omega_k\dot{q}_k(t) + \omega_k^2 q_k(t) = Q_k(t),\ (k = 1, 2, 3\ldots), \tag{13}$$

where ξ_k is the modal damping ratio in the kth mode. In practical application, the number of modes is to be limited to a finite size.

The generalized force (Q_k) of bridge in flexure [36] is given as

$$Q_k(t) = \frac{1}{M_k}\int_0^L f_c(x,t)\varphi_k(x)\mathrm{d}x, \tag{14}$$

where the term M_k is generalized mass in the kth mode [36] given by

$$M_k = \int_0^L m_b \varphi_k^2(x)\mathrm{d}x. \tag{15}$$

Substituting Eq. (4) in Eq. (14) and using property of Dirac delta function, the generalized force in the kth of mode of bridge transverse vibration has been worked out as

$$Q_k(t) = -\frac{1}{M_k}\left[k_t\left\{z_2(t) - \sum_{k=1}^{n_b}\varphi_k(x_c)q_k(t) - h(x_c)\right\}\varphi_k(x_c) + c_t\left\{\dot{z}_2(t) - \sum_{k=1}^{n_b}\varphi_k(x_c)\dot{q}_k(t) - Vh'(x_c)\right\}\varphi_k(x_c) - \{m_w + m_v\}g\varphi_k(x_c),\right. \tag{16}$$

in which g denotes acceleration due to gravity and ($'$) represents space derivative.

2.4 Solution of forward problem

The system Eqs. (1), (2), and (13) are coupled second-order ordinary differential equations. Theoretically, there is infinite number of modes in continuous system. However, for practical purpose, the number of modes has to be truncated to a finite size. Considering first 'n_b' flexural modes of the beam, the number of coupled equations then becomes $n = 2 + n_b$. The system equations can be expressed in matrix notation as

$$[\boldsymbol{M}]\{\boldsymbol{r}(t)\} + [\boldsymbol{C}(t)]\{\dot{\boldsymbol{r}}(t)\} + [\boldsymbol{K}(t)]\{\boldsymbol{r}(t)\} = \{\boldsymbol{F}(t)\}, \tag{17}$$

where $\{\boldsymbol{r}(t)\} = \{z_1(t), z_2(t)\ q_1(t), q_2(t)..., q_n(t)\}^T$ is the response vector, $\{\boldsymbol{F}(t)\}$ is the generalized stochastic force vector. $[\boldsymbol{M}]$, $[\boldsymbol{C}(t)]$, and $[\boldsymbol{K}(t)]$ are the system mass, damping, and stiffness matrix, respectively, of size $(2 + n_b) \times (2 + n_b)$. It may be noted that due to coupling of bridge–vehicle equations, the stiffness and damping matrix becomes time dependent due to change of wheel contact position on the bridge with time. The 'n' system equations are now recast into a $2n$-dimensional first-order state-space form [37] as given below:

$$\{\dot{\boldsymbol{P}}(t)\} + [\boldsymbol{A}(t)]\{\boldsymbol{p}(t)\} = \{\boldsymbol{P}(t)\}, \tag{18}$$

where

$$\{\boldsymbol{p}(t)\} = \begin{Bmatrix}\dot{\boldsymbol{r}}(t)\\ \boldsymbol{r}(t)\end{Bmatrix};\quad [\boldsymbol{A}(t)] = \begin{Bmatrix}\frac{[\boldsymbol{M}]^{-1}[\boldsymbol{C}(t)]}{-[\boldsymbol{I}]} & \frac{[\boldsymbol{M}]^{-1}[\boldsymbol{K}(t)]}{[\boldsymbol{0}]}\end{Bmatrix};\quad \{\boldsymbol{P}(t)\} = \begin{Bmatrix}\frac{[\boldsymbol{M}]^{-1}\{\boldsymbol{F}(t)\}}{\{\boldsymbol{0}\}}\end{Bmatrix}, \tag{19}$$

Here, $\{\boldsymbol{p}(t)\}$ is the state vector, $[\boldsymbol{A}(t)]$ is the state matrix, $[\boldsymbol{I}]$ is an identity matrix, $\{\boldsymbol{P}(t)\}$ is the augmented excitation vector, and [0] is a null vector or matrix. This form is suitable for bridge–vehicle interaction problems, since suspension damping is not small and diagonalization of damping matrix as in case of Rayleigh's damping may not be fully convincing. Let the eigenvalues of the state matrix $[\boldsymbol{A}(t)]$ be α_1, α_2, $\alpha_3...\alpha_{2n}$ and the corresponding complex conjugate eigenvectors be $\{u\}^1$, $\{u\}^2$, $\{u\}^3...\{u\}^{2n}$ Defining modal matrix $[\boldsymbol{U}(t)]$ and using linear transformation $\{\boldsymbol{p}(t)\} = [\boldsymbol{U}(t)]\{\boldsymbol{v}(t)\}$ in Eq. (19), along with orthogonality condition of the complex eigen vectors, the decoupled first-order system is given below:

$$\dot{v}_j(t) + \alpha_j v_j(t) = R_j(t), \qquad j = 1, 2, 3...2n, \tag{20}$$

where

$$R_j = \sum_{s=1}^{n} u'_{js}\sum_{k=1}^{n} m'_{sk}F_k(t), \tag{21}$$

where u'_{js} denotes the elements in the inverse of the matrix $[\boldsymbol{U}(t)]$ and m'_{sk} denotes the elements in the inverse of matrix $[\boldsymbol{M}]$. Using Fourier–Stieltjes transform [37], Eq. (21) can be written in frequency domain as

$$R_j(\omega) = \sum_{s=1}^{n} u'_{js}\sum_{k=1}^{n} m'_{sk}\int_{-\infty}^{\infty}\exp(-i\omega t)\,\mathrm{d}[F_k(\omega)] \tag{22}$$

The general solution of Eq. (20) may now be expressed as

$$v_j(t) = X_{0j}\exp(-\alpha_j t) + \int_{-\infty}^{\infty} H_j(\omega, t)R_j(\omega)\mathrm{d}\omega, \tag{23}$$

where X_{0j} are constants of integration to be determined from the initial conditions. $H_j(\omega,t)$ is the transient frequency response function given by [37]

$$H_j(\omega, t) = \frac{1}{-i\omega + \alpha_j}\exp(-i\omega t). \tag{24}$$

Using Eqs. (22), (24) in Eq. (23) and then utilizing linear transformation of generalized coordinate to physical coordinate, the original response vector $\{\boldsymbol{r}(t)\}$ may be expressed as

$$r_m(t) = \sum_{j=1}^{2n} u_{m+n,j}X_{0j}\exp(-\alpha_j t) + \sum_{j=1}^{2n} u_{m+n,j}\sum_{s=1}^{n} u'_{js}\sum_{k=1}^{n} m'_{sk}\int_{-\infty}^{\infty} F_k(\tau)\left[\frac{1}{2\pi}\int_{-\infty}^{\infty}\frac{\exp[i\omega(\tau - t)]}{-i\omega + \alpha_j}\mathrm{d}\omega\right]\mathrm{d}\tau,$$
$$m = 1, 2, 3......n. \tag{25}$$

The first term of Eq. (25) represents homogeneous solution of the system equation due to initial condition, and the second term of the equation represents particular solution due to imposed dynamic force.

Using Cauchy's residue theorem [38], the general solution of Eq. (25) now can be expressed in compact form

$$r_m(t) = \sum_{j=1}^{2n} u_{m+n,j} X_{0j} \exp(-\alpha_j t) + \sum_{j=1}^{2n} u_{m+n,j} \sum_{s=1}^{n} u'_{js} \sum_{k=1}^{n} m'_{sk} I_{jk}, \quad (26)$$

where

$$I_{jk} = \int_0^t \exp[\alpha_j(\tau - t)] F_k(\tau) \mathrm{d}\tau, \quad (27)$$

in which t is the bridge loading time defined as $t = x/V$, x being the distance traversed by the vehicle at instant t and V being the constant speed of the vehicle. When the vehicle is at the point of exit, then $t = L/V$ where L is the span of the bridge.

Closed-form expressions for the components of the above integral which generates each of the response samples have been developed and given in Appendix-1. The response samples thus form complete ensemble of the process. Averaging across the ensemble at each time step yields mean $\mu_Y(t_k)$ and standard deviation $\sigma_Y(t_k)$ of response process Y.

3 Identification of vehicle parameters

This paper presents the applicability of particle filter to identify the unknown vehicle parameters and provides an estimate of time-dependent moving load on the bridge. The main idea of this method is to estimate the unknown vehicles parameters from the available bridge response measurements. Since both the unknown parameters and the observation data are contaminated by noise, complete information of the parameters is possible if we can construct the probability density function of unknown parameters conditioned on the available bridge response measurement, called posterior probability density function. Vehicle parameters to be identified include sprung mass, unsprung mass, suspension stiffness, suspension damping, tyre stiffness, and tyre damping. Bridge parameters and the vehicle forward velocity are taken to be known.

The basic principle of particle filter method is to represent the required posterior density function of unknown vehicle parameters by a set of random samples (particles) with associated weights, and to compute the estimates based on these samples and weights. As the number of samples becomes very large, this Monte Carlo characterization becomes an equivalent representation of the posterior probability function, and the solution approaches the optimal Bayesian estimate [39]. The system states r_l are assumed to propagate according to the system equation

$$r_{l+1} = g_l(r_l; \Phi_l, \eta_l), \quad l = 0, 1, 2, 3 \ldots N_t, \quad (28)$$

where l represents the discretized time dimension and N_t is the number of time instants considered. $\Phi_l \in R^d$ is a d-dimensional vehicle parameters vector which is considered as constant, $r_l \in R^n$ is a n-dimensional vector denoting the state of the system, and a model noise $\eta_l \in R^m$ is the discretized m-dimensional vector of a sequence of independent and identically distributed random variables. The noise is assumed to be independent of past and current state with known probability density function. $g_l(.)$ is a system transition function.

When the system measurements become available, the system states are related to these measurements via the observation equation given below:

$$Z_l = f_l(r_l; \Phi_l, \zeta_l)\ ; \quad l = 0, 1, 2, 3 \ldots N_t, \quad (29)$$

where $Z_l \in R^p$ is a p-dimensional bridge response measurement vector, a measurement noise $\zeta_l \in R^s$ is a s-dimensional vector of a sequence of independent and identically distributed random variables, and $f_l(.)$ is a nonlinear function that relates the measurements to the system state.

Since, state of the system is dependent on system parameters $\{\boldsymbol{\Phi}\}$, observation equation can be rewritten as

$$Z_l = f_l(\Phi_l, \zeta_l), \quad l = 0, 1, 2, 3 \ldots N_t. \quad (30)$$

Vehicle parameters identification problem can now be considered as being equivalent to the determination of the posterior probability density function $p(\Phi_l|Z_l)$. According to Bayesian theorem, $p(\Phi_l|Z_l)$ can be written as [39]

$$p(\Phi_l|Z_l) = \frac{p(Z_l|\Phi_l)p(\Phi_l)}{\int p(Z_l|\Phi_l)p(\Phi_l)\mathrm{d}\Phi_l}, \quad (31)$$

where $p(\Phi_l|Z_l)$ is the posterior PDF, $p(Z_l|\Phi_l)$ is the likelihood of individual parameters, $p(\Phi)$ is the prior probability density function, and $\int p(Z_l|\Phi_l)p(\Phi_l)\mathrm{d}\Phi_l$ the normalizing parameter, respectively.

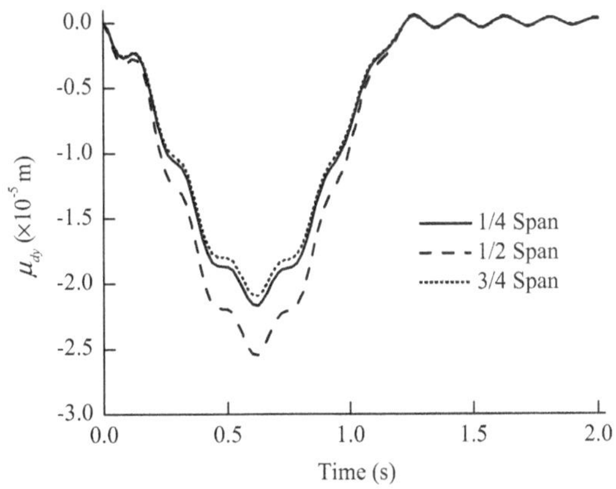

Fig. 2 Mean displacement of bridge at different locations

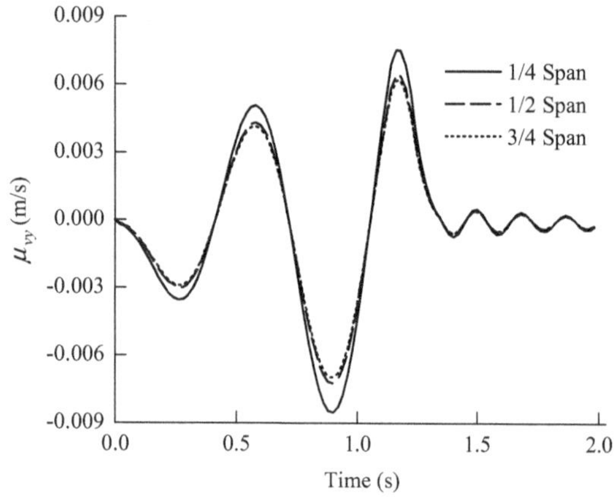

Fig. 3 Mean velocity of bridge at different locations

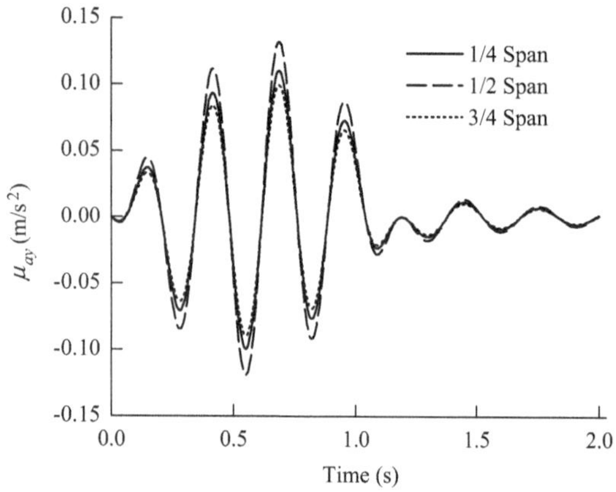

Fig. 4 Mean acceleration of bridge at different locations

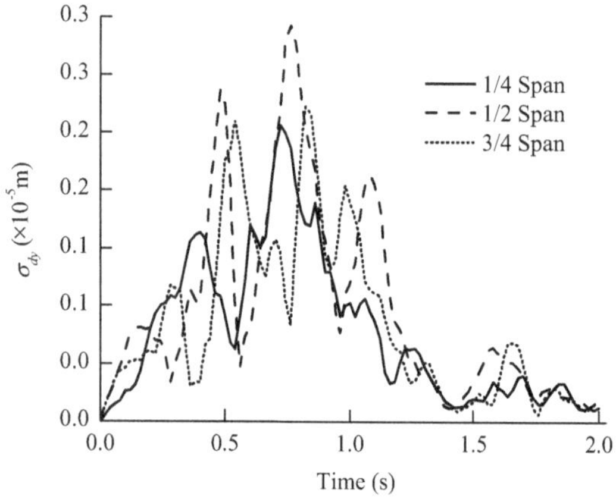

Fig. 5 Standard deviation of bridge displacement response at different locations

Thus knowing posterior PDF $p(\Phi_l\,|Z_l)$ first few moments of the vehicle parameters Φ_l, conditioned on bridge response measurement Z_l, at each time step can be determined. The particle filtering algorithm for identifying vehicle parameters has been implemented by the following principal operations [39]:

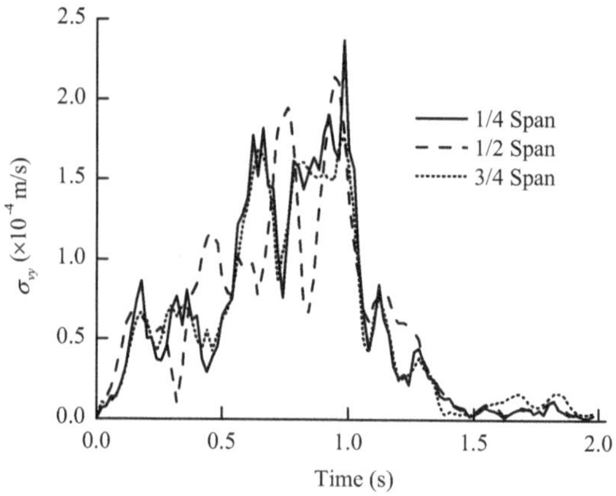

Fig. 6 Standard deviation of bridge velocity response at different locations

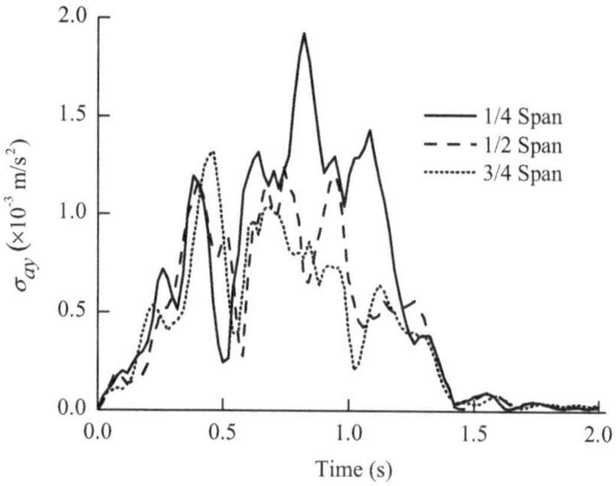

Fig. 7 Standard deviation of bridge acceleration at different locations

(i) *Prediction:* Draw N_p random samples of vehicle parameters $\{\Phi_{0j}\}_{l=1}^{N_l}$ from the assumed PDF $\{p(\Phi_{0l})\}_{l=1}^{N_l}$.

(ii) Forward Solution: Determine bridge response using present closed-form expressions at time step l. If N_l is sufficiently large, these estimates are approximately distributed as $\{p(f_l\,[\Phi_{0l}]|Z_l)\}_{l=1}^{N_l}$.

(iii) *Updating:* Once the measurements are available for different measurement locations ($i = 1,2,3\ldots.N_m$) at time l, evaluate the likelihood corresponding to all the samples. This implies that one needs to evaluate $\{p(Z_l|f_l\,[\Phi_{0l}])\}_{l=1}^{N_l}$.

For the lth measurement, calculate the weighting function as

$$w_l = \frac{p(Z_l|\Phi_{il}^*)}{\sum_{j=1}^{N_p} p(Z_l|\Phi_{il}^*)}. \tag{32}$$

The discrete mass probability function for the next iteration is defined as

$$P(\Phi_{lj} = f_l\,[\Phi_{0j}]) = w_j. \tag{33}$$

Table 1 Comparison of the estimate of gross vehicle weight with published results [41]

Vehicle speed (m/s)	Percentage error in estimation of gross vehicle weight (%)					
	Class B (good condition)		Class C (average condition)		Class D (poor condition)	
	Law et al. [41]	Present study	Law et al. [41]	Present study	Law et al. [41]	Present study
10	11.57	3.18	13.71	5.82	23.49	8.95
15	11.55	4.13	15.35	7.38	25.46	11.51
20	12.46	4.96	16.73	8.94	27.79	13.03
25	13.45	8.55	19.47	13.33	30.01	16.22

(iv) *Resampling*: From the discrete mass distribution function, a new set of N_p samples of Φ_{lj} is generated. This constitutes the posterior estimates of Φ_{lj}. The mean of estimates is obtained by averaging across the ensemble.

Set $l = l+1$ and if $l < N_t$, then go to the step (ii) and repeat other steps, otherwise stop. In this way, the filtering is carried out for the entire available time history of measurements.

4 Results and discussions

In this study, particular attention is given to examine the applicability of particle filter method in identifying vehicle parameters (sprung mass, unsprung mass, suspension stiffness, suspension damping, tyre stiffness, and tyre damping) on measured bridge dynamic response. Since no physical experiments have been undertaken, the measured response samples have been synthetically generated using the present analytical expression with artificial noise added to it to mimic field data. The ends of the bridge are simply supported. The following data for bridge and vehicle have been assumed to simulate measured response:

Bridge span $(L) = 20$ m; mass $(m_b) = 11.15 \times 10$ kg/m; flexural rigidity $(EI) = 1.695 \times 10^{10}$ N-m^2; Speed range in which vehicle movement is considered is 40–80 km/h; Vehicle mass $(m_v) = 18{,}000$ kg; wheel mass $(m_w) = 1{,}500$ kg; suspension stiffness $(k_v) = 3.6 \times 10^7$ N/m; Suspension damping $(c_v) = 7.2 \times 10^4$ N-sec/m; tyre stiffness $(k_t) = 0.9 \times 10^7$ N/m; and tyre damping $(c_t) = 0.7 \times 10^4$ N-sec/m. For modeling deck surface roughness, the values of spectral roughness coefficient (ς) have been taken as 2×10^{-6}–18×10^{-6} m^2/(m/cycle) according to International Organization for Standardization (ISO-8608) specifications for the class of different road conditions [40]. The lower and upper limits of the spatial frequencies of the road profile are taken as $\omega_L = 0.01$ cycle/m and $\omega_U = 3$ cycle/m. The cut-off spatial frequencies are chosen in view of the practical size of tyre. The forward problem has been solved using closed-form solution given in

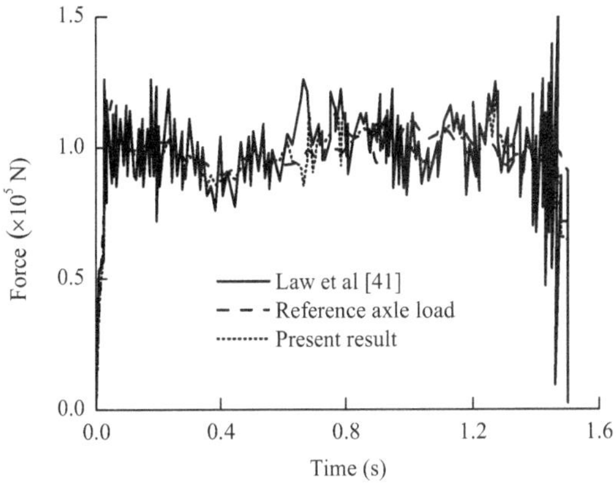

Fig. 8 Comparison of reconstructed dynamic axle load with published result in Ref. [41]

Sect. 2.4, and above data have been employed to generate bridge dynamic responses.

With the assumed parameters as above, time histories of response (displacement, velocity, and acceleration) at three different locations (at $L/4$, $L/2$, and $3L/4$) have been obtained, and the response time series are made corrupted by addition of artificial noise. In generating response samples to mimic measured data, vehicle parameters, bridge roughness profile and vehicle speed, have been taken as known quantities. The focus is on demonstrating whether the simulated (measured) data used as input to the particle filter be able to estimate the values of the vehicle parameters and to what precision. With identified vehicle parameters and known bridge sectional and material properties, state has been estimated and used to reconstruct dynamic tyre force imposed on the bridge during movement of the vehicle.

4.1 Response statistics of bridge

Response statistics have been found by a semi-analytical solution of forward problem which is intended to assist particle filtering algorithm to identify vehicle parameters. We first present the mean displacement, velocity, and acceleration of the bridge at one-fourth, half, and three-

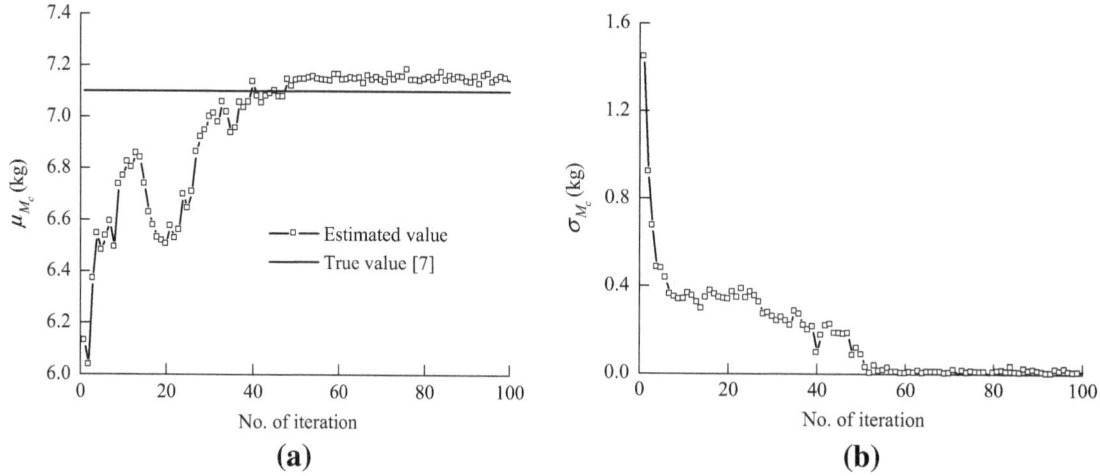

Fig. 9 Progressive estimate of model car mass using experimental data [7] and its comparison with true value

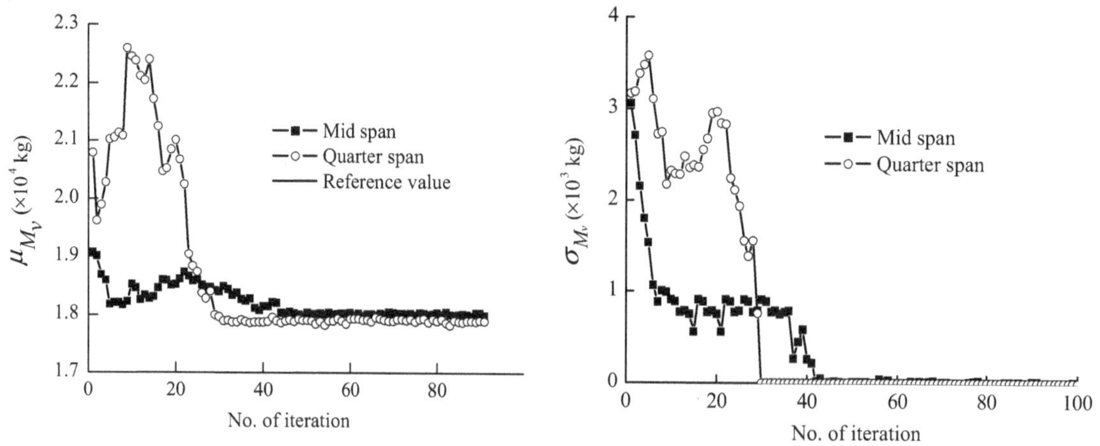

(a) Mean value of M_v

(b) Standard deviation of M_v from acceleration data at different locations

Fig. 10 Progressive estimate of vehicle mass

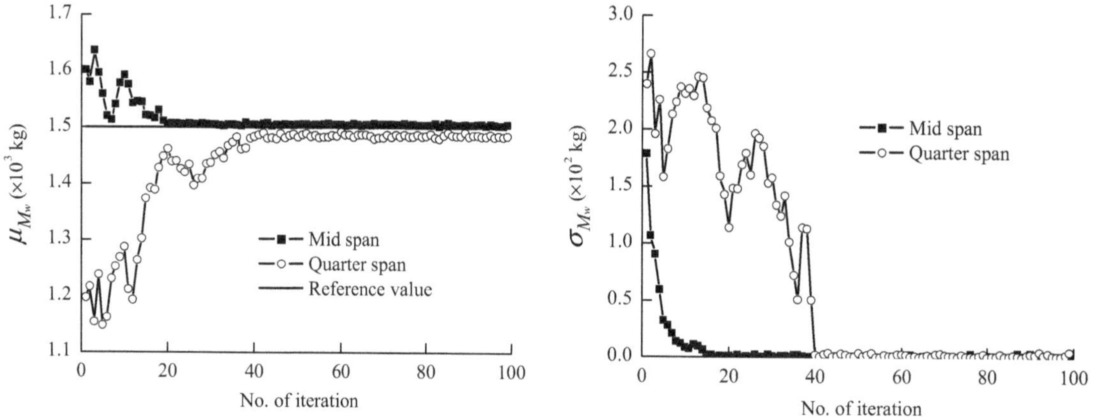

(a) Mean value of M_w

(b) Standard deviation of M_w from acceleration data at different locations

Fig. 11 Progressive estimate of wheel mass

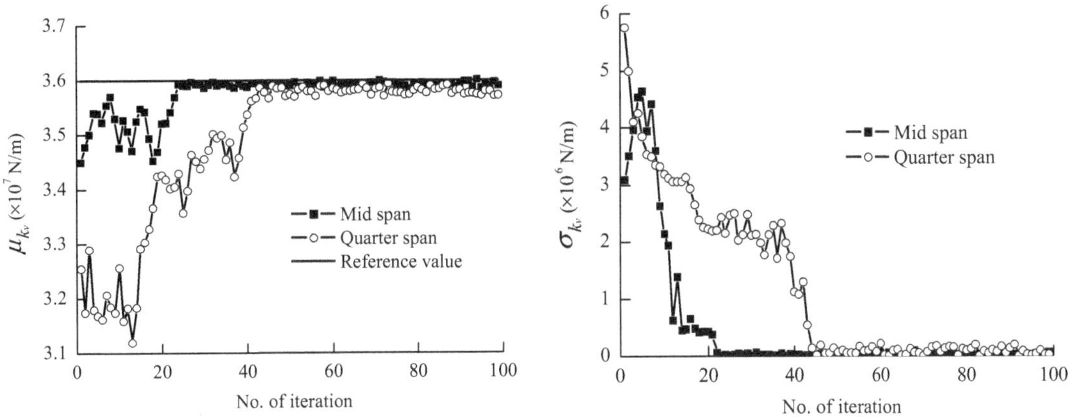

(a) Mean value of k_v

(b) Standard deviation of k_v from acceleration data at different locations

Fig. 12 Progressive estimate of vehicle suspension stiffness

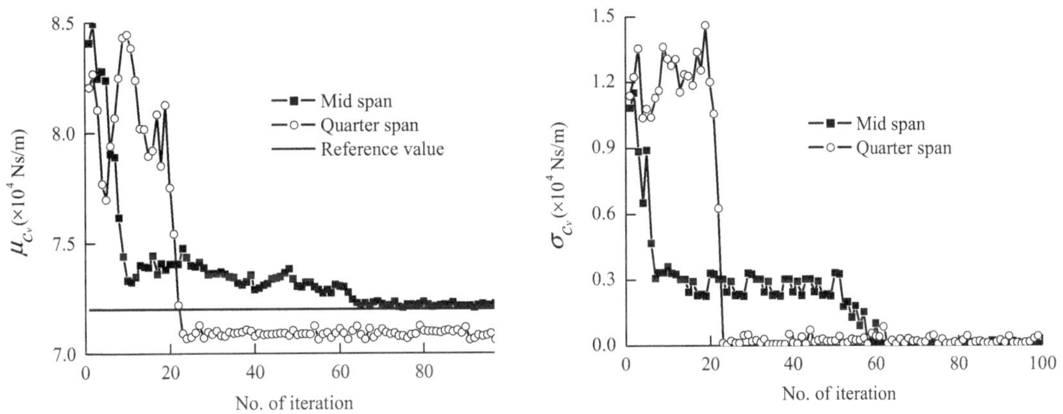

(a) Mean value of c_v

(b) Standard deviation of c_v from acceleration data at different locations

Fig. 13 Progressive estimate of vehicle suspension damping

fourth of span (Figs. 2, 3, and 4). Corresponding standard deviations are shown in Figs. 5, 6, and 7, respectively. Vehicle velocity is taken as 60 km/h (16.67 m/sec). These results are obtained from the present analytical formulations laid down in Sect. 2. No noise has been added at this stage. However, measured data will be considered in the particle filter algorithm by adding different levels of noise in the simulated response sample using closed-form expressions.

The mean quantities of mid-span response is larger compared to other span locations. Due to variation of roughness in random manner, the response magnitudes at other points are also different. The frequency of oscillation for the mean response at different locations shown in Figs. 2, 3, and 4 are not varying as the vehicle wheel is excited by same frequency due to uniform velocity considered in the presentation of the results here. Standard deviation values shown in Figs. 5 and 7 do not show any definite pattern of variation.

4.2 Comparison of vehicle load estimation with published results

Main focus of the present study is to estimate vehicle parameters from bridge dynamic response using particle filtering technique. Before conducting a parametric study for the vehicle parameters to be identified, a comparative study of the results obtained by particle filtering approach with published results obtained by different identification techniques has been carried out.

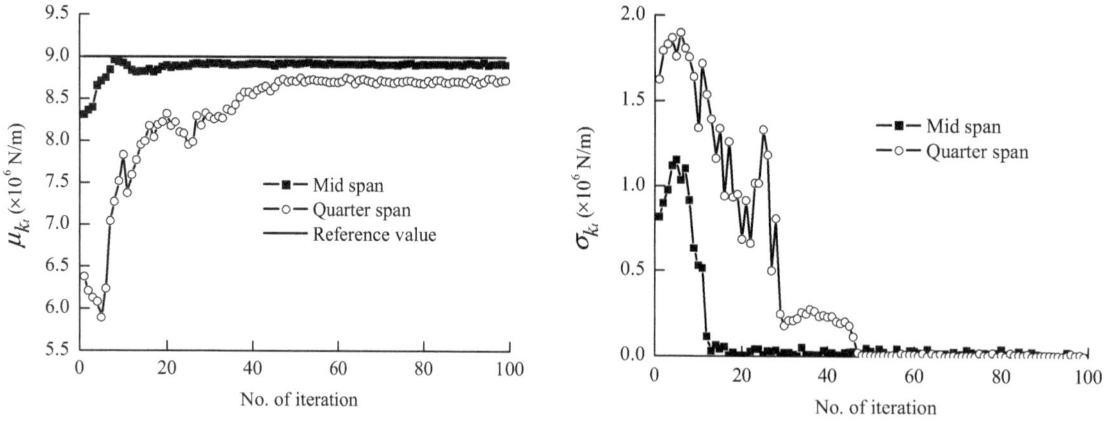

(a) Mean value of k_t

(b) Standard deviation of k_t from acceleration data at different locations

Fig. 14 Progressive estimate of tyre stiffness

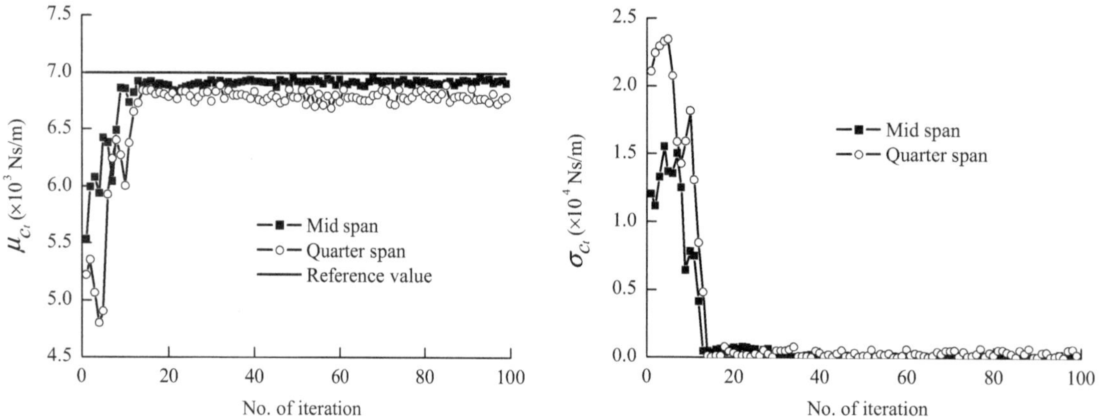

(a) Mean value of c_t

(b) Standard deviation of c_t from acceleration data at different locations

Fig. 15 Progressive estimate of tyre damping

Table 2 Effect of response measurement location on indentified vehicle parameters

Vehicle parameters	% Error		Vehicle parameters	% Error	
	Mid span	Quarter span		Mid span	Quarter span
m_v	5.07	8.53	k_t	4.33	9.35
m_w	2.20	9.29	c_v	5.30	7.69
k_v	4.72	7.17	c_t	6.24	8.78

4.2.1 Comparison with the results of numerical study

First, we present a comparative study with the method proposed by Law et al. [41] based on simulated data to judge the efficiency of particle filter approach for moving load identification. The loading on the bridge pavement used by them was a deterministic harmonic function composed of two different frequencies as given below:

$$f(t) = 121\ [1 - 0.1\sin(10\rho\pi t) + 0.05\sin(50\pi t)]\text{kN}. \tag{34}$$

The result of Ref. [41] was based on interpretive method. The comparison is shown in Table 1. It may be noted that percentage error increases with the vehicle speed. Besides, it has been observed that poor surface condition of the bridge produces more error in the estimation. The same patterns have been found even in the present method. However, particle filter method gives improved estimate of the gross axle load in all cases as compared to the interpretive method. Further, gross axle force time history has been reconstructed from the estimated vehicle parameters by considering same bridge surface condition and vehicle speed as assumed by Law et al. [41]. Comparison of reconstructed force time history with the assumed time varying axle load is shown in Fig. 8. It may be seen that high fluctuation of moving force about the reference loading has been produced at the end of time history in the results obtained by Law et al. [41]. This might be due to the fact that high-frequency component of random unevenness could not be properly filtered out with the sampling frequency considered in the earlier study. However, the high-frequency component could not disturb the estimation of moving load when particle filter technique has been used in the present study. This resulted higher accuracy in parameter estimation.

Table 3 Effect of different noise levels on the estimated vehicle parameters

Vehicle parameters	5 % noise		10 % noise	
	No. of iteration	Error (%)	No. of iteration	Error (%)
m_w	31	5.07	54	8.75
m_v	42	2.20	38	6.04
k_v	28	4.72	53	6.92
k_t	36	4.33	56	9.35
c_t	25	5.31	31	7.74
c_v	63	6.24	39	9.63

4.2.2 *Comparison with the results of experimental study*

The second comparative study has been done with the experimental data from Law et al. [7]. They conducted a laboratory experiment with a model car having a total mass (M_c) 7.1 kg being pulled at a speed of 3.102 m/s over a simply supported beam of span 3.376 m. The model bridge was a flat bar having cross-sectional dimension of 100 mm × 25 mm. The mass and flexural rigidity of the beam were 24.12 kg/m and 63.4 kNm^2, respectively. Other details of experiments can be found in Ref [7]. Accelerometers were mounted on the beam at 1/4, 1/2, and 3/4 span. Sampling frequency was 256 Hz. Acceleration record at 3/4th of the span experimentally obtained in Ref. [7] was

Table 4 Range of mass of vehicle and wheel mass to construct prior PDF

Range	Vehicle mass (m_v) kg		Wheel mass (m_w) kg	
	Case 1	Case 2	Case 1	Case 2
Φ_L	0.4×10^4	1.7×10^4	0.3×10^3	1.4×10^3
Φ_U	3.6×10^4	1.9×10^4	3.5×10^3	1.6×10^3

Table 5 Range of suspension stiffness and tyre stiffness to construct prior PDF

Range	Vehicle suspension stiffness (k_v) N/m		Wheel stiffness (k_t) N/m	
	Case 1	Case 2	Case 1	Case 2
Φ_L	0.3×10^7	2.8×10^7	5.2×10^6	0.6×10^7
Φ_U	8.5×10^7	4.0×10^7	8.5×10^7	1.1×10^8

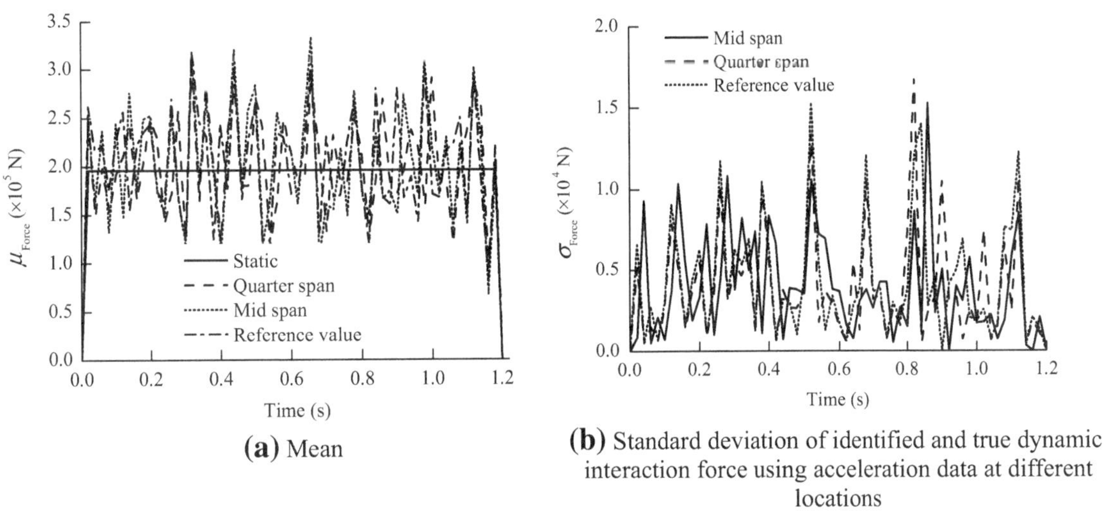

Fig. 16 Comparison of **a** mean and **b** standard deviation of identified and true dynamic interaction force using acceleration data at different locations

Table 6 Range of suspension damping and tyre damping to construct prior PDF

Range	Vehicle suspension damping (c_v) N-s/m		Wheel damping (c_t) N-s/m	
	Case 1	Case 2	Case 1	Case 2
Φ_L	1.5×10^4	6.5×10^4	5.2×10^3	0.5×10^4
Φ_U	12.3×10^4	8.3×10^4	6.4×10^4	0.9×10^4

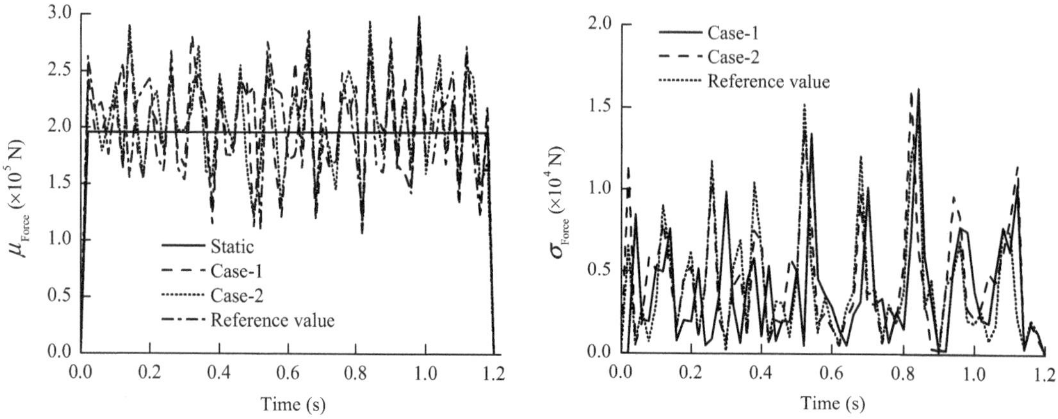

(a) Mean

(b) Standard deviation of identified and true dynamic interaction force for different range of values used to construct prior PDF $p(\Phi_0)$

Fig. 17 Comparison of **a** mean and **b** standard deviation of identified and true dynamic interaction force for different ranges of values used to construct prior PDF $p(\Phi 0)$

taken as measured data in particle filter method for estimating the mass of the model car. The progress of iteration with updated estimation as obtained in the present study is shown in Fig. 9. The result shows that the filtering process converges at the 57th of iteration to a mass value of 7.26 kg. The error has been estimated as 2.25 %. This demonstrates successful application of particle filter method for the estimation of moving mass when experimentally acquired data are utilized in the algorithm.

4.3 Influence of various factors on vehicle parameter identification

The particle filter method is now applied to estimate the unknown vehicle parameters which include sprung mass, unsprung mass, suspension stiffness, suspension damping, tyre mass, stiffness, and damping. Only acceleration response of the bridge has been used as in most of the practical situation, accelerometers are common type of sensors. The analytically computed time history has been contaminated by the addition of artificial noise to mimic field data. The mean and standard deviation values of the vehicle parameters are calculated at each stage of iteration at each time step of the synthetically generated time history. The progress of iteration and its convergence are presented in the subsequent sub-sections taking various factors into considerations.

4.3.1 Effect of bridge response measurement location

The bridge acceleration measurement at different locations along the span has been used as input to the particle filter algorithm. Vehicle parameters as well as dynamic interaction force are estimated. In both the cases, standard deviation approaches very low value at a certain number of iteration which implies that the algorithm has achieved convergence. The progress of estimation of mean and standard deviation has been displayed in the form of graphical plot of estimated parameters versus corresponding number of iterations (Figs. 10, 11, 12, 13, 14, 15). Result shows that measurement taken other than the mid span takes a longer iteration to achieve convergence. Percentage error has been calculated for different measurement inputs as shown in Table 2. The results are obtained when response samples from analytical model are contaminated by 5 % noise. The effect of noise level on the algorithm has been tested by adding 10 % noise also. Table 3 shows the percentage error as well as number of iteration required to converge, when noise level is increased. Result shows that mid-span response gives the

Table 7 Performance of the algorithm at different vehicle speeds (V)

Vehicle parameters	$V = 60$ km/h		$V = 80$ km/h	
	No. of iteration	Error (%)	No. of iteration	Error (%)
m_w	26	1.21	12	8.15
m_v	43	2.20	18	7.04
k_v	35	1.72	22	6.63
k_t	22	3.12	42	3.45
c_t	21	3.72	19	3.93
c_v	65	2.31	18	8.63

Table 8 Identification of mass of vehicle and wheel for different road conditions

Road condition	Mass of vehicle (kg)		Mass of wheel (kg)	
	No. of iteration	Error (%)	No. of iteration	Error (%)
Good	16	3.21	19	2.02
Average	38	3.59	25	3.97
Poor	49	6.06	43	5.95

best estimate for the vehicle parameters as well as dynamic interaction force identification. The reconstructed mean and standard deviation of dynamic force induced in the bridge with identified parameters is shown in Fig. 16. The result reveals that bridge response measurement at mid span gives around 2%–5 % error, while measurement other than mid span leads to 4%–10 % error.

4.3.2 *Effect of assumption of the prior PDF* $p(\Phi_0)$

In the absence of any information about the unknown parameters, it is assumed that the prior PDF $p(\Phi_0)$ is uniformly distributed within a range $[\Phi_L, \Phi_U]$. Two different cases have been considered to specify the range within which random particles are generated assuming uniform probability density function $p(\Phi_0)$.

Case-1: Keeping the lower and the upper limits with large variation from the true value.

Case-2: When the lower and upper limits are not widely apart from the actual value.

The ranges of values of the parameters assumed for the above two cases are mentioned in Tables 4, 5, 6. In these two cases, the number of particles $N_p = 1{,}000$ and artificial noise is taken to be 5 % of the simulated maximum bridge dynamic response. The mean and standard deviation are observed at each stage of iteration and stop when standard deviation becomes less than equal to tolerance. Identified vehicle–bridge interaction force is shown in Fig. 17 simultaneously comparing with the true value of dynamic interaction force. It has been found that a wrong choice of $p(\Phi_0)$ does not necessarily lead to wrong estimates by the particle filter identification method. However, a crude assumption of the prior probability density is found to consume longer time to achieve convergence. Assumption based on the first case of prior density function leads to 3%–5 % error, while the second case assumption gives 2%–4 % error.

Table 9 Identification of stiffness and damping of vehicle suspension for different road conditions

Road condition	Suspension stiffness (N/m)		Suspension damping (Ns/m)	
	No. of iteration	Error (%)	No. of iteration	Error (%)
Good	20	2.16	23	3.95
Average	25	2.72	44	4.31
Poor	43	4.02	72	6.14

Table 10 Identification of stiffness and damping of tyre for different road conditions

Road condition	Tyre stiffness (N/m)		Tyre damping (Ns/m)	
	No. of iteration	Error (%)	No. of iteration	Error (%)
Good	36	1.98	19	2.24
Average	41	2.33	44	4.59
Poor	85	8.02	68	10.27

4.3.3 *Effect of different vehicle velocities*

The identification algorithm has been examined from the measured response for different vehicle speeds over the bridge. The response samples have been generated at 60 km/h (16.67 m/s) and 80 km/h (22.20 m/s) of vehicle speed. The sampling time interval in measured response sample (after adding 5 % artificial noise) has been initially chosen as 0.02 s. This time interval was not satisfactory in case of vehicle moving at higher speed. The reason may be that vehicle leaves the bridge in a shorter time span generating less number of data points that are available in the working of a particle filter algorithm. Number of iteration required to get convergence and percentage error is tabulated in Table 7. It has been found that lower speed gives better estimate but it requires more number of iteration to achieve the convergence.

Table 11 Comparison of computer processing time for estimation of vehicle parameters

Parameters	Sampling frequency (Hz)	Analytically generated samples		Numerically generated samples	
		Processing time (s)	% error	Processing time (s)	% error
m_v	300	20.21	5.98	54.37	8.81
	500	21.47	5.07	73.66	6.22
	700	26.52	3.82	102.88	5.47
m_w	300	13.47	4.09	45.65	6.50
	500	19.79	2.20	77.16	4.13
	700	22.73	1.66	117.67	3.06
k_v	300	16.00	6.21	39.22	8.39
	500	20.63	4.72	63.66	5.64
	700	23.58	2.99	167.18	3.36
c_v	300	22.31	6.54	42.44	5.99
	500	25.26	5.30	115.74	5.85
	700	32.42	5.29	158.82	5.01
k_t	300	20.63	7.51	46.30	9.30
	500	23.16	4.33	57.23	6.07
	700	25.68	4.02	83.59	5.84
c_t	300	11.79	7.25	32.15	10.11
	500	12.63	6.24	52.73	7.79
	700	13.89	2.89	91.31	5.83

4.3.4 Effect of different roughness conditions

Bridge deck surface irregularity has been considered in the identification of vehicle parameters based on ISO specification [40] for different conditions—good, average, and poor. Among the three conditions, results show that good condition gives the best estimate with less number of iterations which is given in Tables 8, 9, 10. This may be attributed to the reason that noise effect in dynamic input for the case of rougher pavement increases, requiring more number of iterations for convergence. This is in conformity with the results obtained when artificial noise level increased from 5 % to 10 % as stated in Sect. 4.3.1.

4.4 Comparison of CPU processing time for estimation of vehicle parameters with numerically generated samples

The efficiency of the proposed analytical approach for the forward scheme in the identification of vehicle parameters has been judged by comparing the CPU processing time when numerically simulated samples are used. For numerical scheme, Newmark's method [42] has been adopted. Three different sampling frequencies 300, 500, and 700 Hz have been used to compare the convergence rate. A personal computer with Inte l(R) Core (TM) i3-2120 CPU 3.3 GHz and 4.0 GB RAM has been employed for all computations. Vehicle speed 60 km/h (16.67 m/s) and poor bridge deck surface condition [40] have been considered for the study. The results are presented in Table 11. It has been found that sampling frequency affects the estimation accuracy for both the method of solutions. Higher sampling frequency leads to lower error in vehicle parameters estimation as shown in Table 11. Since several parameters are estimated simultaneously, the CPU processing time is also different for each parameter. The overall time requirement for the parameter estimation has been found to be decreased by 27 %, while average accuracy has been increased by 24 % due to the use of present semi-analytical method in forward scheme of Particle Filtering approach. This again demonstrates the superiority of particle filter method when combined with analytical method for generations of response samples.

5 Conclusions

In the present study, particle filtering combined with a semi-analytical method has been outlined for the identification of vehicle parameters as it passes over a simply supported bridge. The deck surface roughness has been considered as a non-homogeneous random process in spatial domain. Comparison of the results obtained by present study using particle filter method with published results has shown higher accuracy. The dynamic interaction force time history has been reconstructed with the identified parameters and compared with true value. Effect of different measurement locations along the bridge span and artificially added noise has been investigated. The accuracy of the proposed method has been checked by considering two different cases of prior density function selection. A comparative study of computational time required in particle filter method using present semi-analytical scheme and numerical scheme has been performed. Some of the major findings and recommendations on the applicability of particle filter technique for vehicle parameter identification are given below:

(1) Response measurement location has greater influence on the accuracy and computational time required in application of particle filtering technique. For simply supported single-span bridge like the one being presented, mid-span sensor data may be the better option.

(2) For identification of vehicle parameters with greater accuracy and within short time, response picked up at lower vehicle movement would be preferable for the implementation of particle filtering method.

(3) Rough bridge deck surface would require more time for the convergence of the result. The same conclusion is also valid for increased noise level in response data.

(4) The initial wrong choice of parameters of prior probability density function does not eventually lead to

wrong estimate, except that the convergence time would increase. This would imply that an idealized bridge model amenable for a closed-form solution of the response, in case of linear problem, may save much computational time.

(5) Choice of sampling frequency governs the time of convergence. Higher sampling frequency leads to the increased accuracy at the cost of increased CPU processing time in the estimation of system parameters.

(6) The present semi-analytical method used in generation of response samples brings rapid convergence and higher accuracy as compared to existing numerical methods.

Appendix-1

Expression of the Integral I_{jk} for generating response sample

The deck roughness is given by

$h(x) = h_{\mathrm{m}}(x) + h_{\mathrm{r}}(x)$.

The mean surface profile has been taken as shallow parabolic (a_0 being the rise at the center), the equation of which with respect to one end of the bridge is given by

$$h_{\mathrm{m}}(x) = \frac{4a_0}{L^2} x(L - x). \tag{35}$$

For one trial, a generation of a set of random phase angles $\theta_s (s = 1, 2, \ldots, N)$ is employed to express a Gaussian process as

$$h_{\mathrm{r}}(x) = \sum_{s=1}^{N} A_s \cos(2\pi\,\Omega_s x + \theta_s). \tag{36}$$

Further, $h'(x) = \frac{\mathrm{d}h}{\mathrm{d}x}$ and $\dot{h}(x) = \frac{\mathrm{d}h}{\mathrm{d}t} = V\frac{\mathrm{d}h}{\mathrm{d}x}$ where V is the speed of vehicle.

The vector $\{\boldsymbol{F}(t)\}$ needed to perform the integration is given below:

$$F_j(t) = 0 \, for j = 1 = k_t h(Vt) + c_t\, V\frac{\mathrm{d}h(x)}{\mathrm{d}x}\bigg|_{x=Vt} \quad \text{for} \quad j = 2$$
$$= k_t \varphi_j(Vt) h(Vt) + c_t\, \varphi_j(Vt) V\frac{\mathrm{d}h(x)}{\mathrm{d}x}\bigg|_{x=Vt}$$
$$- (m_w + m_v) g \varphi_j(x)\big|_{x=Vt} \quad \text{for} \quad j = 3, 4, \ldots, n_b. \tag{37}$$

The components of I_{jk} are given below for systematic computation

For $k = 1$, $I_{jk} = 0$.

Let us take $A_s = \sqrt{2S(V\Omega_s)^{-2}}$ and

$$B_s = 2\pi\left\{\Omega_{\mathrm{L}} + \frac{(s - 0.5)(\Omega_{\mathrm{U}} - \Omega_{\mathrm{L}})}{N}\right\}. \tag{38}$$

Now we write for $k = 2$, $I_{jk} = I_{j21} + I_{j22}$.

The components $I_{\mathrm{j}21}$ and $I_{\mathrm{j}22}$ are obtained as

$$I_{j21} = k_t\Bigg[\sum_{s=1}^{N} A_s \frac{\exp(-\alpha_j t)}{\alpha_j^2 + (B_s V)^2}\{-\alpha_j \cos(\theta_s) - B_s V \sin(\theta_s)$$
$$+ \exp(\alpha_j t)[\alpha_j \cos(B_s Vt + \theta_s) + B_s V \cos(B_s Vt + \theta_s)]\}$$
$$+ \frac{4a_0 V}{\alpha_j^3 L^2}\{\alpha_j L(\alpha_j t + \exp(-\alpha_j t) - 1)\}$$
$$- \frac{V}{\alpha_j^3 L^2}\{\alpha_j t(\alpha_j t - 2) - 2\exp(-\alpha_j t) + 2)\}\Bigg], \tag{39}$$

$$I_{j22} = c_t\Bigg[\sum_{s=1}^{N} VA_s B_s \frac{\exp(-\alpha_j t)}{\alpha_j^2 + (B_s V)^2}\{-\alpha_j \sin(\theta_s) - B_s V \cos(\theta_s)$$
$$+ \exp(\alpha_j t)[\alpha_j \sin(B_s Vt + \theta_s) - B_s V \sin(B_s Vt + \theta_s)]\}$$
$$- \frac{4a_0 V \exp(-\alpha_j t)}{\alpha_j^3 L^2}\{2V[\exp(-\alpha_j t)(\alpha_j t - 1) + 1]\}$$
$$- \frac{1}{\alpha_j L}\{\exp(\alpha_j t) - 2\}\Bigg], \tag{40}$$

For $k = 3, 4, \ldots, n$, the integral is split up into five parts as follows:

$I_{jk} = I_{jk1} + I_{jk2} + I_{jk3} + I_{jk4} + I_{jk5}$,

$Q_s = B_s V$; $R_k = (k - 2)\pi V/L$,

$$I_{jk1} = \frac{1}{2} k_t \sum_{s=1}^{N} A_s$$
$$\Bigg[\frac{2R_k \exp(-\alpha_j t)\{(\alpha_j^2 - Q_s^2 + R_k^2)\cos(\theta_s) + 2\alpha_j Q_s \sin(\theta_s)\}}{\{\alpha_j^2 + (Q_s - R_k)^2\}\{\alpha_j^2 + (Q_s + R_k)^2\}}$$
$$- \frac{\alpha_j \sin\{(Q_s - R_k)t + \theta_s\} + (R_k - Q_s)\cos\{(Q_s - R_k)t + \theta_s\}}{\{\alpha_j^2 + (Q_s - R_k)^2\}}$$
$$+ \frac{\alpha_j \sin\{(Q_s + R_k)t + \theta_s\} - (R_k + Q_s)\cos\{(Q_s + R_k)t + \theta_s\}}{\{\alpha_j^2 + (Q_s + R_k)^2\}}\Bigg], \tag{41}$$

$$I_{jk2} = k_t \frac{2a_0 V \exp(-\alpha_j t)}{L^2(\alpha_j^2 + Q_s^2)^3}$$
$$\Big[\Big\{L(\alpha_j^2 + Q_s^2)\Big[\exp(\alpha_j t)\sin(Q_s t)\{Q_s^2(\alpha_j t + 1) + \alpha_j^2(\alpha_j t - 1)\Big\}\Big]$$
$$- Q_s \exp(\alpha_j t)\cos(Q_s t)[Q_s^2 t + \alpha_j(\alpha_j t - 2)] - 2\alpha_j Q_s\}$$
$$+ V(2Q_s^3 + Q_s \exp(\alpha_j t))\cos(Q_s t)$$
$$\{Q_s^4 t^2 + 2Q_s^2(\alpha_j^2 t^2 - 2\alpha_j t - 1) + \alpha_j^2(\alpha_j^2 t^2 - 4\alpha_j t + 6)\}$$
$$- \exp(\alpha_j t)\sin(Q_s t)\{Q_s^4(\alpha_j t + 2) + \alpha_j Q_s^2(\alpha_j^2 t^2 - 3)$$
$$+ \alpha_j^3(\alpha_j^2 t^2 - 2\alpha_j t + 2)\} - 6\alpha_j^2 Q_s\Big]. \tag{42}$$

$$I_{jk3} = \frac{1}{2} c_t \sum_{s=1}^{N} A_s Q_s$$
$$\left\{ -\frac{2R_k \exp(-\alpha_j t)\{(\alpha_j^2 - Q_s^2 + R_k^2)\sin(\theta_s) - 2\alpha_j Q_s \cos(\theta_s)\}}{\{\alpha_j^2 + (Q_s - R)^2\}\{\alpha_j^2 + (Q_s + R_k)^2\}} \right.$$
$$-\frac{\alpha_j \cos\{(Q_s - R_k)t + \theta_s\} + (R_k - Q_s)\sin\{(Q_s - R_k)t + \theta_s\}}{\{\alpha_j^2 + (Q_s - R_k)^2\}}$$
$$\left. + \frac{\alpha_j \cos\{(Q_s + R_k)t + \theta_s\} + (R_k + Q_s)\sin\{(Q_s + R_k)t + \theta_s\}}{\{\alpha_j^2 + (Q_s + R_k)^2\}} \right\}. \tag{43}$$

$$I_{jk4} = c_t \frac{2a_0 V \exp(-\alpha_j t)}{L^2(\alpha_j^2 + V^2)^3} \{\exp(\alpha_j t)\cos(Q_s t) \times [Q_s^2(2\alpha_j tV - \alpha_j L + 2V) + \alpha_j^2(2\alpha_j tV - \alpha_j L - 2V)] - Q_s \exp(\alpha_j t)\sin(Q_s t)[Q_s^2(L - Vt) + \alpha_j(\alpha_j L - 2\alpha_j tV + 4V)]\}. \tag{44}$$

$$I_{jk5} = -\frac{\exp(-\alpha_j t) g L(m_v + m_w)}{L^2\alpha_j^2 + k^2\pi^2 V^2} \times [R_k L + \exp(-\alpha_j t)\{R_k L \cos(R_k t) + L\alpha_j \sin(R_k t)\}]. \tag{45}$$

References

1. Moses F (1979) Weigh-in-motion system using instrumented bridges. ASCE Trans Eng J 105:233–249
2. Clayton A, Peter R (1990) Truck weights as a function of regulatory limits. Can J Civil Eng 17:45–54
3. Green MF, Cebon D (1997) Dynamic interaction between heavy vehicle and highway bridges. Comput Struct 62:253–264
4. Yang YB, Yau JD (1997) Vehicle-bridge interaction element for dynamic analysis. J Struct Eng ASCE 123(11):1512–1518
5. Yu L, Chan THT (2007) Recent research on identification of moving loads on bridges. J Sound Vib 305:3–21
6. Connor CO, Chan THT (1988) Dynamic wheel loads from bridge strains. J Struct Eng ASCE 114(8):1703–1723
7. Law SS, Chan THT, Zeng QH (1997) Moving force identification: a time domain method. J Sound Vib 201(1):1–22
8. Chan THT, Law SS, Yung TH, Yuan XR (1999) An interpretive method for moving force identification. J Sound Vib 219(3):503–524
9. Chan TH, Yu TL, Law SS (2000) Comparative studies on moving force identification from bridge strains in laboratory. J Sound Vib 235(1):87–104
10. Law SS, Fang YL (2001) Moving force identification: optimal state estimation approach. J Sound Vib 239(2):233–254
11. Law SS, Chan THT, Zeng QH (1997) Moving force identification: a time domain method. J Sound Vib 201(1):1–22
12. Law SS, Chan THT, Zeng QH (1999) Moving force identification: a frequency and time domain analysis. J Dyn Meas and Control ASME 12:394–401
13. Chan THT, Law SS, Yung TH, Yuan XR (1999) An interpretive method for moving force identification. J Sound Vib 219(3):503–524
14. Zhu XQ, Law SS (2000) Study on different beam models in moving force identification. J Sound Vib 234(4):661–679
15. Wu SQ, Law SS (2010) Moving load identification based on stochastic finite element model. Eng Struct 32:1016–1027
16. Wu SQ, Law SS (2011) Vehicle axle load identification on bridge deck with irregular road surface profile. Eng Struct 33:591–601
17. Ghanem R, Shinozuka M (1995) Structural system identification-I, theory. J Eng Mech ASCE 121(2):255–264
18. Shinozuka M, Ghanem R (1995) Structural system identification-II, experimental verification. J Eng Mech ASCE 121(2):265–273
19. Deng L, Cai CS (2009) Identification of parameters of vehicles moving on bridges. Eng Struct 31(2474):2485
20. Kalman RE (1960) A new approach to linear filtering and prediction method. J Basic Eng ASME 191(82D):35–42
21. Ying DH, Bin CAO, Ping YY (2010) Application of particle filter for target tracking in wireless sensor networks. Int Conf Commun Mob Comput IEEE 3:504–508
22. Ristic B, Arulampalam S, Gordon N (2004) Beyond the Kalman Filter: particle filter for tracking application. Artech House Publishers, London
23. Qian SS, Stow CA, Borsuk ME (2003) On Monte Carlo methods for Bayesian inference. Ecol Model 159:269–277
24. Weerts AH, Ghada YH, Serafy E (2006) Particle filtering and ensemble Kalman filtering for state updating with hydrological conceptual rainfall-runoff models. Water Resour Res 42(9):1–17
25. Schon T, Gustafsson F, Nordlund PJ (2005) Marginalized particle filters for mixed/nonlinear state space models. IEEE Trans Signal Process 53(7):2279–2288
26. Chatzi EN, Smyth AW (2009) The unscented Kalman filter and particle filter and particle filter methods for non linear structural system identification with non collocated heterogeneous sensing. Struct Control Health Monit 16(1):99–123
27. Nasrellah HA, Manohar CS (2011) Finite element based Monte Carlo filters for structural system identification. Probab Eng Mech 26(2):294–307
28. Pokale B, Gupta S (2014) Damage estimation in vibrating beams from time domain experimental measurements. Arch Appl Mech.
29. Ching J, Beck JL, Porter KA (2006) Bayesian state and parameter estimation of uncertain dynamical systems. Probab Eng Mech 21:81–96
30. Nasrellah HA, Manohar CS (2010) A particle filtering approach for structural system identification in vehicle–structure interaction problems. J Sound Vib 329:1289–1309
31. Green MF, Cebon D, Cole JD (1995) Effects of vehicle suspension design on dynamics of highway bridges. J Struct Eng ASCE 121(2):272–282
32. Fryba, L (1996) Dynamics of Railway Bridge. Thomas Telford, London
33. Shinozuka M (1971) Simulation of multivariate and multidimensional random process. J Acoust Soc Am 49:357–367
34. Huang DZ, Wang TL (1992) Impact analysis of cable stayed bridges. Comp Struct 43(5):897–908
35. Yin X, Fang Z, Cai CS, Deng L (2010) Non-stationary random vibration of bridges under vehicles with variable speed. Eng Struct 32(8):2166–2174
36. Meirovitch L (1986) Elements of vibration analysis. Mc. Graw Hill Book Co., New York
37. Nigam NC, Narayanan S (1994) Introduction to random vibrations. Narosa Publishing House, Delhi
38. Potter MC, Goldberg J (1991) Mathematical methods. Prent Hall of India, Delhi
39. Arulampalam MS, Maske IS, Gordon N, Clapp T (2002) A tutorial on particle filters for online nonlinear/non-Gaussian Bayesian tracking. Trans Signal Process IEEE 50(2):174–188
40. ISO 8606:1995. Mechanical vibration-Road surface profiles-reporting measured data
41. Law SS, Bu JQ, Zhu XQ, Chan SL (2004) Vehicle axle loads identification using finite element method. Eng Struct 26:1143–1153
42. Collatz L (1966) The numerical treatment of differential equations. Springer-Verlag, New York

A multi-scale point of view on the structure–property relationships of A15 superconductors

Yanlong Ding · Min Pan · Shuiquan Deng · Yong Zhao

Abstract Nb_3Sn and other A15 members have been widely applied in nuclear power, nuclear magnetic resonance, and high-energy particle accelerators for their high critical current density (J_c) and upper critical field (B_{c2}). There have been comprehensive and intensive studies on the applications, the fundamental lattice dynamic and electronic properties, etc., of A15 superconductors. Various reviews on the preparations, structures, and properties have already been written in the last few years. Nevertheless, on account of the large amount of existing facts and views, a coherent view on the relations between the structures and properties has not appeared to unify the facts. This article sketches a multi-scale point of view on the relations between the multi-scale structures and the corresponding properties.

Keywords Multi-scale · A15 compounds · Structure · Properties

1 Introduction

The first A15 compound, V_3Si [1], was discovered by Hardy and Hulm in 1953. Since then a large number of compounds in this family have been prepared and characterized [2–5]. Among the important discoveries, the superconducting properties of some members such as Nb_3Sn, Nb_3Ge, etc., have received increasing attention. Although many other high T_c superconductors have also been discovered in the last few years, they have never received large-scale applications due to their various technical problems. In fact, nowadays—in almost everywhere—a high magnetic field is necessary, so A15 superconductors are used. With the pressing energy problem, Nb_3Al and Nb_3Sn have been suggested as the superconductors for creating the high magnetic field to control the plasma in the International thermonuclear experimental reactor (ITER). In addition, the large amount of need for superconductors in the superconducting maglev (SCMaglev) [6] vehicle has also stimulated studies for developing new techniques for manufacturing stable and durable superconductors. In this regard, the conventional A15 superconductors have their own advantages. The main advantages of A15 superconductors are their overall best merit factors. For example, Nb_3Al has a very high upper critical field (B_{c2}), $\sim$30 T at 4.2 K, and very good mechanical property which is important in actual applications. As the studies in this field are rather comprehensive and intensive, a large amount of facts and knowledge have been accumulated, ranging from microscopic electronic structures to macroscopic industry preparations. As diversified views exist for many experimental results at different length and energy scales, it is obvious that a unified and multi-scale point of view is necessary to summarize and analyze the existing results and conclusions. This work attempts to sketch such a frame to understand the relationship between the structures and properties of A15 compounds.

2 Atomic scales

2.1 Crystal structure

The A15 compounds with the chemical formula A_3B (the space group $Pm\bar{3}n$) have the same crystal structure as

Y. Ding · M. Pan · S. Deng (✉) · Y. Zhao
Key Laboratory of Magnetic Levitation Technologies and Maglev Trains, Ministry of Education of China, and Superconductivity and New Energy R&D Center, Southwest Jiaotong University, Chengdu 610031, China
e-mail: s.deng@swjtu.edu.cn

shown in Fig. 1. The structure belongs to a simple cubic crystal system. Each unit cell contains six A-type atoms which lie on the surface of the unit cell and form chains along the axis directions, respectively. On each chain, the A-type atoms distribute equidistantly. The two B-type atoms occupy the corners and the center of a unit cell.

In principle, a crystallographic structure at the atomic scale is decided by the structure of subatomic scale, namely the electronic structure. In the case of A15 compounds, although the electronic structure and phonon spectrum have been intensively studied, a direct link between the electronic structure and crystallographic structure has not been established. The structures at this two length scales are usually studied separately.

2.2 Phenomena related to the structure at atomic scale

Some properties of a material are highly related to the structure at the atomic scale. The studies on the defects and their influences are crucial for practical applications of a material. Sweedler et al. [7] have studied the neutron irradiation effect on the superconducting properties of A15 compounds. As shown in Fig. 2, the superconducting transition temperature (T_c) is dramatically reduced with the irradiation of neutron. Although such kind of phenomenological relation as shown in Fig. 2 is established, the structural damages due to irradiation have remained unknown. Brown et al. [8] have also studied the influence of neutron irradiation on the critical current density of Nb_3Sn (Fig. 3). As can be seen from Fig. 3, the critical currents are strongly affected by the field and irradiation dose. At high field, the change rate of critical current ($\Delta J_c/J_{c0} \equiv (J_c - J_{c0})/J_{c0}$, where J_{c0} is the unirradiated J_c) increases with the increase of the irradiation dose. At low field, the change rate of critical current decreases with the increase of the irradiation dose.

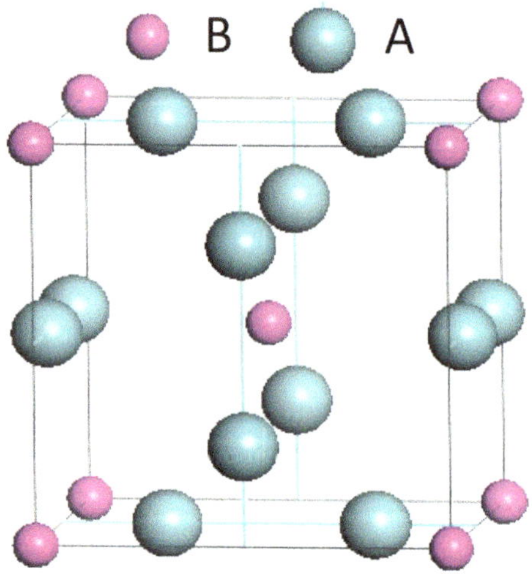

Fig. 1 The unit cell structure of A15 compound, A_3B

As the structural details after the irradiation have remained unknown, many experimentally established relations, as shown in Figs. 2 and 3, have not been well understood. In some studies, the reduction of the transition temperature (T_c) of A15 superconductors is attributed to the static disorder. It is thought that the high-energy irradiation induces "antisite defects" (A-type atom occupies the site of B-type atom and vice versa) [9]. These defects reduce the Bragg-Williams long-range-order parameter S [10] of the A15 superconductors. A possible consequence of such homogeneous defects is the reduction of the density of states at the Fermi level, DOS (E_F). Other possible effects of irradiation are the creation of disordered micro-regions, i.e., the so-called inhomogeneous defects, or the static displacement of atoms. Such defects may also cause the decrease of T_c by the proximity effect [11]. Although a wealth of experimental data has been obtained by various experimental means, the connection between the properties and the defected structures at the atomic scale has not been established due to lack of both the knowledge of the structures and the reliable microscopic models.

3 Electronic structure

Although in principle it is the structure at smaller scale which decides that at larger scale, the actual studies are performed in a reversed direction. This is because that humankind lives in a macroscopic world and thus our knowledge on the nature is established from macroscopic scale down to the microscopic scale in one direction and up to the cosmological scale in the other direction. The case of A15 compounds follows the above general principle. Because of the identical crystallographic structure, all A15 compounds have very similar electronic structures which differ only in details resulting from either different A- or B-type atoms. In Fig. 4, we have shown the electronic density of states (DOS) and the band structure of Nb_3Al as

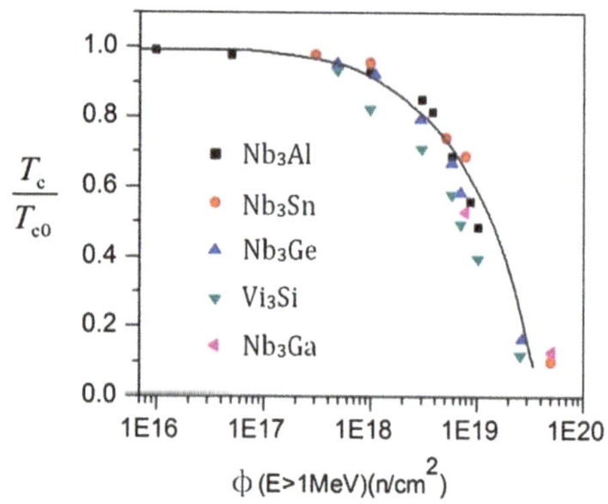

Fig. 2 T_c as a function of neutron fluence ($E \leq 1$ MeV, $T_{irr} \leq 150$ °C)

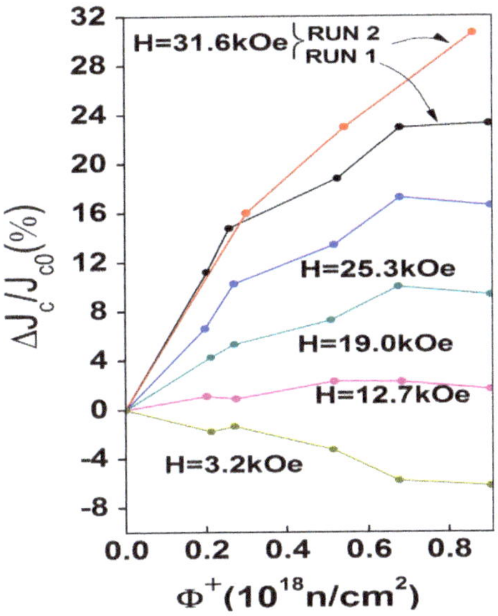

Fig. 3 The change rate of critical current as a function of field and irradiation dose

an example. The DOS has a sharp peak near the Fermi level. The states around the Fermi level are mainly contributed by the Nb *d* states. The lowest peak at about −5.6 eV is contributed mainly by the *s* states of B-type atoms [12]. These states form the lowest dispersion-less non-bonding energy bands. The bands above the Fermi level are mainly contributed by the *p* states of the B atoms. The coexistence of the very flat bands and steep bands around the Fermi level indicates that the A15 superconductors are essentially of "flat/steep" band superconductors [13]. The consequence of this latest observation [14] is worth of further studies.

4 Properties related to mesoscopic structure

For practical applications, the critical current density (J_c) and the upper critical field (H_{c2}) are crucial parameters. There have been a number of established experimental facts that indicate the notable dependence of J_c on the grain size of the sample [15–17]. It has also been found that the pinning force shows also some dependence on the grain size and grain boundary. In addition, the critical superconducting properties have also been found to be affected by the stacking faults [18]. It should be noted that knowledge of the grain size, grain boundary, etc., provides structural information possibly ranging from nano to mesoscopic length scale. Because of the inadequacy of the structural information and the lack of reliable model at these length scales, the explanations on the phenomenological structure–properties relations as established by experiments are not conclusive [18–20].

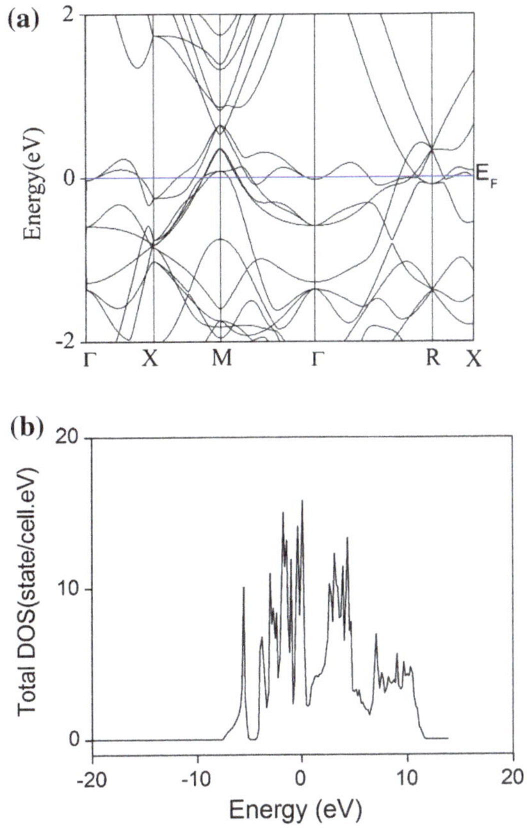

Fig. 4 **a** The band structure of Nb_3Al. **b** The total density of states of Nb_3Al

5 Macroscopic structure and properties

It is evident that to apply a stress on a material, its macrostructure will be changed in the form of strain, deformation, or cracks, etc. [21, 22]. For A15 superconductors, indirect indicators for such changes are superconducting critical parameters such as J_c, H_{c2}, and T_c. Figure 5 shows the effect of uniaxial strain on the upper critical field of A15 superconductors [23]. As shown in Fig. 5, although the crystal structures, electronic structures, etc., of A15 compounds are very similar, their critical field and critical current density have very different responses on the stress. Taylor and Hampshire [24] and Markiewicz [25] have attributed the effects of the uniaxial strain on the superconducting critical parameters to the changes of the phonon spectrum, rather than to the electronic density of states. In contrast, other researchers [26, 27] have attributed these effects mainly to the electronic properties. The problem of such studies is that the direct effect of the macroscopically applied stress has not been known with regard to the structures at macroscopic, mesoscopic, nanoscopic, and atomic length scales. Such information has essentially been taken as assumptions in the above-mentioned studies. A recent work [27] which has taken electronic and phonon

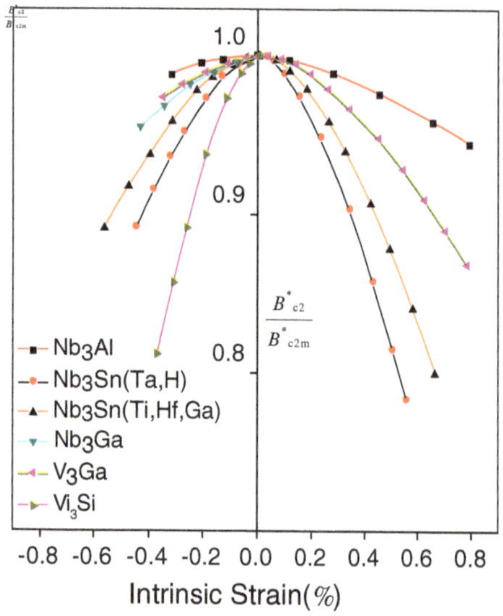

Fig. 5 Effect of uniaxial strain on the upper critical field of practical A15 superconductors

contributions into account made a new attempt to understand the stress effect. However, it did not consider the multi-scale structures at all.

6 Discussion and conclusions

Through surveying both the experimental and theoretical results on A15 compounds, we have provided a new multi-scale point of view on the structure–properties relationships. It is obvious that there have been many phenomenological relations established. However, in most cases they are not well understood. The reason is that, in many cases, the passage from one smaller scale to larger scale is necessary, which remains a challenge in science. For A15 compounds, the question that which and how some information are transmitted from one scale to the other one has remained unclear. We suggest that the density of states may be the key parameter that survives the renormalization processes from smaller scale to the larger scale. This needs to be studied by more experimental and theoretical work. Superconductors have already been applied to many aspects of our modern society. A large-scale application to the high-speed train can be expected with the maturity of SCMaglev technique, which provides the fastest ground transportation system and saves more energy in comparison with the conventional Maglev system. Because of these large-scale applications, studies on superconductivity are both scientifically and economically crucially important.

Acknowledgments This work is financially supported by the Science Foundation for International Cooperation of Sichuan Province (2014HH0016), the Fundamental Research Funds for the Central Universities (SWJTU2014: A0920502051113-10000), National Magnetic Confinement Fusion Science Program (2011GB112001).

References

1. Hardy G, Hulm J (1953) Superconducting silicides and germanides. Phys Rev 89(4):883–884
2. Weger M, Goldberg B (1973) Lattice and electronic properties of the β-tungstens. Solid State Phys 28:1–177
3. Dew-Hughes D (1975) A-15 compounds: a review. Cryogenics 15(8):435–454
4. Godeke A (2006) A general scaling relation for the critical current density in Nb_3Sn. Supercond Sci Technol 19(10):100–116
5. Glowacki BA (1999) Niobium aluminide as a source of high-current superconductors. Intermetallics 7:117–140
6. Central Japan Railway Company Annual Report (2012), pp. 23–25
7. Sweedler AR, Cox DE, Newkirk L (1975) Neutron irradiation effects in superconducting A-15 compounds. J Electron Mater 4(5):883–889
8. Brown BS, Blewitt TH, Wozniak DG et al (1975) Critical current changes in Nb_3Sn irradiated with fast neutrons at 6 K. J Appl Phys 46(12):5163–5168
9. Proceedings of the International Discussion Meeting on Radiation Effects on Superconductors, J Nucl Mat 1978, 72: 1–300
10. van Reuth EC, Waterstrat RM (1968) Atomic ordering in binary A15-type phases. Acta Cryst B24:168–196
11. Pande CS (1977) Effect of nuclear irradiation on the superconducting transition temperatures of A-15 materials. Solid State Commun 24(3):241–245
12. Paduani C (2007) Electronic structure of the A_3B compounds: A = Nb: B = Al, Ga, Ge, Sn and In. Braz J Phys 37(3B):1073–1079
13. Deng S, Simon A, Köhier J (1998) Superconductivity and chemical bonding in mercury. Angew Chem Int Ed 37(5):640–642
14. Ding YL, Deng SQ, Zhao Y The electronic structure of Nb3Al/Nb3Sn, a new test case for flat/steep band model of superconductivity. J Mod Transp, see this issue for the article of Ding et al.
15. Scanlan RM, Fietz WA, Koch EF (1975) Flux pinning centers in superconducting Nb3Sn. J Appl Phys 46(5):2244–2250
16. Shaw BJ (1976) Grain size and film thickness of Nb3Sn formed by solid-state diffusion in the range 650–800 °C. J Appl Phys 47(5):2143–2146
17. Wadayama Y, Suzuki T, Aihara K et al (1991) Microstructural characterization of Nb_3Al ultrafine multifilamentary superconducting wire by analytical electron microscopy. J Appl Phys 59(4):473–476
18. Banno N, Takeuchi T, Tsuchiya K (2012) Possible pining centers in transformation-processed Nb_3Al superconductors. IEEE Trans Appl Supercond 22(3):6001504
19. Takeuchi T, Tsuchiya K, Saeda M (2010) Electron backscatter diffraction analysis of Nb_3Al multifilamentary strands prepared by rapid heating, quenching and transformation annealing. Supercond Sci Technol 23:125001–125013
20. Takeuchi T, Tsuchiya K, Saeda M et al (2011) Electron-backscatter diffraction analysis of Nb_3Sn and Nb_3Al multifilaments. IEEE Trans on Appl Supercond 21(3):2541–2545

21. Savitski EM, Bychkova MI, Baron VV et al (1976) Lattice instability of compound Nb_3Sn in strained state. Phys Status Solidif A 37(2):165–167
22. Matthew CJ, Peter JL, David CL (2003) The influence of Nb_3Sn strand geometry on filament breakage under bend strain as revealed by metallography. Supercond Sci Technol 16:1005–1011
23. Ekin JW (1984) Strain effects in superconducting compounds. Adv Cryo Eng 30:823–836
24. Taylor DM, Hampshire DP (2005) The scaling law for the strain dependence of the critical current density in Nb3Sn superconducting wires. Supercond Sci Technol 18:241–251
25. Markiewicz WD (2004) Invariant formulation of the strain dependence or the critical temperature T_c of Nb_3Sn in a three term approximation. Cryogenics 44(11):895–908
26. Lim KC, Thompson JD, Webb GW (1983) Electronic density of states and T_c in Nb_3Sn under pressure. Phys Rev B 27(5):2781–2787
27. Marzi GD, Morici L, Muzzi L et al (2013) Train sensitivity and superconducting properties of Nb_3Sn from first principles calculations. J Phys 25(13):135702–135709

30

Comparative study on mode split discrete choice models

Xianlong Chen · Xiaoqian Liu · Fazhi Li

Abstract Discrete choice model acts as one of the most important tools for studies involving mode split in the context of transport demand forecast. As different types of discrete choice models display their merits and restrictions diversely, how to properly select the specific type among discrete choice models for realistic application still remains to be a tough problem. In this article, five typical discrete choice models for transport mode split are, respectively, discussed, which includes multinomial logit model, nested logit model (NL), heteroscedastic extreme value model, multinominal probit model and mixed multinomial logit model (MMNL). The theoretical basis and application attributes of these five models are especially analysed with great attention, and they are also applied to a realistic intercity case of mode split forecast, which results indicating that NL model does well in accommodating similarity and heterogeneity across alternatives, while MMNL model serves as the most effective method for mode choice prediction since it shows the highest reliability with the least significant prediction errors and even outperforms the other four models in solving the heterogeneity and similarity problems. This study indicates that conclusions derived from a single discrete choice model are not reliable, and it is better to choose the proper model based on its characteristics.

Keywords Discrete choice model · Mode split · NL · MMNL · HEV · MNP

1 Introduction

A good understanding on the travellers' mode choice behaviours serves as one of the prerequisites for passenger transport policy-making. Being important tools of travellers' mode choice behaviour studies, discrete choice models are widely used both in theory and practice within transportation planning field.

In the context of transportation, travellers tend to choose the transport mode which fits them best, and in the meanwhile, transportation means tend to 'choose' the most capable travellers as well. As a result, the decision making process of travellers' mode choice is influenced by the attributes of transportation means as well as the internal factors (individual attributes) of the travellers themselves, such as economic capability, personal preference, etc. Since several factors affect the description and the prediction accuracy of the mode choice behaviours, selecting a suitable discrete choice model with good interpretation ability appears to be very critical.

The discrete choice model used in early times is the multinomial logit model (MNL). The rigid assumption that the utility random terms of alternative parts satisfy independent identical distribution (IID) conditions makes MNL simple in calculation as well as provides MNL with independence of irrelevant alternatives (IIA) property, which

X. Chen (✉)
Guangzhou Transport Planning and Research Institute, 11th Floor, No. 80 of Jixiang Road, Guangzhou 510030, China
e-mail: harleych@163.com

X. Liu
Transportation Planning and Design Division, Architectural Design & Research Institute of Tongji University (Group) Co., Ltd., Shanghai 200092, China

F. Li
Guangzhou Transport Planning and Research Institute, Guangzhou 510030, China

weakens its ability of recurring the actual choice behaviours [1].

The IIA property of MNL model stems from the rigid assumption that the utility random terms of alternatives parts have totally independent structures. To relax the MNL model's IIA property, and in the meanwhile, keep its calculation convenience, the researchers gradually relax the restrictions on the assumption of utility random terms structures and successively explore and develop several MNL-based models, which are more capable of recurring decision-makers' choice behaviours, such as nested logit model (NL) [2], generalised extreme value model [3, 4], heteroscedastic extreme value model (HEV) [5], mixed multinomial logit model (MMNL) [6], etc. Many scholars [7–10] have analysed the travel mode of urban commuters. Schmidt and Strauss [11] and Boskin [12] analysed occupational choice among multiple alternatives. Rossi and Allenby [13] studied consumer brand choices in a repeated choice (panel data) model. Train [14] studied the choice of electricity supplier by a sample of California electricity customers. Hensher et al. [15] analysed choices of automobile models by a sample of consumers that offered a hypothetical menu of features. In each of these cases, there is a single decision among two or more alternatives. Faced with so many discrete choice models, how to choose an appropriate one to simulate travel behavior is still a rather difficult problem. In this article, we focused on transport mode choice behaviour modelling and made a comparison between five typical discrete choice models and discussed the rules for choosing the optimal discrete choice model.

2 MNL model and its application restrictions

Stemming from psychology and economics, discrete choice theory has become a mainstream since 1980s. Most discrete choice theory studies are grounded on the utility functions, which are expressed as

$$\boldsymbol{U}_{nj} = \boldsymbol{V}_{nj} + \boldsymbol{\varepsilon}_{nj} = \gamma_n' \boldsymbol{Z}_{nj} + \boldsymbol{\varepsilon}_{nj} = \beta_n' \boldsymbol{x}_{nj} + \alpha_n' \boldsymbol{y}_{nj} + \boldsymbol{\varepsilon}_{nj}, \quad (1)$$

where $\boldsymbol{U}_{nj}$ denotes the utility that the decision-maker n associates with alternative j; $\boldsymbol{V}_{nj}$ denotes measurable utility; ε_{nj} is the error term (immeasurable utility), γ_n' is the parameter vector of decision-maker n, $\boldsymbol{Z}_{nj}$, is the observed variable; $\boldsymbol{x}_{nj}$ is the individual attributes vector of decision-maker n, $\boldsymbol{y}_{nj}$ is attributes vector of alternative j; $n \in N$, N denotes the amount of decision-makers; $j \in J$, J denotes the amount of alternatives.

In the modelling process of individual choice models, assume that the consumers are rational choosers; therefore, the probability that individual n selects programme i is

$$\begin{aligned} \boldsymbol{P}_{ni} &= \boldsymbol{P}(\boldsymbol{V}_{ni} + \boldsymbol{\varepsilon}_{ni} > \boldsymbol{V}_{nj} + \boldsymbol{\varepsilon}_{nj}, \ \forall j), \quad i,j \in J, \\ &= \boldsymbol{P}(\boldsymbol{\varepsilon}_{nj} < \boldsymbol{\varepsilon}_{ni} + \boldsymbol{V}_{ni} - \boldsymbol{V}_{nj}, \ \forall j \neq i). \end{aligned} \quad (2)$$

Equation (2) assumes that the error term $\boldsymbol{\varepsilon}_{nj}$ satisfies independent and identically distributed (IID) assumption and subjects to type I extreme value distribution:

$$f(\varepsilon_{nj}) = \lambda e^{-\lambda(\varepsilon_{nj} - \eta_{nj})} \exp\left[-e^{-\lambda(\varepsilon_{nj} - \eta_{nj})}\right], \quad (3)$$

where η_{nj} and λ denote the location parameter and dispersion parameter, respectively. The variance of this distribution is $\pi^2/6\lambda^2$. Let $\lambda = 1$ and $\eta_{nj} = 0$, the MNL choice probability model takes the form of

$$\boldsymbol{P}_{ni} = \frac{e^{\boldsymbol{V}_{ni}}}{\sum_{j=1}^{J} e^{\boldsymbol{V}_{nj}}}. \quad (4)$$

In terms of measurable utility, it is usually defined as linear-in-parameter specification, i.e. $\boldsymbol{V}_{ni} = \beta_n \boldsymbol{x}_{ni} + \alpha \boldsymbol{y}_{ni}$. Thus, the probability that individual n chooses alternative i can be expressed as:

$$\boldsymbol{P}_{ni} = \frac{e^{\beta_n \boldsymbol{x}_{ni} + \alpha \boldsymbol{y}_{ni}}}{\sum_{j=1}^{J} e^{\beta_n \boldsymbol{x}_{nj} + \alpha \boldsymbol{y}_{nj}}}. \quad (5)$$

The following equation is produced by a simple derivation from Eq. (5) and shows the IIA property of multinomial logit model:

$$\frac{\boldsymbol{P}_{ni}}{\boldsymbol{P}_{nj}} = \frac{\exp(\beta_n \boldsymbol{x}_{ni} + \alpha \boldsymbol{y}_{ni})}{\exp(\beta_n \boldsymbol{x}_{nj} + \alpha \boldsymbol{y}_{nj})}, \quad i \neq j, \ i, \ j \in J. \quad (6)$$

Equation (6) means that among all the alternative sets, the ratio of choice probabilities of any two of the decision-maker's alternatives only associates with the utilities of these two alternatives but has nothing to do with the utilities of any other alternatives.

On the other hand, if we assume that the decision individual n makes is affected by his personal attributes $\boldsymbol{x}_{ni}$, then MNL model in Eq. (5) can be rewritten as:

$$\boldsymbol{P}_{ni} = \frac{e^{\alpha \boldsymbol{y}_{ni}}}{\sum_{j=1}^{J} e^{\alpha \boldsymbol{y}_{nj}}}. \quad (7)$$

Equation (7) is the conditional logit model. The feature of this model lies in that all the decisions are merely dependent on the attributes of the chosen alternatives $(\boldsymbol{y}_j)$ but irrelevant to decision-makers' attributes $(\boldsymbol{X}_n)$. No matter how many sets of alternatives exist, only one group of parameters needs to be estimated because of the assumption that the influences on individual utility of every choice set are identical. If there are many alternatives, then the conditional logit model can be served as a better choice for modelling and is also simpler to be compared with other models.

In practice, multinominal logit model (MNL) has to satisfy IIA property, which means that, all the alternatives are independent with each other and the ratio of choice probabilities only associates with the utilities of the given alternatives and is irrelevant to the utilities of any other alternatives. IIA property is generated from IID assumption of error term (The error term satisfies independent and identical type I extreme value distribution). IID constrains cannot be guaranteed if heterogeneity and similarity problems remain, which may result in the wrong statistic inference.

Except for IIA property restrictions, MNL model has two shortcomings in application. One is its incapability of handling with random preference discrepancies, the other is its incapability of finding out the correlative factors with panel data. Two merits of MNL are that it has a close-formed structure, and its parameters can be easily estimated.

Some scholars hold that IIA property is totally reasonable in terms of model theory. McFadden and Domencich [16] found that although the IIA restrictions result in a value loss of those studies on MNL, the deviations caused by IIA are owning to study objects, not the theory itself. They believed that IIA property is tenable in homogeneous populations. Ben-Akiva and Lerman [17] further pointed out that although IIA does not fit the whole populations, it does exist in homogeneous populations, such that the reliability of IIA property relies on whether the populations significantly show their heterogeneity. MNL model has the best performance to explain discrete choice behavior if the populations' heterogeneity is not significant.

3 Improvement and development of discrete choice models

Heterogeneity and similarity problems are directly related to the assumption of error terms in model. Observed samples and alternatives will cause the error terms. Therefore, we can consider heterogeneity and similarity problems from the perspective of these two factors. Viewing the heterogeneity issues from the perspective of samples means that the decision-maker holds different viewpoints towards specific transport modes in his mode choice behaviour, which can be called divergent tastes in individuals, or individual heterogeneity. Individual heterogeneity mainly comes from preference heterogeneity and response heterogeneity. The former includes the observed and unobserved effects that individual socioeconomic characteristics put on transport mode choice, and the latter refers to individual evaluation discrepancy on level of service across transport modes, which brings observed and unobserved effects as well.

Similarity across alternatives refers to the situation that similarity issues arise because of spatial or time autocorrelation during the survey process of samples (e.g. repeatedly investigations on the same respondent or samples across sampling objects are self-correlated because of adjacent zone effects, etc.) [5]. Heterogeneity and similarity are prone to biassed parameter evaluation or even overestimation on the effects of some specific factors.

When it comes to examining heterogeneity and similarity attributes from the perspective of alternatives, we need to consider whether alternatives share IIA property. If the alternatives appear to be dependent or heterogeneous, there may be similarity and heterogeneity problems among alternatives, which are called alternatives similarity and alternatives heterogeneity, respectively.

In the recent three decades, many models were developed to alleviate the heterogeneity and similarity problems, including NL model [17], HEV model [5], mixed multinominal logit (MMNL) model [10, 15], and multinominal probit (MNP) model [6].

3.1 NL model

NL model introduces the concept of nest layers, in which similar alternatives are put in the same nest layer. Assuming that, the error terms across alternatives in the same nest layer are independently and identically type I extreme value distributed, and the error terms across alternatives which belong to different nest layers are different. Here, we take two-layer nest structure as an example. Suppose that, there are M nests in the model and J_m alternatives in the mth nest layer. Alternative i is one of the alternatives in the mth nest layer, and thus, P_j^i is the probability that decision-maker chooses alternative i:

$$P_j^i = P_m^j \cdot P_j(i|m), \tag{8}$$

where P_m^j represents the marginal probability that decision-maker chooses nest layer m,

$$P_m^j = \frac{\exp(\mu_m \cdot I_m)}{\sum_{k=1}^{M} \exp(\mu_k \cdot I_k)}, P_j(i|m);$$

denotes the conditional probability that alternative $\underline{i}$ of nest layer m is chosen,

$$P_j(i|m) = \frac{\exp(V_i/\mu_m)}{\exp(I_m)};$$

I_m is inclusive value, which means comprehensive utility of nest layer $m, I_m = \ln \sum_{j\in J_m} \exp(V_j/\mu_m)$; μ_m is the parameter of inclusive value, which explains the similarity degree of the alternatives in nests. The estimated inclusive value parameter must subject to $0 \le \mu_m \le 1$, i.e. the principle of utility maximisation. When $\mu_m = 1$, NL model is simplified as MNL model. The more μ_m approaches to zero, the higher the correlation degree among alternatives is.

In order to make all the alternatives independent with each other, NL model imposes all the correlated alternatives on the same independent nest layer and makes use of inclusive value to represent the common utility of these alternatives, and then builds models with other independent alternatives. The NL model is good at solving the similarity problems among alternatives. However, its disadvantages are also evident. First, it has to be assigned a fixed nest layer structure; Second, it is not able to accommodate the situation that all the error terms correlate with each other at the same time; third, decision procedures should be supposed to satisfy the continuity condition; Forth, each alternative is restricted to appear in only one nest.

3.2 HEV model

HEV model is put forward by Bhat [18]. This model allows that alternative ε_{nj} satisfies independent non-uniform type I extreme value distribution, which means that each alternative has its own variance, and the variances may be same or not, but the covariance of different alternatives is zero. The probability that individual n chooses alternative i is

$$P_i = \int_{w=-\infty}^{w=+\infty} \prod_{j \in C, j \neq i} \Psi\left[\frac{V_i - V_j + \theta_i w}{\theta_j}\right] \psi(w)\mathrm{d}w, \tag{9}$$

where $\Psi(\cdot)$ and $\psi(\cdot)$ are cumulative distribution function and probability density function of type I extreme value distribution, respectively; C is the choice set; $\mathrm{var}(\varepsilon_i) = \pi^2 \cdot \theta_i^2/6$, $w = \varepsilon_{ni}/\theta_i$; θ_i represents heterogeneity parameter of alternatives, and it reflects the degree of uncertain factors, namely the weight of uncertain factors. Different alternatives have different effects on the whole utility. Increasing θ_i will decrease the unit variation that observed utility brings to choice probability.

HEV model allows variance discrepancy (the variance can be identical to each other or not) among error terms across alternatives by introducing scale factor into the expression of error terms, and the covariance across different alternatives is zero. HEV model is only able to handle with the heterogeneity problems among alternatives. During the model application process, large deviation may happen if similarity problems among alternatives simultaneously exist.

3.3 MNP model

Daganzo [6] proposed that the MNP model can be derived if assuming that random error terms follow normal distribution in Eq. (2). MNP model allows the situation that not all the random error terms are independent and identical with each other. It is the most generalised model as it fully reflects the realistic choice behaviours. The MNP model can be expressed as:

$$P_{ni} = \int I(\varepsilon_{nj} < \varepsilon_{ni} + V_{ni} - V_{nj},\ \forall j \neq i)\ \phi(\varepsilon_n)\ \mathrm{d}\varepsilon_n, \tag{10}$$

where $I(\cdot)$ is the index function, $I(\cdot) = 1$ means that the decision-maker has the one with max utility chosen, otherwise is zero; $\phi(\varepsilon_n)$ obeys multivariate normal distribution,

$$\phi(\varepsilon_n) = \frac{\exp\left(-\frac{1}{2}\varepsilon_n' \Omega^{-1} \varepsilon_n\right)}{(2\pi)^{\frac{J}{2}}|\Omega|^{\frac{1}{2}}},\ \varepsilon_n' = (\varepsilon_{n1}, \varepsilon_{n2}, \cdots, \varepsilon_{nJ}),$$

expectation $E(\boldsymbol{\varepsilon}_n) = \mathbf{0}$, and Ω denote covariance matrix.

Equation (10) has very complex integral components. When there are more than four alternatives within the choice set, it is difficult to estimate parameters.

MNP model is free of the MNL model's three restrictions. It is capable of handling with heteroscedasticity problems, defining error structures of any types as well as dealing with error terms related to time series by using panel data. The only limitation of MNP model is that all the error terms of utility functions must normally distribute. In most cases, assuming that the random terms satisfy normal distribution seems to be proper, but in some cases, this assumption may lead to unconventional prediction results. The most well-known example is about the price variable coefficient, the density distribution of which ought to only appear in the side of distribution greater than zero. Besides, MNP model appears to be much complicated in finding its parameters.

3.4 MMNL model

MMNL model based upon the assumption that decision-makers show different preferences. It assumes that marginal utility obeys Gumbel distribution, and the probability of MMNL has to be obtained by integrating the parameters of MNL model. The probability that decision-maker n chooses alternative i is:

$$P_{ni} = \int \boldsymbol{L}_{ni}(\gamma) g(\gamma)\mathrm{d}\gamma = \int \boldsymbol{L}_{ni}(\gamma) g(\gamma|\theta)\mathrm{d}\gamma, \tag{11}$$

where $\boldsymbol{L}_{ni}(\gamma)$ is the multinomial logit choice probability along with specific parameter vectors,

$$\boldsymbol{L}_{ni}(\gamma) = \frac{\exp[\mathbf{V}_{ni}(\gamma)]}{\sum_{j=1}^{J} \exp[\mathbf{V}_{nj}(\gamma)]}. \tag{12}$$

$g(\gamma) = g(\gamma|\boldsymbol{\theta})$ represents probability density function, $\boldsymbol{\theta}$ denotes deep parameter vector, which include mean value, variance or covariance, etc., and $V_{ni}(\gamma)$ is measurable utility. If the utility is linearly combined, i.e. $V_{ni}(\gamma) = \gamma' Z_{ni}$, then the choice probability of MMNL can be expressed as below:

$$P_{ni} = \int \frac{\exp(\gamma' Z_{ni})}{\sum_{j=1}^{J} \exp(\gamma' Z_{nj})} g(\gamma) \mathrm{d}\gamma. \tag{13}$$

The choice probability of MMNL model relies on the distribution form of γ. $g(\gamma)$ is normally distributed or log normally distributed [19].

In the utility function of MMNL model, except for observed non-random terms and error terms, unobserved random terms are also involved. The correlation, heterogeneity and individual preference heterogeneity of alternatives need to be considered with these random terms. MMNL is able to deal with heterogeneity and similarity problems simultaneously. Thus, the assumption of MMNL model is most practical and performs best in interpreting preference behaviours.

If the utility of MMNL model is set as linear combination, then MNL turns out to be the special case of MMNL model. The merits of MMNL model lie in that preference discrepancy among individuals is allowed, the correlation among different trips of the same consumer can be described, and it can approach to the estimated results of any other random utility models. The demerit of MMNL model is its complex computing process.

Note that, the parameter estimation methods of MNL, NL, HEV, MMNL and MNP models are not totally the same. Generally, the parameters of MNL, NL and HEV models can be identified and obtained by the maximum likelihood estimation method while the unknown parameters of MMNL model and MNP model can only be estimated by the maximum simulated likelihood method.

By analysing discrete choice model on error terms' assumptions from two dimensions, which are independent and identical, five introduced models can be classified as shown in Table 1. Each model has its own merits and demerits, and in application, desirable results can be obtained if they are well combined.

4 Illustrations

The source data used for the comparison of these five discrete choice models were drawn from the questionnaire survey on transport mode (car, train, bus and air) choice behaviours from 210 commuters between Sydney and Melbourne [12]. The main variables include:

TTME Terminal time, The TTME for car is zero (min)
INVT In-vehicle time (min)
GC Generalised cost
HINC Household income

The utility function to be estimated is constructed as

$$U_{nj} = \alpha_{\text{air}} d_{i,\text{air}} + \alpha_{\text{train}} d_{i,\text{train}} + \alpha_{\text{bus}} d_{i,\text{bus}} + \gamma_G GC_{ij} + \gamma_T TTME_{ij} + \gamma_H d_{i,\text{air}} HINC_i + \varepsilon_{ij}, \tag{14}$$

where for each j, ε_{ij} has the same independent, type 1 extreme value distribution,

$$F_{\varepsilon}(\varepsilon_{ij}) = \exp\left(-\exp(-\varepsilon_{ij})\right),$$

which has standard deviation $\pi^2/6$. $d_{i,m}$ is the binary variable which indicates if individual i made choice $m, m =$ air, train, bus, car. α_m is an estimate parameter for mode m.

We take the car mode as the basic alternative to construct MNL model and estimate parameters by maximum likelihood estimation method. The parameter values of universal set and restricted set are shown in Table 2, where restricted set means that the set excludes the air mode.

The calculation results show that as Hausman test value HM = 33.3363 is greater than $\chi^2_{0.05} = 9.488$, which indicates that IIA assumption of the MNL model is not proper, and there exist heterogeneity and similarity problems. The train, bus and car modes can all be used as the standard basic group, except for the air mode that will result in non-identified parameters,

The tree-like NL model is shown in Fig. 1.

After testing the estimated results of NL model we have $10.945 > \chi^2_{0.05} = 5.99$, which rejects IIA assumption. It is shown that the NL model outperforms the MNL model in terms of interpreting choice behaviours.

Table 1 Discrete choice models' classification

Error term assumption	Independent	Dependent (Similarity)
Identical	MNL	NL
Non-identical (Heterogeneity)	HEV	MMNL, MNP

Table 2 Results for MNL calibration

Estimated parameters	Universal set	Restricted set
α_{air}	5.207	–
α_{train}	3.869	4.464
α_{bus}	3.163	3.105
γ_G	−0.015	−0.0639
γ_T	−0.096	−0.0699
γ_H	0.013	–
Log likelihood at $\gamma = 0$	−291.1218	
Log likelihood at convergence	−199.1284	

It is possible that heterogeneity exists across alternatives. Here, we try to introduce scale parameters into the error terms across alternatives to make error terms unequal and alternatives heterogeneous. At least, one of the alternative scale parameter has to be fixed in HEV model. For the convenience of model comparison, car mode is set to be the basic alternative, and its scale parameter is assumed to be 1. The estimated scale parameter values for other three modes are: $\alpha_{air} = 0.2485$, $\alpha_{train} = 0.2595$, $\alpha_{bus} = 0.6065$. In Bhat's empirical study [5], HEV model has better interpretation ability over NL model and MNL model. The example below (Table 3) indicates that HEV model does out-perform MNL model on interpretation of choice behaviours but this doesn't mean it is better than NL model.

The parameter value of car mode is set as zero in the parameter estimation of MNP model, and the results are shown in Table 3. The MNP model does not enhance the interpretation ability of choice behaviours. This is because, some error terms of utility function do not distribute normally.

The MMNL model was built on the basis of the MNL model under the universal set mentioned above. The parameters estimated by maximum simulated likelihood estimation method are listed in Table 4 (classified as independent random parameters and correlated random parameters). It is shown that the MMNL model has the best performance in interpretation among all the models.

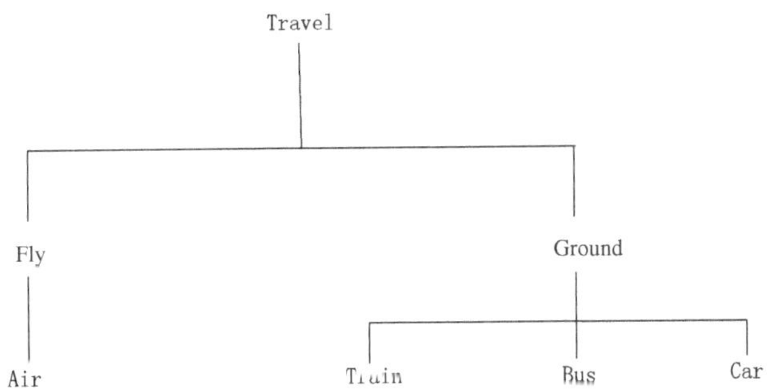

Fig. 1 Nested logit model structure

Table 3 Calibration results for NL, HEV and MNP models

Estimated parameters	NL Model	HEV Model	MNP Model
α_{air}	6.062	7.833	1.358
α_{train}	4.096	6.866	4.298
α_{bus}	5.065	7.172	3.609
γ_G	−0.032	−0.052	−0.035
γ_T	−0.112	−0.197	−0.077
γ_H	0.015	0.040	0.059
Log likelihood at convergence	−193.6561	−195.6605	−196.9244

Table 4 Calibration results for the MMNL model

Estimated parameters	Independent random parameters		Correlated random parameters	
	Estimate	t-stat	Estimate	t-stat
α_{air}	9.41	5.83	10.8	3.8
α_{train}	9.55	5.75	10.7	3.6
α_{bus}	8.59	5.69	9.7	3.7
γ_G	−0.03	−3.35	−4.02	1.9
γ_T	−0.21	−5.96	−13.4	3.9
γ_H	0.059	2.50	5.5	2.0
Chol(*GC*, *GC*)			3.00	1.3
Chol(*TTME*, *TTME*)	0.13	3.85	3.86	0.4
Chol(*GC*, *TTME*)			7.70	2.0
Log likelihood at convergence	−178.810		−176.816	

Chol denotes the parameter value derived by Cholesky decomposition of variance–covariance matrix

5 Conclusions

(1) MMNL model can be the first option when the parameter distribution is available because it performs best in interpretation.
(2) MNP model has its natural defect that all its error terms of utility functions should be normally distributed, which leads to a poor interpretation performance in practical application. Of all the test models, MNP shows the poorest performance in interpretation.
(3) The prediction accuracy of NL model depends on the given behaviour structure of decision-makers. If the decision-making procedures are unknown, then it will turn out to be very difficult to construct choice structure, and it has great influences on final model results if the decision structure is built with considerable mistakes.
(4) Illustration analysis indicates that the HEV model has a better interpretation ability in behavioral choice than the MNL model, but worse than NL model.

Acknowledgments This study was supported by the Science & Technology pillar project (No. 0556) of Guangzhou, China in 2011 for the study on railway transportation passenger flow forecast in the context of intercity and city integration based on low carbon.

References

1. Fuss MA, McFadden D (1978) Production economics: a dual approach to theory and applications. North-Holland Pub. Co, Amsterdam
2. McFadden D, Cowing TG (1984) Microeconomic modeling and policy analysis: studies in residential energy demand. Academic Press, Orlando
3. McFadden D (1976) The theory and practice of disaggregate demand forecasting for various modes of urban transportation. Dept. of Transportation, Office of the Secretary, Washington
4. Hensher DA, Swait JD, Louviere JJ (2000) Stated choice methods: analysis and applications. Cambridge University Press, Cambridge
5. Bhat CR (1998) Accommodating flexible substitution patterns in multi-dimensional choice modeling: formulation and application to travel mode and departure time choice. Transp Res B 32(7):455–466.
6. Daganzo CF (1979) Multinomial probit: the theory and its application to demand forecasting. Academic Press, New York
7. Hensher D (1986) Sequential and full information maximum likelihood estimation of a nested logit model. Rev Econ Stat 68(4):657–667
8. Hensher D (1991) Efficient estimation of hierarchical logit mode choice models. J Jpn Soc Civil Eng 425(IV-14):117–128
9. McFadden D (1974) The measurement of urban travel demand. J Public Econ 3:303–328
10. Greene WH (2012) NLOGIT Version 5 reference guide. Econometric Software, New York
11. Schmidt P, Strauss R (1975) The prediction of occupation using multinomial logit models. Int Econ Rev 16:471–486
12. Boskin M (1974) A conditional logit model of occupational choice. J Polit Econ 82:389–398
13. Rossi P, Allenby G (2003) Bayesian statistics and marketing. Mark Sci 22:304–328
14. Train K (2009) Discrete choice methods with simulation, 2nd edn. Cambridge University Press, New York
15. Hensher DA, Rose JM, Greene WH (2006) Applied choice analysis. Cambridge University Press, Cambridge
16. McFadden D, Domencich TA (1975) Urban travel demand: a behavioral analysis: a Charles River Associates research study. North-Holland Pub. Co., Amsterdam
17. Ben-Akiva ME, Lerman SR (1985) Discrete choice analysis: theory and application to travel demand. MIT, Cambridge
18. Bhat CR (2001) Quasi-random maximum simulated likelihood estimation of the mixed multinomial logit model. Transport Res B 35(7):677–693.
19. Hensher DA, Greene WH (2003) The mixed logit model: the state of practice. Transport 30(2):133–176

Permissions

All chapters in this book were first published in JMT, by Springer; hereby published with permission under the Creative Commons Attribution License or equivalent. Every chapter published in this book has been scrutinized by our experts. Their significance has been extensively debated. The topics covered herein carry significant findings which will fuel the growth of the discipline. They may even be implemented as practical applications or may be referred to as a beginning point for another development.

The contributors of this book come from diverse backgrounds, making this book a truly international effort. This book will bring forth new frontiers with its revolutionizing research information and detailed analysis of the nascent developments around the world.

We would like to thank all the contributing authors for lending their expertise to make the book truly unique. They have played a crucial role in the development of this book. Without their invaluable contributions this book wouldn't have been possible. They have made vital efforts to compile up to date information on the varied aspects of this subject to make this book a valuable addition to the collection of many professionals and students.

This book was conceptualized with the vision of imparting up-to-date information and advanced data in this field. To ensure the same, a matchless editorial board was set up. Every individual on the board went through rigorous rounds of assessment to prove their worth. After which they invested a large part of their time researching and compiling the most relevant data for our readers.

The editorial board has been involved in producing this book since its inception. They have spent rigorous hours researching and exploring the diverse topics which have resulted in the successful publishing of this book. They have passed on their knowledge of decades through this book. To expedite this challenging task, the publisher supported the team at every step. A small team of assistant editors was also appointed to further simplify the editing procedure and attain best results for the readers.

Apart from the editorial board, the designing team has also invested a significant amount of their time in understanding the subject and creating the most relevant covers. They scrutinized every image to scout for the most suitable representation of the subject and create an appropriate cover for the book.

The publishing team has been an ardent support to the editorial, designing and production team. Their endless efforts to recruit the best for this project, has resulted in the accomplishment of this book. They are a veteran in the field of academics and their pool of knowledge is as vast as their experience in printing. Their expertise and guidance has proved useful at every step. Their uncompromising quality standards have made this book an exceptional effort. Their encouragement from time to time has been an inspiration for everyone.

The publisher and the editorial board hope that this book will prove to be a valuable piece of knowledge for researchers, students, practitioners and scholars across the globe.

List of Contributors

Hongsheng Qiu
School of Transportation, Wuhan University of Technology, Wuhan 430063, China

Ximing Tan
School of Transportation, Wuhan University of Technology, Wuhan 430063, China

Shu Shi
School of Transportation, Wuhan University of Technology, Wuhan 430063, China

Heng Zhang
School of Transportation, Wuhan University of Technology, Wuhan 430063, China

Jing YANG
Institute of Mobile Communications, Southwest Jiaotong University, Chengdu 610031, China

Pingzhi FAN
Institute of Mobile Communications, Southwest Jiaotong University, Chengdu 610031, China

D. C. Watson Jr.
289 North Lake Lulu Drive, Winter Haven, FL 33880, USA

A. Al-Kaisy
Western Transportation Institute, Montana State University, 213 Cobleigh Hall, Bozeman, MT 59717, USA

N. D. Anderson
16304 South Yancey Lane, Spangle, WA 99031, USA

Sunao KUNIMATSU
Institute for Geo-Resources and Environment, National Institute of Advanced Industrial Science and Technology, 1-1-1 Higashi, Tsukuba, Ibaraki, 305-8567, Japan

Yoshihiro HIRAO
Kobayashi Institute of Physical Research, 3-20-41 Higashi-Motomachi, Kokubunji, Tokyo, 185-0020, Japan

Yasutoshi KITAMURA
Construction Engineering Research Institute Foundation, 1-3-10 Turukabuto, Nada-ku, Kobe, Hyogo, 657-0011, Japan

Li WANG
MOE Key Laboratory of High-Speed Railway Engineering, Southwest Jiaotong University, Chengdu 610031, China

Ping WANG
MOE Key Laboratory of High-Speed Railway Engineering, Southwest Jiaotong University, Chengdu 610031, China

Shunxi QUAN
China Railway Siyuan Survey and Design Group Co. Ltd., Wuhan 430063, China

Rong CHEN
MOE Key Laboratory of High-Speed Railway Engineering, Southwest Jiaotong University, Chengdu 610031, China

Yangzhou Chen
College of Electronic Information and Control Engineering, Beijing University of Technology, Beijing 100124, China

Hong Bi
College of Electronic Information and Control Engineering, Beijing University of Technology, Beijing 100124, China

Defu Zhang
College of Electronic Information and Control Engineering, Beijing University of Technology, Beijing 100124, China

Zhuoxi Song
College of Electronic Information and Control Engineering, Beijing University of Technology, Beijing 100124, China

Xi CAI
Institute of Mobile Communications, Southwest Jiaotong University, Chengdu 610031, China

Pingzhi FAN
Institute of Mobile Communications, Southwest Jiaotong University, Chengdu 610031, China

Qingchun CHEN
Institute of Mobile Communications, Southwest Jiaotong University, Chengdu 610031, China

Tengfei XU
School of Civil Engineering, Southwest Jiaotong University, Chengdu 610031, China

Tianyu XIANG
School of Civil Engineering, Southwest Jiaotong University, Chengdu 610031, China

Yulin ZHAN
School of Civil Engineering, Southwest Jiaotong University, Chengdu 610031, China

Renda ZHAO
School of Civil Engineering, Southwest Jiaotong University, Chengdu 610031, China

Tong CHEN
School of Electrical & Electronic Engineering, Newcastle University, Newcastle upon Tyne NE1 7RU, UK

Kanapathippillai CUMANAN
School of Electrical & Electronic Engineering, Newcastle University, Newcastle upon Tyne NE1 7RU, UK

Zhiguo DING
School of Electrical & Electronic Engineering, Newcastle University, Newcastle upon Tyne NE1 7RU, UK

Guiyun TIAN
School of Electrical & Electronic Engineering, Newcastle University, Newcastle upon Tyne NE1 7RU, UK

Xianfu Feng
Superconductivity and New Energy R & D Center, Southwest Jiaotong University, Chengdu 610031, China

Shuping Tian
Institute of Nuclear Science and Technology, Sichuan University, Chengdu 610064, China

Xinkui He
Superconductivity and New Energy R & D Center, Southwest Jiaotong University, Chengdu 610031, China

Fujun Gou
Institute of Nuclear Science and Technology, Sichuan University, Chengdu 610064, China

Shuiquan Deng
Superconductivity and New Energy R & D Center, Southwest Jiaotong University, Chengdu 610031, China

Yong Zhao
Superconductivity and New Energy R & D Center, Southwest Jiaotong University, Chengdu 610031, China

Peter Dely
Computer Science Department, Karlstad University, 65188 Karlstad, Sweden

Andreas Kassler
Computer Science Department, Karlstad University, 65188 Karlstad, Sweden

Lawrence Chow
Department of Management Science and Engineering, Stanford University, Stanford, CA 94305, USA

Nicholas Bambos
Department of Management Science and Engineering, Stanford University, Stanford, CA 94305, USA

Nico Bayer
Telekom Innovation Laboratories, Deutsche Telekom, Berlin, Germany

Hans Einsiedler
Telekom Innovation Laboratories, Deutsche Telekom, Berlin, Germany

Christoph Peylo
Telekom Innovation Laboratories, Deutsche Telekom, Berlin, Germany

Daniel Mellado
The Charles III University of Madrid, Madrid, Spain

Miguel Sanchez
The Charles III University of Madrid, Madrid, Spain

Ming FEI
Institute of Mobile Communications, Southwest Jiaotong University, Chengdu 610031, China

Pingzhi FAN
Institute of Mobile Communications, Southwest Jiaotong University, Chengdu 610031, China

Yanlong Ding
Key Laboratory of Magnetic Levitation Technologies and Maglev Trains, Ministry of Education of China, and Superconductivity and New Energy R&D Center, Southwest Jiaotong University, Chengdu 610031, China

Shuiquan Deng
Key Laboratory of Magnetic Levitation Technologies and Maglev Trains, Ministry of Education of China, and Superconductivity and New Energy R&D Center, Southwest Jiaotong University, Chengdu 610031, China

Yong Zhao
Key Laboratory of Magnetic Levitation Technologies and Maglev Trains, Ministry of Education of China, and Superconductivity and New Energy R&D Center, Southwest Jiaotong University, Chengdu 610031, China

Chuan Xiao
School of Civil Engineering, Southwest Jiaotong University, P.O. Box 520, No. 111 First North Section, 2nd Ring Road, Chengdu 610031, Sichuan, China

Tianqing Ling
School of Civil Engineering and Architecture, Chongqing Jiaotong University, Chongqing 400074, China

Yanjun Qiu
School of Civil Engineering, Southwest Jiaotong University, P.O. Box 520, No. 111 First North Section, 2nd Ring Road, Chengdu 610031, Sichuan, China

Saiji FUKADA
School of Environmental Design, Kanazawa University, Kakuma-Machi, Kanazawa City, Ishikawa Pref., 920-1192, Japan

Hirokazu HAMA
Fuji Engineering Co., Ltd., 5-5-28 Higashimikuni, Yodogawa-Ku, Osaka City, Osaka Pref., 532-0002, Japan

Kimihisa USUI
Fuji Engineering Co., Ltd., 5-5-28 Higashimikuni, Yodogawa-Ku, Osaka City, Osaka Pref., 532-0002, Japan

Peng Chen
Key Laboratory of Magnetic Levitation Technologies and Maglev Trains, Ministry of Education of China, and Superconductivity and New Energy R&D Center, Southwest Jiaotong University, Chengdu 610031, China

Dajin Zhou
Key Laboratory of Magnetic Levitation Technologies and Maglev Trains, Ministry of Education of China, and Superconductivity and New Energy R&D Center, Southwest Jiaotong University, Chengdu 610031, China

Pingyuan Li
Key Laboratory of Magnetic Levitation Technologies and Maglev Trains, Ministry of Education of China, and Superconductivity and New Energy R&D Center, Southwest Jiaotong University, Chengdu 610031, China

Yajing Cui
Key Laboratory of Magnetic Levitation Technologies and Maglev Trains, Ministry of Education of China, and Superconductivity and New Energy R&D Center, Southwest Jiaotong University, Chengdu 610031, China

Yongliang Chen
Key Laboratory of Magnetic Levitation Technologies and Maglev Trains, Ministry of Education of China, and Superconductivity and New Energy R&D Center, Southwest Jiaotong University, Chengdu 610031, China

Hui Wang
Institute of Railway and Urban Mass Transit, Tongji University, Shanghai 201804, China

Gang Shen
Institute of Railway and Urban Mass Transit, Tongji University, Shanghai 201804, China

Jinsong Zhou
Institute of Railway and Urban Mass Transit, Tongji University, Shanghai 201804, China

Yuming Gong
Superconductivity and New Energy R&D Center, Southwest Jiaotong University, Chengdu 610031, China

Gang Liang
Superconductivity and New Energy R&D Center, Southwest Jiaotong University, Chengdu 610031, China

Lifeng Zhao
Superconductivity and New Energy R&D Center, Southwest Jiaotong University, Chengdu 610031, China

Yong Zhang
Superconductivity and New Energy R&D Center, Southwest Jiaotong University, Chengdu 610031, China

Yong Zhao
Superconductivity and New Energy R&D Center, Southwest Jiaotong University, Chengdu 610031, China

Xuyong Chen
Beijing Xinli Machinery Co, Ltd, Beijing 100039, China

Hyuk-Jae Roh
City of Regina, Regina, SK S4P 3C8, Canada

Prasanta K. Sahu
National Institute of Construction Management and Research, Pune 411045, India

Ata M. Khan
Department of Civil and Environmental Engineering, Carleton University, Ottawa, ON K1S 5B6, Canada

Satish Sharma
Environmental Systems Engineering, University of Regina, Regina, SK S4S 0A2, Canada

Xin LI
Institute of Mobile Communications, Southwest Jiaotong University, Chengdu 610031, China

Pingzhi FAN
Institute of Mobile Communications, Southwest Jiaotong University, Chengdu 610031, China

Jayant P. Sangole
Department of Civil Engineering, Indian Institute of Technology Bombay, Powai, Mumbai 400076, India

Gopal R. Patil
Department of Civil Engineering, Indian Institute of Technology Bombay, Powai, Mumbai 400076, India

Hong Zhang
Superconductivity and New Energy R&D Center, Southwest Jiaotong University, Chengdu 610031, China

Yong Zhao
Superconductivity and New Energy R&D Center, Southwest Jiaotong University, Chengdu 610031, China

Wentao Wang
Superconductivity and New Energy R&D Center, Southwest Jiaotong University, Chengdu 610031, China

Min Pan
Superconductivity and New Energy R&D Center, Southwest Jiaotong University, Chengdu 610031, China

Ming Lei
Superconductivity and New Energy R&D Center, Southwest Jiaotong University, Chengdu 610031, China

Dehua Xu
Key Laboratory of Magnetic Levitation Technologies and Maglev Trains, Ministry of Education of China, and Superconductivity and New Energy R&D Center, Southwest Jiaotong University, Chengdu 610064, China
School of Physical Science and Technology, Southwest Jiaotong University, Chengdu 610064, China

Xinkui He
Key Laboratory of Magnetic Levitation Technologies and Maglev Trains, Ministry of Education of China, and Superconductivity and New Energy R&D Center, Southwest Jiaotong University, Chengdu 610064, China

Shuiquan Deng
Key Laboratory of Magnetic Levitation Technologies and Maglev Trains, Ministry of Education of China, and Superconductivity and New Energy R&D Center, Southwest Jiaotong University, Chengdu 610064, China

Yong Zhao
Key Laboratory of Magnetic Levitation Technologies and Maglev Trains, Ministry of Education of China, and Superconductivity and New Energy R&D Center, Southwest Jiaotong University, Chengdu 610064, China

Weichen XIANG
School of Electrical and Electronic Engineering, Newcastle University, NE1 7RU, UK

Stéphane Le Goff
School of Electrical and Electronic Engineering, Newcastle University, NE1 7RU, UK

Zhiguo DING
School of Electrical and Electronic Engineering, Newcastle University, NE1 7RU, UK

Guangbing Luo
State Key Laboratory of Traction Power, Southwest Jiaotong University, Chengdu, China

Jing Zeng
State Key Laboratory of Traction Power, Southwest Jiaotong University, Chengdu, China

Qunsheng Wang
State Key Laboratory of Traction Power, Southwest Jiaotong University, Chengdu, China

Zhantao Wei
Key Laboratory of Magnetic Levitation Technologies and Maglev Trains, Ministry of Education of China, Chengdu 610031, China

Xinsheng Yang
Key Laboratory of Magnetic Levitation Technologies and Maglev Trains, Ministry of Education of China, Chengdu 610031, China
Superconductivity and New Energy R&D Center, Southwest Jiaotong University, Chengdu 610031, China

Li Lv
Key Laboratory of Magnetic Levitation Technologies and Maglev Trains, Ministry of Education of China, Chengdu 610031, China
Superconductivity and New Energy R&D Center, Southwest Jiaotong University, Chengdu 610031, China

Min Zhang
Key Laboratory of Magnetic Levitation Technologies and Maglev Trains, Ministry of Education of China, Chengdu 610031, China
Superconductivity and New Energy R&D Center, Southwest Jiaotong University, Chengdu 610031, China

Yong Zhang
Key Laboratory of Magnetic Levitation Technologies and Maglev Trains, Ministry of Education of China, Chengdu 610031, China
Superconductivity and New Energy R&D Center, Southwest Jiaotong University, Chengdu 610031, China

Zhao Pingrui
MOE Key Laboratory of High-speed Railway Engineering, Southwest Jiaotong University, Chengdu 610031, China

Liu Xueyi
MOE Key Laboratory of High-speed Railway Engineering, Southwest Jiaotong University, Chengdu 610031, China

Liu Guan
MOE Key Laboratory of High-speed Railway Engineering, Southwest Jiaotong University, Chengdu 610031, China

Sachin Sudhakar Harak
Vibration and Noise Control Lab., Mechanical and Industrial Engineering Department, Indian Institute of Technology, Roorkee 247667, Uttarakhand, India

Satish Chand Sharma
Vibration and Noise Control Lab., Mechanical and Industrial Engineering Department, Indian Institute of Technology, Roorkee 247667, Uttarakhand, India

Suraj Prakash Harsha
Vibration and Noise Control Lab., Mechanical and Industrial Engineering Department, Indian Institute of Technology, Roorkee 247667, Uttarakhand, India

Hai ZUO
National Key Laboratory of Science and Technology on Communications, University of Electronic Science and Technology of China, Chengdu 610054, China

Xia LEI
National Key Laboratory of Science and Technology on Communications, University of Electronic Science and Technology of China, Chengdu 610054, China

Rongzhen LE
National Key Laboratory of Science and Technology on Communications, University of Electronic Science and Technology of China, Chengdu 610054, China

Maozhu JIN
Business School, Sichuan University, Chengdu 610065, China

Yamin SONG
National Key Laboratory of Science and Technology on Communications, University of Electronic Science and Technology of China, Chengdu 610054, China

Xia LEI
National Key Laboratory of Science and Technology on Communications, University of Electronic Science and Technology of China, Chengdu 610054, China

Zhaofu KONG
National Key Laboratory of Science and Technology on Communications, University of Electronic Science and Technology of China, Chengdu 610054, China

Binhong DONG
National Key Laboratory of Science and Technology on Communications, University of Electronic Science and Technology of China, Chengdu 610054, China

Zhao Liu
Key Laboratory of Magnetic Levitation Technologies and Maglev Trains, Ministry of Education of China, and Superconductivity and New Energy R&D Center, Southwest Jiaotong University, Chengdu 610031, People's Republic of China

Yongliang Chen
Key Laboratory of Magnetic Levitation Technologies and Maglev Trains, Ministry of Education of China, and Superconductivity and New Energy R&D Center, Southwest Jiaotong University, Chengdu 610031, People's Republic of China

Lupeng Du
Key Laboratory of Magnetic Levitation Technologies and Maglev Trains, Ministry of Education of China, and Superconductivity and New Energy R&D Center, Southwest Jiaotong University, Chengdu 610031, People's Republic of China

Pingyuan Li
Key Laboratory of Magnetic Levitation Technologies and Maglev Trains, Ministry of Education of China, and Superconductivity and New Energy R&D Center, Southwest Jiaotong University, Chengdu 610031, People's Republic of China

Yajing Cui
Key Laboratory of Magnetic Levitation Technologies and Maglev Trains, Ministry of Education of China, and Superconductivity and New Energy R&D Center, Southwest Jiaotong University, Chengdu 610031, People's Republic of China

Xifeng Pan
Key Laboratory of Magnetic Levitation Technologies and Maglev Trains, Ministry of Education of China, and Superconductivity and New Energy R&D Center, Southwest Jiaotong University, Chengdu 610031, People's Republic of China

Guo Yan
Key Laboratory of Magnetic Levitation Technologies and Maglev Trains, Ministry of Education of China, and Superconductivity and New Energy R&D Center, Southwest Jiaotong University, Chengdu 610031, People's Republic of China

R. Lalthlamuana
Department of Civil Engineering, Indian Institute of Technology Guwahati, Guwahati 781039, India

S. Talukdar
Department of Civil Engineering, Indian Institute of Technology Guwahati, Guwahati 781039, India

Yanlong Ding
Key Laboratory of Magnetic Levitation Technologies and Maglev Trains, Ministry of Education of China, and uperconductivity and New Energy R&D Center, Southwest Jiaotong University, Chengdu 610031, China

Min Pan
Key Laboratory of Magnetic Levitation Technologies and Maglev Trains, Ministry of Education of China, and uperconductivity and New Energy R&D Center, Southwest Jiaotong University, Chengdu 610031, China

Shuiquan Deng
Key Laboratory of Magnetic Levitation Technologies and Maglev Trains, Ministry of Education of China, and uperconductivity and New Energy R&D Center, Southwest Jiaotong University, Chengdu 610031, China

Yong Zhao
Key Laboratory of Magnetic Levitation Technologies and Maglev Trains, Ministry of Education of China, and uperconductivity and New Energy R&D Center, Southwest Jiaotong University, Chengdu 610031, China

Xianlong Chen
Guangzhou Transport Planning and Research Institute, 11th Floor, No. 80 of Jixiang Road, Guangzhou 510030, China

Xiaoqian Liu
Transportation Planning and Design Division, Architectural Design & Research Institute of Tongji University (Group) Co., Ltd., Shanghai 200092, China

Fazhi Li
Transportation Planning and Design Division, Architectural Design & Research Institute of Tongji University (Group) Co., Ltd., Shanghai 200092, China